China as an Innovation Nation

Edited by
Yu Zhou, William Lazonick,
and Yifei Sun

OXFORD
UNIVERSITY PRESS

OXFORD
UNIVERSITY PRESS

Great Clarendon Street, Oxford, OX2 6DP,
United Kingdom

Oxford University Press is a department of the University of Oxford.
It furthers the University's objective of excellence in research, scholarship,
and education by publishing worldwide. Oxford is a registered trade mark of
Oxford University Press in the UK and in certain other countries

First published 2016

Published in the United States of America by Oxford University Press
198 Madison Avenue, New York, NY 10016, United States of America

British Library Cataloguing in Publication Data
Data available

ISBN 978–0–19–875356–8

China as an Innovation Nation

Acknowledgments

This project has developed over several years. It would not have reached the completion without the help of many people along the way. The editors would like to thank David Musson at Oxford University Press for his support of this project and Clare Kennedy at OUP for coordination with us throughout. Zoe Tucker did an excellent job of copyediting all of the chapter manuscripts before they were sent to the publisher. Vaishnavi Ananthasubramanyam diligently oversaw the publication process for Oxford University Press. Ashley Harding of Ace Creative (www.acecreativce.biz) brought imagination and relevance to the creation of the book cover.

The editors' research for this volume was funded by the Ford Foundation Project on Financial Institutions for Innovation and Development, directed by William Lazonick, who particularly acknowledges the collaboration with Dr. Rongping Mu, Director of the Institute of Policy and Management, Chinese Academy of Sciences, in the sponsoring and running of the conference on financial institutions for innovation and development held in Beijing on October 17-18, 2013 (www.fiid.org). The Institute for New Economic Thinking also funded some of Lazonick's research on China in cross-national comparison. The National Science Foundation and Vassar College funded Yu Zhou and Yifei Sun's research and the workshop of the chapter contributors at Vassar College, April 2014.

The preparation of this manuscript also involved the assistance of many students. Students or former students at Vassar College, Eroll Kuhn, Mercedes Arndt, Xiyang Wang, and Cathy Zhuang assisted Yu Zhou in various stages of the book manuscript. Dongxu Li, Qiaoling Ma, and Xiahui Xia, all graduate students in the Regional Economic and Social Development program at the University of Massachusetts Lowell, worked as research assistants for William Lazonick. Last but not least, the editors would also like to acknowledge the hard work and cooperation of all the chapter contributors throughout the process.

Contents

List of Figures ix
List of Tables xiii
List of Contributors xv

1. Introduction: China's Transformation to an Innovation Nation 1
William Lazonick, Yu Zhou, and Yifei Sun

2. Evolution of Chinese State Policies on Innovation 33
Yu Zhou and Xielin Liu

3. Venture Capital in China 68
Jun Zhang

4. China's Mechanical Engineering Industry: Offering the Potential for Indigenous Innovation? 98
Ingo Liefner and Gang Zeng

5. Chinese Indigenous Innovation in the Car Sector: Being Integrated or Being the Integrator 133
Kaidong Feng

6. High Speed Rail Development in China: A Case Study of State-Guided Technology Transfer 165
Rongfang (Rachel) Liu, Liu (Willow) Lv, and Zhaodong (Tony) Huang

7. State, Market, and Business Enterprise: Development of the Chinese Integrated Circuit Foundries 192
Yin Li

8. Catching Up and Developing Innovation Capabilities in China's Telecommunication Equipment Industry 215
Peilei Fan and Xudong Gao

9. State, Multinational Corporations, Returnees, and Development of China's Integrated Circuit (IC) Design Industry 240
Yifei Sun, Zi Xue, and Debin Du

10. The Evolution of China's Mobile Phone Industry and Good-Enough Innovation 261
Shin-Horng Chen and Pei-Chang Wen

11. The Development of China's Wind Power Technology Sector: Characterizing National Policy Support, Technology Acquisition, and Technological Learning 283
Joanna I. Lewis

12. The Rise of the Chinese Solar Photovoltaic Industry: Firms, Governments, and Global Competition 306
Matthew Hopkins and Yin Li

Acronym Glossary 333
Index 339

List of Figures

2.1. Main Administrative Bodies of China's S&T System 35
2.2. Ratio of Expenditure on Assimilation to Importation of Technology in Different Ownership Enterprises 43
2.3. Funds Controlled by Ministry of Science and Technology 46
2.4. Total Value of Industrial Output of Different Ownership Enterprises (above the designated size) (2002–12) 56
2.5. Number of Scientists and Engineers in Different Ownership Enterprises 57
2.6. Amount of R&D in Different Ownership Enterprises 57
2.7. Number of Research Institutions in Different Ownership Enterprises 58
2.8. Average Amount of Funding of Every Institution of Different Ownership Enterprises 59
2.9. Number of Invention Patent Applications of Different Ownership Enterprises 59
3.1. Spatial Distribution of VC Offices and Investments in 2008 72
3.2. Growth of Venture Capital Investments in China: 1992–2013 73
3.3. The Annual Growth of Newly Raised VC Funds: 2002–13 74
3.4. Number of IPOs of VC-Backed Firms in Different Stock Markets: 2006–12 79
3.5. The Distribution of VC Investment in Leading Cities/Regions by 2013 81
3.6. The Number of Inter-Firm Coinvestment Ties Aggregated at the Urban Scale by 2008 84
3.7. Sectoral Structure of VC-Backed Startups in Top Three Centers (2008) 87
3.8. Sectoral Distribution of VC Investments: Percent of No. of Deals (1999–2013) 88
4.1. The Growth of China's Mechanical Engineering Industry (1952–77 and 1978–2011) 101
4.2. The Spatial Distribution of China's Mechanical Engineering Industry 104
4.3. The Spatial Distribution of China's National Mechanical Engineering Research Laboratories 106
4.4. Yizumi's and Propower's Different Pathways towards Indigenous Innovation 124

List of Figures

5.1. Linear Bottom-Up Model of Technological Learning 141
5.2. Output of the Chinese Car Sector 150
6.1. Historical Development of High Speed Rail in China 167
6.2. Speed Increases for National Rail Networks 170
6.3. Mode Shares in China 176
6.4. Capital Investment in Rail Infrastructure 177
6.5. Historical Railway Track Kilometers in China 182
8.1. China's Telecom Infrastructure 219
8.2. China's Mobile Phone Production (in billion units) 220
9.1. Number of IC Design Enterprises in China (1990–2011) 244
9.2. Revenues of China's IC Design Sector (2000–11) 245
10.1. Key Milestones in the Evolution of China's Mobile Phone and Mobile Communications Sectors 265
10.2. The Three-Level Model for Standards and Innovation in ICT 266
10.3. A Market-Share Comparison of Leading Smartphone Brands in the Global and Chinese Markets, 2012 269
10.4. The New Ecosystem and Industrial Transformation of Smartphones and Services in China 271
10.5. Value Chain Coverage of the Chinese Key Players under the Different Models in China 272
10.6. Industrial Standards and the New Ecosystem of Smartphones and Services in China 276
10.7. The Market Share of Mobile Operating Systems in China (2009–12) 277
11.1. Global Wind Power Installations: Leading Countries 285
11.2. Top 10 Wind Turbine Manufacturers in 2012 285
11.3. China's National Wind Power Support System and Conventional Technology Development Processes 287
11.4. Chinese Market Shares of Chinese and Foreign Wind Turbine Manufacturers (2004–12) 295
11.5. Wind Turbine Market Shares in the Chinese Market (2012) 296
11.6. Maximum Wind Turbine Size of Foreign and Domestic Chinese Firms Compared 300
11.7. Total Patent Cooperation Treaty Wind Energy Patents (1999–2011) 301
11.8. Wind Energy Patenting: Global Trends 302
11.9. Wind Turbine Prices and Future Predictions 303
12.1. Producers of Solar PV by Percent of Annual Total, Selected Nations (1995–2012) 312
12.2. Reported Raw Silicon Production Capacity, Selected Firms (2006–13) 315

12.3. Reported Manufacturing Capacity, Selected Firms (2005–13) 316

12.4. Solar PV Cell and Module Prices (1989–2013) 317

12.5. China's Annual Capacity Additions (2006–13) 326

List of Tables

2.1. Breakdown of Investment in Research Activities after Implementation of Indigenous Innovation Strategy by Country (%) 48
2.2. Percent of Intramural Expenditure on R&D 48
3.1. Zero2IPO's Annual Top 20 VC Investors 2012 74
3.2. Types of Limited Partners as Sources of China's Domestic VC Funds: 2012 80
4.1. China's Trade and Unit Values for 8428: Lifting, Handling, Loading Machinery (2012) 108
5.1. Imported Cars in Early Half of 1980s (unit: set) 137
5.2. Production Localization Rate of Components of Santana (Shanghai-Volkswagen) 141
5.3. Initial Expenditures of TMFT Practices 142
5.4. Product Sequencing of the TMFT JVs (up to 2009) 149
5.5. Newly Launched Car Models in the Chinese Market 151
5.6. Chery's Cooperation with AVL in Development of Engines (2002–8) 156
5.7. Important External Technical Projects of Chery in 2005 156
6.1. Main Events during the HSR Exploration Phase 167
6.2. Timeline for the Maglev and HSR Comparison 169
6.3. Imported HSR Technology and Their Chinese Partners 173
6.4. Second Generation HSR Models and their Key Features 175
6.5. Highway Comparison 176
6.6. Land Use Planning along Beijing–Shanghai HSR 180
9.1. Growth of China's IC Industry (2000–11) 245
9.2. Top 10 Chinese IC Design Firms in 2011 246
9.3. China's Growth in Electronic and Information Technology Products 249
11.1. Technology Transfer Models and Sources of Leading Chinese Wind Turbine Manufacturers 297
12.1. A Styled Description of the Solar PV Value Chain 314
12.2. IPO, Follow-On, and Debt Financing for Selected Companies 322
12.3. Percent of Total Revenues Derived from Europe (2004–13) 323

List of Contributors

Shin-Horng Chen, Chung-Hua Institution for Economic Research, Taiwan

Debin Du, School of Urban and Regional Sciences, East China Normal University, China

Peilei Fan, School of Planning, Design, & Construction, Michigan State University, USA

Kaidong Feng, School of Government, Peking University, China

Xudong Gao, School of Economics and Management (SEM), Tsinghua University, China

Matthew Hopkins, Center for Industrial Competitiveness, University of Massachusetts Lowell, USA

Zhaodong (Tony) Hang, Ningbo University, China

William Lazonick, Center for Industrial Competitiveness, University of Massachusetts Lowell, USA

Joanna I. Lewis, Edmund A. Walsh School of Foreign Service, Georgetown University, USA

Yin Li, School of Public Policy, Georgia Institute of Technology, USA

Ingo Liefner, Department of Geography, Justus Liebig University Giessen, Germany

Xielin Liu, School of Economics and Management, University of Chinese Academy of Sciences, China

Rongfang (Rachel) Liu, New Jersey Institute of Technology, USA

Liu (Willow) Lv, New Jersey Institute of Technology, USA

Yifei Sun, Department of Geography, California State University Northridge, USA

Pei-Chang Wen, Chung-Hua Institution for Economic Research, Taiwan

Zi Xue, Shanghai Semiconductor Industry Association, China. Mr. Xue unfortunately passed away in 2015

Gang Zeng, School of Urban and Regional Sciences, East China Normal University, China

Jun Zhang, Department of Geography and Planning, University of Toronto, Canada

Yu Zhou, Department of Earth Science and Geography, Vassar College, USA

1

Introduction

China's Transformation to an Innovation Nation

William Lazonick, Yu Zhou, and Yifei Sun

China's Rapid Growth

In the last three decades, China has transformed itself from one of the poorest countries to the world's second largest economy. In the process, hundreds of millions of people have left behind lives in poverty. Especially over the past decade, and at an accelerating rate, a Chinese middle class has emerged.

Much of China's growth since the late 1970s has been based on massive investments in human knowledge and physical infrastructure. In 1980, 33.1 percent of the population had no schooling; in 2010 only 6.6 percent did. Meanwhile the average years of schooling among this population increased from 3.87 to 7.12 (Barro and Lee 2000). The Chinese government has invested in road, rail, and air transportation networks, a national telecommunications system, new sources of energy, and massive amounts of industrial materials such as steel, all of which have underpinned China's industrialization process.[1]

In the end, however, all of these national investments in physical and human capital cannot be sustained unless business enterprises make use of them to produce goods and services that buyers need or want at prices that they are able or willing to pay. The business enterprises that succeed in national and global competition are those that have developed the capabilities to produce higher quality, lower cost goods and services than other firms in their industries. The name for the process that generates a higher quality, lower cost product than was previously available is *innovation*. The markets for

[1] See key statistics on trends and international comparisons of China's demography, infrastructure, knowledge base, industry, and trade, at <http://fiid.org/?page_id=1572>.

these innovative products may be internal to China, with the nation's rapidly growing middle class creating vast opportunities for selling these goods and services. Or the markets for these products may be global, in which case there is an opportunity for Chinese business enterprises to advance in terms of the quality of products that they can supply, often moving up global value chains through upgrading their productive capabilities.

Even with over three decades of sustained rapid growth, per capita incomes in China remain that of a developing economy. Large sections of China's population still live in poverty, and many aspiring young people have been unable to achieve their full potential because of limited education and employment opportunities. With massive investments in physical infrastructure and human knowledge in place, the only way for China to sustain its growth is by becoming an "innovation nation." The purpose of this volume is to assess China's transition to innovation-nation status.

China's future growth path is not just of concern to the Chinese people. The development of the world's most populous nation has been carried out in a highly globalized economy with countries around the world being affected in many ways by the rate and direction of China's growth. Some nations export vast amounts of goods, including natural resources, to China and the health of their economies has become dependent on China's continued growth. Under various arrangements, many multinational companies have a large and growing presence in China, producing for the burgeoning Chinese domestic markets or engaging in value-added production of components or end products for global markets. And increasingly, just as has happened in the cases of Japanese and Korean development, companies around the world have to be concerned about the emergence in China of indigenous companies that through investment in productive capabilities can compete globally in even the most sophisticated technology industries.

The Chinese government's stated goals are for China to join the top rank of "innovative nations" by 2020, and become a world-leading technology power by the mid-21st century (Li 2012). Central to this vision is Zizhu Chuangxin (自主创新), translated as "indigenous innovation," a strategy that was formally articulated in the Medium- and Long-Term Plan for Science and Technology in 2006.[2] Zizhu literally means self-directing. It stresses autonomy and strategic control at the national government and domestic enterprise levels, involving organization building, technology selection and transfer, and innovative learning. This current policy represents a correction from the expectation of the 1980s and 1990s that, through technology transfer, a complete embrace of globalization would automatically result in industrial

[2] See China State Council (2006).

upgrading and technological progress in China. Zizhu Chuangxin also signals the increased involvement of the Chinese state in the innovation processes.

Are these goals realistic? One cannot answer this question simply by looking at government policy or macro-economic indicators such as GDP growth or changes in the balance of trade. Nor can we make this judgment by looking at science and technology (S&T) indicators such as patenting, R&D expenditures, and China's STEM (science, technology, engineering, mathematics) labor supply. These indicators can be useful, but they must be analyzed in the context of evolving technologies, markets, and competitors of the particular industries, and in some cases the particular firms, that develop and utilize innovative capabilities. In the attempt to generate the higher quality, lower cost products that represent innovation, industries vary dramatically in terms of the organizations that must be transformed, production technologies that must be continuously upgraded, and the product markets that must be accessed. China also has to compete with the productive capabilities of other national industries for global and domestic market shares. And within a national industry there will be particular firms, with unique competitive capabilities, that emerge as leaders in global competition. The analysis of China as an innovation nation must therefore delve into the conditions for dynamic industrial sectors to continuously generate high-quality, low-cost products across a range of industries, while paying particular attention to the strategies and structures of the leading business enterprises and industrial districts within those sectors.

This volume provides studies of a range of industries of importance to China's future as an innovation nation, along with analyses of the evolving roles of investment by government agencies and business interests in the process. Each of the chapters has been written by one or more leading academic experts, recruited by the volume editors not only for their deep knowledge of the industries concerned but also for their insights into the role of industrial innovation in the larger process of economic development. The volume includes traditional industries such as mechanical engineering, railroads, and automobiles; rapidly evolving and internationally highly integrated industries such as information and communication technology (ICT); and newly emerging sectors such as wind and solar energy. The industries included in this volume are not exhaustive; for example, we did not include a study of the newly emerging biotech industry. But the sum total of the studies provided in this volume is, we think, a big step (if not a great leap) forward in our understanding of the industrial foundations of China's attempt to become an innovation nation.

With these industry case studies taken together, the book attempts to understand China's growth path in terms of the conditions, characteristics, and impacts of technological innovation over the past decades and into the future. Specifically, this volume is motivated by the following larger questions.

- What is the state and potential of China's indigenous innovation in important industrial sectors?
- How important is innovation to the sustainability of Chinese growth and national competitiveness?
- How do China's innovation paths differ from those of advanced or other newly industrialized countries?
- What are some of the key social conditions and characteristics that underpin the paths of Chinese innovation?
- What are the social implications of Chinese innovation for the stability of economic growth, the equity of income distribution, the social wellbeing of the Chinese people, and China's contribution to global wellbeing?

The volume sheds light on these questions. Definitive answers are not possible because the paths of innovation are long and inherently uncertain. The state of Chinese innovation is diverse across industries and enterprises and fluid over time. In each sector, we observe continued co-evolution of state policy, market demand, and technology development. The strategies and structures of individual companies and industrial ecosystems are changing rapidly. In almost all sectors, the gaps between Chinese indigenous companies and the global lead firms are shrinking, but at varying rates. Chinese governments and businesses are engaging in a variety of experiments in corporate governance, business models, employment management, and financial arrangements. Rather than viewing the Chinese path of innovation as a top-down movement, mostly powered by state-owned enterprises (SOEs), we are struck by the diversity of the types of business enterprises and the diversity of innovative experience across different industrial sectors. China's innovation path is being shaped by both top-down initiatives and bottom-up strategies, building productive capabilities from both technology transfer from abroad and indigenous investment. In our view, China has great potential to shape its institutions and organizations to be an innovation nation. But it still has a long way to go in developing and utilizing its innovative capabilities to achieve higher living standards, environmental sustainability, and social equality.

What is an Innovation Nation? Lessons from Japan and the United States

It is generally assumed that a nation needs innovation to prosper. Why? "Innovation" signifies that a national economy has acquired the capability to produce "higher quality" products than it was previously capable of producing. For any product, there are myriad dimensions of quality. Take a passenger car as an example, a product that, as in many of the world's most advanced economies, has been strategically important for the economic growth of China over the past 15 years or so. In the passenger car industry, "high quality" may mean that a car is safe (high-quality brakes, high-quality tires, seat belts, airbags, injury-proof, etc.), fuel-efficient, and environmentally friendly—dimensions of quality that are of public concern and are hence often subject to regulation. It may also mean that the car is rust-resistant, air-conditioned, roomy, stylish, comfortable, etc.—dimensions of quality that will be left to consumer choice. But it costs money to build quality into cars, and different types of government regulators and car buyers may register very different views about what "high quality" means and how much they are willing to pay for it.

Most nations have some car-producing capacity but few nations have the capability to produce high-quality cars. The "innovation nation" question is whether a national car industry can transform from producing low-quality cars to high-quality cars on a scale that has a significant impact on the nation's economic growth. In the age of globalization, such high-quality cars also have to be competitive on global markets.

The dramatic development of the Japanese economy in the last half of the twentieth century demonstrated the possibility of transforming national industries into world-leading producers of sophisticated manufactured goods, including cars. Coming into the 1970s, after almost two decades of high-speed growth, Japan was still known in the West as a nation that produced low-quality goods. But by the 1980s the Japanese had become renowned for their high-quality production of automobiles, consumer electronics, memory chips, machine tools, and steel. The passenger car industry was at the center of the transformation of Japan from a relatively poor nation into a relatively rich nation within a few decades. A short review of this transformation process illustrates the social conditions of innovative enterprise that a nation such as China must put in place.

From the late 1950s, Japanese carmakers, including Toyota, Nissan, and Honda, had been trying to sell their small ("compact") imported cars in the United States, gradually cutting into the leading market share of the Volkswagen Beetle. The inexpensive, fuel-efficient Japanese cars made some progress from the late 1960s, and then attracted the attention of a growing proportion of US consumers during the oil crisis of 1973–4, when prices at the gas pump quadrupled. But it was only from the last half of the 1970s, as Japanese cars became recognized as high-quality as well as low-cost, that Japanese car exports to the United States entered into sustained growth.

Most observers of the car industry in the early 1970s attributed Japanese competitive advantage to the low wages and long working hours of its labor force as well as a favorable exchange rate. Yet in the last half of the 1970s the Japanese compact cars attained a reputation for being very high quality, especially given their relatively low cost. Indeed during the 1980s, as Japanese wages rose rapidly and the Japanese yen strengthened, the Japanese car producers continued to gain market share in global competition. Japanese car producers also started to make massive investments in manufacturing plants in the United States as well as in Europe. Thus, besides exporting its high-quality products to the West, Japan also began exporting its management methods. Meanwhile the Japanese car producers transitioned from competitive advantage in compact cars to leadership in producing the whole range of vehicles, and by 1989 Toyota with Lexus and Nissan with Infiniti were able to compete with the high-quality German car producers, Mercedes-Benz and BMW, at the top of the price range of mass-produced luxury cars.

How did the Japanese manage to upgrade the quality of their cars in the 1970s and 1980s? The answers to this question provide us with insights into what China must now do to become an innovation nation. Japanese carmakers, led by Toyota, had three "social conditions of innovative enterprise" working on their behalf, that permitted them to produce cars that were higher quality than their competitors at lower unit costs, even when the advantages of low wages, long working hours, and a weak currency had disappeared (Lazonick 2007: 21–69; 2010a: 317–49; 2010a: 675–702).

- The first condition of innovative enterprise is "strategic control." Japanese executives with both the abilities and incentives to build world-class car producers controlled the allocation of resources in the Japanese companies. Their abilities came from their careers as professional managers, often with engineering backgrounds and their incentives came from the expectation of career progress within the company, rather than remuneration that depended on the company's stock price as increasingly became the case in the United States. Indeed from the 1950s, Japanese companies constructed a system of stable shareholding, also known as cross-shareholding, that protected their corporate treasuries from being looted by outside shareholders.
- The second condition is "organizational integration." The range of Japanese management methods that from the 1980s became famous in the West from just-in-time inventory systems to continuous improvement (kaizen) to total quality control entailed the integration of the skills and efforts of shop-floor workers with managerial (professional, technical, administrative) personnel. This organizational integration enabled the collective and cumulative learning processes that resulted in high-quality cars and large market shares. Supporting this organizational integration was the Japanese norm of permanent employment, also known as lifetime employment.
- The third condition is "financial commitment." The Japanese companies retained corporate profits and reinvested them in the physical and human

> capital that would ultimately, but with great uncertainty, enable these companies to generate higher quality products at lower unit costs than their competitors. These investments were not just in plant and equipment. The Japanese institution of permanent employment turned labor into a massive fixed cost that then required the attainment of large market shares to transform these high fixed costs into low unit costs. Central to this process of transformation of the high fixed costs of human capital into low unit costs was the combination of sustained organizational learning and hard, steady work. In the immediate postwar decades, Japanese car companies faced financial shortages, but were able to leverage their retained earnings with loans from Japanese banks under what became known as the main-bank system. By the 1980s the most successful Japanese car companies, along with companies in many other industrial sectors, had become so competitive on global markets that they were awash with cash (and were no longer dependent on bank loans), even as their employees enjoyed soaring living standards as they shared in their companies' gains from innovative enterprise.

Given the phenomenal growth of the Chinese car industry over the past 15 years or so, there are many lessons from the Japanese experience for understanding the possibilities and problems of China as an innovation nation through the development of this particular industry. China's manufacturing capacity of cars has increased rapidly. In 1998 just over half a million cars were produced in China, out of world production of 37.2 million (OICA 2013). Fifteen years later, in 2013, China's production of cars was 18.1 million, or 27.6 percent of the world total of 65.5 million. In units, in 2013 China's production just surpassed the *combined* total of Japan (8.2 million), Germany (5.4 million), and the United States (4.4 million). Right behind the United States was South Korea with 4.1 million cars produced, a big leap from 1.6 million 15 years earlier, and another success story that, like that of Japan, can be explained by the "social conditions of innovative enterprise" framework.

Enabling the growth of the Chinese automobile industry has been the domestic market populated by expanding numbers of upper middle-class households who now have the incomes to afford to buy cars. Of the 18.1 million cars produced in China in 2013, less than 600,000 were exported, mainly to low-income nations (almost 20 percent of these exports went to Algeria, China's largest foreign car market) (China Auto Web 2013). While China's total car output in 2013 was 2.2 times that of Japan's, China's total population is ten times that of its Asian neighbor. While the Chinese car industry cannot yet match the quality of Japanese, German, American, and Korean cars, the quality of Chinese cars has already surpassed those produced by the industries of India (3.1 million cars) and Russia (2.9 million), which have far longer histories of making cars. In comparative-historical perspective, the Indian and Russian car industries have lacked the organizational learning at the enterprise level that car companies operating in China have been acquiring over the past three decades.

The Chinese market is dominated by the foreign brands of cars produced through joint ventures (JVs) with Chinese SOEs, with relatively limited technological contribution from the indigenous carmakers. As Kaidong Feng details in his chapter (Chapter 5) on the automobile industry, under a policy called "Trading Market for Technology" (TMFT), launched in the first half of the 1980s, the Chinese state has permitted multinational companies (MNCs) to engage in JVs with SOEs based at the municipal or provincial levels. Through these JVs, China has been able to transfer automobile manufacturing from abroad. Now a number of these JVs are the leading producers of cars in China.

Under TMFT, however, the development of capabilities within the JVs has been constrained by the dependence of the Chinese SOEs on the foreign partners to provide them with the knowledge that is needed to produce cars that would be considered high quality in the Chinese market. This limitation of TMFT has thus translated into the lack of strategic control by the Chinese partners in the JVs (see more details in Chapters 2 and 5).

Feng argues that the exercise of strategic control by Chinese automobile firms is a critical next step for Chinese innovation in the automobile industry. Now there are a number of indigenous firms in the Chinese car industry that have gained significant market shares, with the aspiration of becoming leading global competitors in the next decade or two. In 2013 three indigenous companies—Geely with 554,000 cars produced in China, BYD with 511,000, and Chery with 459,000—ranked nos. 13, 14, and 15 respectively in Chinese car production. Together these three companies had just under 8 percent of Chinese car production. Geely's acquisition of Volvo has transformed the company into a small global competitor, but to date very few Geely cars have been exported from China.

Competition within the Chinese car market has intensified in recent years, with slower growth and a crowded field. Newer indigenous companies face difficult challenges in producing cars that are comparable in quality with those of the established carmakers. Geely CEO An Conghui was quoted as saying: "Chinese can not only make cars, but are also capable of making good cars. We are utilizing the very best designers the automotive world offers to create a new global design language" (Ying 2014). No one can say whether Geely's strategy will be successful, or the extent to which, over any given timeframe, it will be able to reap a significantly larger share of the Chinese car market. But we can say that the innovative strategies of indigenous companies like Geely, BYD, and Chery will be central to the competitive dynamics that will over the next decade or two bring increasingly higher quality, lower cost cars to China.

The biggest lesson that we can draw from this comparative experience of the Chinese automobile industry is that the building of an innovation nation is a long process of collective and cumulative learning, with no certainty of success. At the enterprise level, strategic control is essential to put in place the organizational learning processes that can generate higher quality products. It is then imperative for the innovative company to attain a large extent of the market to drive down unit costs.

Investments in organization and technology in advance of product revenues, which is inherent in the innovation process, require what many have called "patient capital"; i.e. financial commitment. The extent of financial commitment required to sustain the innovation process depends on not only the size of the investment as a point in time but also the duration of time from when investments in the innovation process are made until, through the creation of competitive products, market sales can provide financial returns.

What are the sources of this financial commitment? The answer to this question varies dramatically depending on the capital requirements of the particular industry in question. The Chinese state is well known for investing in long-term infrastructure projects. The Chinese government financed the building of the modern railroad system, including high-speed rail (see Chapter 6); a highway system; a telecommunication system; and an electric power system. Through SOEs, the Chinese government has also overseen the financing of a steel industry that in 2014 produced 49.5 percent of the world's crude steel.[3] In effect, steel has also functioned as an infrastructural input without which China's construction and transportation booms could not have taken place. The car industry, for example, could not have experienced its spectacular growth in China if it had had to wait for the necessary indigenous steel capacity to be put in place.

China used to have a highly centralized financial system with the state-owned banks being the only source of finance capital. In recent years, however, much of the decision-making for allocating financial resources to China's industrial development has occurred at the local government level, as detailed in Chapter 2. Even then, central government financial agencies have played key roles. The capital available to local governments has not been constrained by the current taxpaying capacity in the local region. Rather the ability of local governments to finance industrial development has been supported by loans from the China Development Bank, with local land, often purchased from peasants, as collateral (Sanderson and Forsythe 2013). The

[3] The shares of the next largest crude-steel producers were EU2-8, 10.2%; Japan, 6.2%; and the USA, 5.3% (World Steel Association 2014).

local government can also compete for subsidies from the central government to help fund strategic industries, as documented in more detail in the chapters on semiconductors and clean technology (Chapters 7, 11, and 12).

The sources of capital for innovation are becoming more diversified. Chinese companies have made creative use of foreign stock markets to raise capital. In the 1980s and 1990s, a number of SOEs and non-SOEs were able to list on the Hong Kong stock exchange to gain direct access to foreign exchange. Over the past decade, as is shown in this volume's chapters on venture capital (Chapter 3) and the solar panel industry (Chapter 12), young Chinese companies have been able to raise significant amounts of capital on the NASDAQ stock exchange or the New York Stock Exchange (NYSE) in the United States. In 2011, there were 179 Chinese firms listed on NASDAQ (up from 41 in 2006), 84 on NYSE (up from 23), 43 on AMEX (8), 549 on the Hong Kong Stock Exchange (333), 182 on Singapore Stock Exchange (117), 81 on London AIM (47), and 141 on others (29) (Pan and Brooker 2014). This stock-market financing has been highly speculative, and if and when, as happened in 2011 in the solar panel industry, the Chinese companies go bankrupt, the productive capacity remains in China but foreign speculators are left holding valueless shares.

More recently, as is detailed in the chapter on venture capital (Chapter 3), startup companies in China have had access to US-style venture capital in China and abroad. But, as was the case in the United States, venture capital in China has emerged as a source of finance at a relatively late stage of industrial development, after governments and large corporations have made the massive investments in physical and human capital that make startups possible (Lazonick 2009a). In the United States from the late 1970s, lobbying by Silicon Valley interests, taken to the national level, convinced the US Congress that the attraction of venture capital to support innovation required dramatically lower tax rates, ignoring the fact that government and business investments in the previous generation had made new high-tech ventures possible. When companies such as Cisco Systems, Microsoft, Intel, Apple, and Amgen, among many others, have transformed themselves from small new ventures to enormous going concerns, their executives have viewed their profits as "returns" to shareholders, in some cases expending more than 100 percent of corporate profits over the past decade on stock buybacks even while paying ample dividends. Yet public shareholders have made *insignificant* investments in the productive assets of these companies, raising the question of why they are entitled to such high levels of rewards (Lazonick 2009a, b, 2014a, 2015; Hopkins and Lazonick 2014). In other cases, especially in the US biopharmaceutical industry where it can take at least a decade and $1 billion to develop an approved medical drug, venture-backed companies have been able to list on the stock market, and financial interests have been able to make vast amounts of money, even when no product is produced (Lazonick and Tulum 2011).

Since the 1990s, China has attempted to replicate the US "New Economy business model" through a combination of foreign direct investment, global value chains, Chinese high-tech returnees from the United States and elsewhere, the establishment of various domestic stock exchanges, and high-profile initial public offerings in the United States such as that of Alibaba (see Chapters 2 and 3). At this stage, it is fair to say that venture capital development supplements the weakness of China's state-owned banking system in supporting the small private startups. China's variant of the internet sector has been a focus of the venture capital boom in China.

In the long run, however, there is a distinct danger that Chinese business executives, along with the Chinese public, might become enamored with the speculative stock market as the primary institution for financing innovation. Unless China, like Japan, develops its own institutions to prevent financial interests from capturing the lion's share of the gains from innovative enterprise, China runs the risk of having its financial economy dominate its productive economy as has happened in the United States, with, as the results, an extreme concentration of income among the richest households and the stifling of a prosperous middle class.[4]

Building Chinese Institutions for an Innovation Nation

If China is to become an innovation nation, the Chinese government will have to pay attention to three types of national institutions—governance institutions, employment institutions, and investment institutions—that can support or undermine the social conditions of innovative enterprise. Through the evolution and operation of these institutions, China can learn how to develop and utilize its indigenous innovation capacity in relation to the global production systems in which its industry is deeply intertwined. Let us consider each of these institutional types in turn.

Governance Institutions

A critically important characteristic of China's development since the late 1970s has been the willingness to encourage the establishment of business enterprises with strategic autonomy from state control. The former Soviet Union did not permit such a strategy and to this day has not spawned even one important global manufacturing enterprise. In contrast, the Chinese state,

[4] For the US case, see Lazonick (forthcoming).

in a business sense, let a hundred flowers bloom to the point that China today permits the coexistence of a variety of corporate governance regimes. Some Chinese high-tech firms such as Lenovo and ZTE were originally spun off from Chinese research institutes or state-owned enterprises. Others such as Alibaba, Huawei, Tencent, and Xiaomi were private startups. Even with this autonomy, however, Chinese business continues to rely heavily on state investments in physical infrastructure and human capital as foundations for business investment in the value-creating capabilities that can bring innovative products to the market. Along with state support come industrial policies, financial incentives, and legal regulations and restrictions that may enable or proscribe business activity.

There has been a heated debate in the West about the rise of state capitalism as opposed to liberal capitalism, with China being the prime example of the former.[5] The definitions of state capitalism vary, but the most common view is that it is characterized by the state use of its power to establish and promote favored enterprises, most likely SOEs, to become national champions in the market (Bremmer 2009). This volume provides a reality check on this perspective on state capitalism as the central feature of the Chinese economy, with every chapter in the volume addressing the issue in one way or another of the role of the Chinese state in innovation. The critical insight from the studies in this volume is that, far from the overarching Leviathan image of Chinese state capitalism, the roles of the Chinese state are extensive and multifaceted, but entail reflexive and collaborative interactions with business enterprises, many of them non-SOEs. Chapter 2 shows that while the Chinese state has indeed maintained a consistent vision of seeking technological leadership and autonomy from the West, it has changed its policies and practices dramatically in the last 60 years, responding to domestic economic reform agendas, global relations, and technological transformations.

In the 1980s and 1990s, the Chinese government set up advanced-technology enterprises through SOEs or JVs in strategic sectors such as automobiles (Chapter 5) and semiconductor fabrication (Chapter 7), in line with the state capitalism model. However, with the exception of China Railway Corporation (formerly Ministry of Railway), these companies, with strategic control in the hands of the state, did not do well in keeping up with rapid technological change. As Chapter 6 on high-speed rail makes clear, this particular sector lends itself to a state capitalism model because of the need for central planning of the system, enormous financial commitment, and the absence of international competition.

[5] See discussion on "The Rise of State Capitalism," *The Economist*, Jan. 21, 2012, at <http://www.economist.com/node/21543160>. For a rebuttal on China as the model of state capitalism, see Lardy (2014).

The various chapters in this volume demonstrate that not all industries are created equal. They can differ dramatically in terms of technologies, markets, and competitors. In industrial sectors in which decentralized decision-making is more important for transforming technologies and accessing markets, financial commitment less severe, and international competition more intense, the Chinese state has deployed a growing list of flexible instruments to promote innovation by diverse actors from SOEs to private companies. These instruments include direct R&D finance to state-owned research institutes or enterprises, preferential finance and tax policies for strategic industries, public procurement, and domestic technological standards.

The roles of central and local governments have expanded and contracted at different stages of the growth of enterprises or industries, as exemplified by the cases of Huawei and ZTE (Zhongxing Telecommunication Equipment Corporation) in the communication equipment sector (Chapter 8). In the railroad (Chapter 2) and semiconductor fabrication (Chapter 7) industries, state direct investment in SOEs or JVs remains central. In the clean-tech industries (Chapters 11 and 12), governmental subsidies, supportive pricing, and regulatory policies such as feed-in tariffs have been critical mechanisms for government influence. In the ICT sectors, public procurement and industrial technological standards have been immensely influential. In all sectors, low-cost land and subsidized rent in industrial districts and science parks, preferential tax rates for innovative startups, as well as local government loans, have been used to support a variety of enterprises.

There are questions, however, whether such local support is serving to preserve local industrial capacity as sources of tax revenue and employment or to encourage the development of higher quality products. Of particular concern is the tendency of governments, especially at the local level, to protect established enterprises through bank loans and subsidies regardless of their performance, while discouraging the emergence of new competitors. In the automobile industry, for example, for decades the state established regulatory barriers to prevent new entrants into the industry that could challenge the JVs set up under TMFT (Chapter 5). These barriers disappeared, however, after China joined the World Trade Organization (WTO) in 2001 (Chapter 5), and, as we have seen, a number of indigenous companies have become serious competitors in the automobile industry.

Overall, the drivers of China's innovation efforts are both top-down and bottom-up, with shifting combinations depending on the nature of the industries and extent of global integration. Top-down state investment has played critical roles in building China's infrastructure and human resources, which directly affect technology markets and business enterprises. China's state governance regime over the management of innovation processes is still evolving through trial and error in response to feedback from industry. For China to become an innovation nation, it needs a governance regime that ensures the complementarity of government and business investment in productive capabilities. As such complementarities are being negotiated between assorted governmental and business players, the emerging governance institutions must guard against "rent-seeking" tendencies, based on value extraction in excess of contributions to value creation, by powerful parties within the governmental and business entities. Each of the chapters in this volume provides examples of these problems.

Stock markets are prime institutions that can enable excessive value extraction (Lazonick 2013). The growth of stock markets and shareholding companies are new phenomena in China since the 1990s, modeled after similar institutions in the United States. It should not be assumed, however, that the publicly listed business enterprise represents the corporate governance model best suited to becoming an innovation nation. In the United States, for well over a century, the advantage of the publicly listed corporation for investing in innovation has been the separation of share ownership from managerial control, which is the fundamental role that the stock market plays. Contrary to conventional wisdom, the separation of ownership from control in the United States did not occur because companies had to raise money from the stock market to enable the growth of the firm. Rather the constraint on the growth of the firm was managerial capability (Lazonick 2014a). By listing on the stock market, the original owner-entrepreneurs and their financial supporters could exit from their investments, leaving salaried managers who had helped make the company a success in positions of strategic control. In the current environment, however, with their compensation mainly in the forms of stock options and stock awards, top executives of US companies have used their positions of strategic control to engage in massive value extraction that benefits themselves. Research by Lazonick and colleagues has shown that this value extraction undermines innovation.[6]

Given the interconnections of the US and Chinese economies, combined with the growing influence of the stock market in China, there is a distinct possibility that Chinese executives will be drawn to the highly financialized business model that now dominates in the United States. If China is to remain on the path to becoming an innovation nation, Chinese government policy must ensure that corporate profits are returned to those taxpayers, workers, and financiers who have *actually invested in productive capabilities* rather than to financial interests who simply buy and sell corporate shares. We should also note that China's most successful high-tech company, Huawei Technologies, is 100 percent employee-owned, and is not listed on any stock market.

[6] See the research at <www.theAIRnet.org>.

Huawei's history to the present exemplifies the roles of strategic control, organizational integration, and financial commitment in the operation and performance of a business enterprise.[7] In the years after the company's establishment in 1987, in the face of intense competition, Huawei's founder, Ren Zhengfei, pursued a strategy of making the company the leading indigenous producer of high-quality telecommunication switches. The company first manufactured equipment for service providers in China's rural areas, which were neglected by MNCs and SOEs. This strategy provided Huawei with not only its first customers but also opportunities for organizational learning. In an industry with about 200 competitors that were seeking to supply equipment to China's burgeoning telecommunications networks, Ren sought to attract, retain, and motivate personnel by sharing ownership of the company with employees. Known collectively as The Union, Huawei employees now own 98.6 percent of Huawei's shares and participate in the election of a two-tier governing board. Ren owns only 1.4 percent of the shares, but he remains chairman of the company and has veto power in decision-making. Employee ownership has provided an institutional foundation for the organizational integration of Huawei's personnel into the company's organizational learning processes while helping to finance the growth of the firm in the 1990s because employees took their raises in the form of shares. From the mid-1990s, Huawei was able to tap local government financing, and from the late 1990s the company secured loans from the China Development Bank to finance Huawei's customers to purchase its equipment. Because Huawei is not listed on a stock exchange, it can use its cash flow as it sees fit for developing the company's competitive capabilities without interference from outside shareholders.

Huawei's experience is a path-dependent process that is difficult to replicate, especially in an era in which speculative stock markets make founders of young companies billionaires when they do initial public offerings (IPOs) at home and abroad. While the ability to do a quick IPO certainly attracts venture capital to startups, it also creates the danger that financial interests both inside and outside the company might have more of an interest in extracting value than creating value and, as has been evident in the case of the United States, use the ideology of "maximizing shareholder value" to reap where they have not sown (Lazonick 2014a, b, c, and forthcoming).

As China seeks to determine the system of corporate governance that will enable it to transform itself into an innovation nation, it needs to understand the structures of strategic control that have enabled its most successful companies to achieve sustained competitive advantage. At the same time, it needs to recognize the ways in which, for publicly listed corporations in the West,

[7] Currently, Kaidong Feng, William Lazonick, and Yin Li are researching and writing a history of Huawei Technologies from the perspective of the theory of innovative enterprise.

and particularly in the United States, the ideology that companies should be run to "maximize shareholder value" has been destructive of innovative enterprise.[8] Rules and regulations concerning corporate governance must recognize that innovation is inherently an uncertain process in which decision-makers must have both the abilities and incentives to invest in learning organizations; hence the centrality of human capital for both strategy and learning in the innovation process. At the same time, a system of corporate governance that promotes innovation must ensure that strategic decision-makers within the firm recognize the multi-dimensional roles of the society in which the firm is embedded in contributing to the innovation process, and advocate for a system of corporate resource allocation that returns a fair share of corporate profits to its employees in the forms of employment stability and wage increases while, through the tax system, reproducing the ability of the state to support the next generation of innovation just as these companies have been supported in the past.

It may be that the leaders of some of the most successful Chinese companies that have been listed on the stock market might have found ways to deal with the problem of predatory shareholders. For example, China's Alibaba, whose $25 billion IPO on the New York Stock Exchange in September 2014 is the biggest IPO in history, has a partnership structure that, like Google and Facebook in Silicon Valley, permits widespread public shareholding while leaving strategic control over corporate allocation decisions in the hands of the founding partners. Although shareholder-value activists rail against dual shares and other modes of keeping outside shareholders at bay, it gives those insiders who exercise strategic control the power to ensure that value creation takes precedence over value extraction and that those who engage in value extraction are the parties who contributed to value creation. In shaping its own corporate governance institutions, the Chinese government should be aware of the ways in which the relation between value creation and value extraction can support or undermine the achievement of stable and equitable economic growth. It needs to be wary of those financial interests and business academics who argue that the highly financialized US new economy business model represents the foundation for a prosperous society.

Employment Institutions

Innovation results from a learning process. The learning that enables a company to transform technologies and access markets to generate higher quality,

[8] See Lazonick (2015). For Lazonick's recent critique of the shareholder-oriented principles of corporate governance promulgated by OECD, see his comments at <http://www.oecd.org/daf/ca/publiccommentsreceivedonthe2014reviewoftheoecdprinciplesofcorporategovernance.htm>.

lower cost products is both collective and cumulative, and hence organizational rather than individual. Learning is collective because large numbers of people in a hierarchical and functional division of labor must engage in interactive learning. Learning is also cumulative because the knowledge that the learning organization developed yesterday provides an indispensable foundation for what that organization is capable of learning today. The most intense and coordinated learning processes tend to occur within particular business enterprises, but collective and cumulative learning can also occur in industrial districts, of which China's most well-known is Zhongguancun in Beijing (Zhou 2008b). Our volume also studies other industrial districts such as mobile phones in Shenzhen (Chapter 10) and IC design in Shanghai (Chapter 9). The mobility of labor from established companies to startups and, in the case of expatriates, from foreign nations back to China are also important for launching or sustaining an innovation process. But that process will only be successful if key contributors remain committed to the collective and cumulative learning processes in the new business organizations that they join.

Indeed, the hypermobility of high-tech labor from one firm to another can undermine the organizational learning that is the essence of the innovation process. By the same token, inflexible employment relations, such as the "Iron Rice Bowl" that used to predominate in Chinese SOEs can severely restrict the ability of a business enterprise to adapt to changes in the market, technological, and competitive conditions that characterize its industry. The mobility of labor between firms and between government, business, and civil society organizations is essential to a society that promotes individual freedom. But an innovation nation requires employment relations that enable and motivate freely mobile labor to remain committed to organizational learning processes for sustained periods of time. At the same time, an innovation nation must have policies that ensure the preservation and, if possible, enhancement of human capital for workers with long years of work experience later in their careers. In recent research, mainly focused on the United States, Lazonick and his colleagues have highlighted the importance for an innovation nation of employment institutions that support "collective and cumulative careers" over the four or five decades that individuals seek to remain in the labor force as productive employees (Lazonick et al. 2014: 51–4).

Compared with Japan and the United States, China has a very fluid labor market, with the exception of employment in SOEs. The intense competition for highly skilled workers means most high-tech companies, even those owned by the state, typical suffer from hypermobility of labor (Zhou 2008b). An inability to retain employees long-term thus represents a significant challenge for Chinese high-tech enterprises to preserve and enhance their human capital. As China concludes its fourth decade of sustained growth since the Economic

Reforms of 1978, government policy at both the local and central levels should consider how employment institutions can be structured to support collective and cumulative careers in a world of highly mobile labor.

As labor costs have risen steeply in China since 2005, it is increasingly imperative for Chinese enterprises to retain their experienced workers. For example, China's most highly profiled nation-wide talent search program, One-Thousand-Talent, promises attractive funding for elite researchers from abroad, but lacks similar inducements that target experienced domestic talent (Qiu 2009). The stock-option culture that has become synonymous with Silicon Valley and to which many Chinese high-tech workers have been exposed during periods in the United States tends to foster individual labor mobility at the expense of collective and cumulative learning (Lazonick et al., 2014). As companies in China become enmeshed in this stock-based labor-market competition, they will need the government to step in to help stem the hypermobility of labor, by for example placing restrictions on exercising stock options or higher taxes on stock-based pay. Thus far, the Chinese government has paid scant attention to the formulation of institutional arrangements to encourage collective and cumulative learning or to discourage the hyper-mobility of high-tech labor.

To be sure, in certain periods and in certain sectors, a high level of labor mobility has been beneficial to industrial clusters. Many chapters of this volume document the emergence of industrial ecosystems, much of which were created through spin-off of key engineers or business managers from the existing enterprises. The ever-evolving divisions of labor and modularized value chains have been the prominent characteristics of the contemporary production system, knitted through inter-firm networks of supply chains and producer services. Since China hosts a significant portion of the global production system, working at different levels in ever-evolving value chains, it draws labor and other resources from within and outside China. The transfer of people with collective and cumulative learning from JVs to indigenous Chinese companies is evident in the automobile industry (Chapter 5), where experienced engineers from JVs have been the backbone for the newer indigenous automakers. Similar cases are also common in the machinery equipment industry (Chapter 4), with the chapter focusing on two companies founded by former employees of a related enterprise or a research institute.

International labor mobility has been important to the accumulation of productive capabilities in China. Returnees from the United States were China's first generation of venture capitalists, and they help to train the newer generation of venture capitalists from China (Chapter 3). Returnees have also powered China's IC foundry and design industries (Chapters 7 and 9). The positive effects of collective and cumulative learning within a specific industrial district are also evident in the cellphone industry. The cluster in

Shenzhen hosted a large number of *Shanzhai* cellphone makers during the 2G era. With the arrival of 3G technology, some of the cellphone makers used their previous expertise to set up companies that upgraded to become important brand-name smartphone makers. In general, at this stage of its development, in the highly dynamic electronics industries in which China is most integrated with global markets, Chinese enterprises tend to prioritize speed of adaptation to changing market opportunities over securing the long-term commitment of employees (Chapter 10). Yet, as a subject for further study, we would hypothesize that, given the collective and cumulative character of the innovation process, even in highly dynamic sectors, it will be those enterprises and those districts that are in the forefront in organizational learning that will be best positioned for competitive advantage.

The Chinese system of higher education has had to adapt to the rapidly changing requirements of China as an innovation nation. Since the 1990s, China's higher education system has expanded at an unprecedented speed. Enrollment increased by close to seven times between 1998 and 2010 (Li 2012). The *New York Times* reported that, in 1996, one in six Chinese 17 year olds graduated from high school, the same proportion as in the United States in 1919. By 2013, however, three in five Chinese 17 year olds graduated from high school, matching the United States in the 1950s. There are concerns about the preparedness of these high school and college graduates in the wake of such a dramatic surge, and whether the Chinese market can absorb the more highly educated population (Bradsher 2013). There has also been a longstanding criticism of the Chinese education system for its suppression of creativity and the failure of its curriculum to respond to changing socioeconomic needs (Zhao 2014). However, these are also common complaints in Japan and South Korea, but these limitations have not prevented those countries from transforming themselves into innovation nations. Since 2000, there has been a sharp increase in Chinese students studying abroad and since 2007 also a similar increase of overseas students returning to China (Zhou and Hsu 2010). Chinese nationals in 2010 represented 25 percent of foreign doctorate recipients in science and engineering fields from American universities, more than double the proportion from India, the second largest national group (National Science Foundation 2012). With growing and diversifying channels for obtaining a higher education, both in China and abroad, China's more highly educated workforce constitutes a foundation of human capital for China to transform into an innovation nation.

Investment Institutions

Innovation requires financial commitment, or what is sometimes called "patient capital." Money needs to be tied up in productive resources until a business enterprise can develop and utilize those resources to generate competitive products that can then generate financial returns. Since innovation is uncertain there is no guarantee that these returns will be forthcoming. Since innovation is collective, the productive resources in which this money is invested include human capital as well as physical capital. The investments in human capital will include the costs of training and retaining people, only some of which are captured in R&D expenditures.[9] And it is because of the cumulative character of the innovation process that financial *commitment* is needed; the innovative enterprise requires sustained access to finance until it can transform investments in productive resources into the high-quality, low-cost products that can generate financial returns.

Some of the chapters in this volume, including those on venture capital (Chapter 3), integrated circuit fabrication (Chapter 7), and solar panels offer substantial new information and insights (Chapter 12), but much research on the sources of financial commitment in Chinese development remains to be done. Clearly, by investing in a wide variety of infrastructure projects, the Chinese government has enabled the rapid economic growth of the past three and a half decades. But, as emphasized earlier, all of this investment in infrastructure would have simply entailed high fixed costs and would not have been translated into sustained economic growth without business enterprises that were able and willing to make use of these infrastructure investments by producing competitive products. For example, the massive investment in highways by the Chinese government would be largely wasted without the mass production and sale of cars and trucks in China to drive on them. But without these prior investments in highways, the usefulness of these cars and trucks to most (potential) buyers would be vastly reduced.

In the financing of infrastructure projects, China's high-speed rail system represents one of the nation's most impressive achievements. Massive loans from the state banks were used to construct over 13,000 km of high-speed railroad tracks from scratch in a decade. China Railway Corporation (formerly Ministry of Railway) imported and developed high-speed train technology and built the largest high-speed networks in the world, an impossible feat for any business enterprise that would need to generate profits over a reasonable time horizon to survive (Chapter 6).

[9] These investments in human capital represent fixed costs to the firm. On a company's financial statements, R&D expenditures appear as current costs in the profit and loss accounts. More generally, companies do not count investments in human capital as an asset on their balance sheets because the people in whom this human capital has been invested cannot be owned. Nevertheless, in terms of the economics of the innovation process, this human capital is an asset that constitutes a fixed cost that can only be transformed into revenues by the production and sale of competitive products.

The dedicated passenger rail system liberated the capacity of the overburdened existing rail system to better accommodate freight shipments, with the two systems working together vastly speeding up the movement of goods and people, and thus transforming China's economic geography. A World Bank study done in 2014 found that the number of China's high-speed rail passengers grew five-fold from 128 million in 2008 to 672 million in 2013, with 530 million of the passengers in 2013 using dedicated passenger lines. In 2013, the passenger-km on high-speed rail was slightly more than the rest of the world combined (Ollivier et al. 2012). Such infrastructure development, financed almost entirely by government loans, means that China Railway Corporation will be heavily in debt well into the future. But the infrastructure investment, which only the government could have financed, has provided Chinese households, businesses, and government agencies with a public good that has profound positive implications for China's urban and industrial development.

As a general rule, it can be argued that the Chinese state has taken control over investment and production through SOEs in those industrial sectors in which massive infrastructure investments must be made to enable Chinese development but in which rapid innovative responses to global competition are not required. Thus the Chinese government has taken direct control over investment in the steel industry, to the point where China now accounts for almost half of the world's crude steel production, to ensure that there would be sufficient indigenous capacity to support the development of the construction and transportation (particularly rail and automobile) industries. In the ICT industries, the Chinese government used SOEs to fund and operate the service-provider networks but encouraged competition among non-SOEs to develop indigenous ICT equipment. In the critical integrated-circuit fabrication industry, the Chinese government initially favored SOEs but since about 2000 came to see that non-SOEs, formed by returnees, of which SMIC (Semiconductor Manufacturing International Corporation) and Grace are the most important examples, were much more capable of making the strategic decisions necessary to remain close to the technological frontier in an industry in which rapid change in process technology is the norm. Yet, as shown by Yin Li in Chapter 7, the extremely high fixed costs of investing in new-generation IC foundries has meant that the Chinese government has been compelled to help finance these non-SOEs if it wants China to retain the possibility of becoming a world leader in this crucial industry.

We have already discussed the relatively new role of venture capital in the Chinese economy, as well as the potential for this source of finance to become "impatient" if it looks for quick returns on a speculative stock market in advance of generating competitive products. The development of venture capital in China has complemented a weak apparatus of state financial institutions for supporting startups. Given that this venture-capital system was established only recently, most of China's successful technology enterprises to date have not primarily relied on this mode of finance. A prime example, already discussed, is Huawei Technologies with its employee stock ownership plan (ESOP), put in place at the beginning of the 1990s. Huawei's shares have never been listed on the stock market, and the company does not pay dividends to its shareholders. Indeed, one can argue that by protecting the company from the value-extracting pressures of a public stock-market listing, Huawei's ESOP has funded the growth of the firm. Once it was clear in the late 1990s that Huawei had emerged as the strongest indigenous innovator among Chinese communication equipment firms, the China Development Bank extended large low-cost loans to Huawei to help fund its expansion as a global company (*Bloomberg News*, 2011). Despite its exclusion from the US market on national security grounds, in 2013 65 percent of Huawei's markets were outside China (although from 2012 to 2013 China was Huawei's fastest growing market) (*Huawei Annual Report*, 2013: 23). In 2014 Huawei surpassed Sweden's Ericsson as the world's larger vendor of equipment to carrier networks, the most technologically sophisticated segment of the communication equipment industry.

In principle, the stock market can serve as a source of finance for investment in productive capabilities. In practice, however, it has become in many places a mode of extracting value from the productive economy. If China wants to become an innovation nation, it needs to learn the lessons of Huawei in the ongoing structuring of its investment institutions, and recognize the significant dangers of a stock-market-oriented economy in empowering financial interests over productive interests, with a failure to become an innovation nation as the likely result. Chinese companies that list on the stock market, as high-tech companies with growth potential, will need to find ways the ensure that the listing supports rather than undermines the value-creation process, with government policies that regulate financial markets supporting this developmental objective.

Managing Indigenous Innovation in a Global Production System

A crucial difference between the rapid industrial growth of China and the earlier transformations of Japan and the United States is the ever-tightening articulation of global production networks (GPNs) today. In the post-World War II decades, national production systems were still the norm, so the rise of leading national companies usually meant the corresponding rise of the entire supply chain within the country. From the late 1960s, however, a transformation in vertical production relations began to take place, with

GPNs becoming the norm. China's rapid growth from the 1980s coincided with the expansion of GPNs, especially in the ICT industries, with Chinese producers becoming integral to their development.

A GPN is "the globally organized nexus of interconnected functions and operations by firms and non-firm institutions through which goods and services are produced and distributed" (Coe et al. 2004: 471). Under such a system, production processes have become modularized and distributed across different countries. Large transnational corporations, usually located in advanced countries, which Ernst and Kim call "flagship" corporations, take on the strategic, technological, and organizational leadership in GPNs (ibid. 2002). As they state: "The main purpose of these networks is to provide the flagship with quick and low-cost access to resources, capabilities and knowledge that are complementary to its core competencies" (Ernst and Kim 2002: 1420). The 2013 World Investment Report estimated that some 80 percent of international trade is organized through GPNs (Yeung and Coe 2015).

GPNs provide China with opportunities to manufacture for the world market, as Chinese companies or foreign subsidiaries become subcontractors of leading corporations in the advanced economies. But GPNs also make it harder to achieve strategic control of technology development because of the interdependent character of the networks. For example, China policymakers initially devised TMFT strategy in the automobile industry (Chapter 5) and integrated-circuit industry (Chapter 7) to transfer technology from abroad to localized production. But the modularization of global production means that control over the core intellectual property rights or crucial technology components can be separated from the particular stages of manufacturing that take place in China. GPNs have also created higher entry barriers for certain segments of production. For example, several global giants in the semiconductor fabrication industry supply the entire world market. Their economies of scale and control over technology standards make it very difficult for latecomers to break in. Examples include Qualcomm and Samsung in IC chips for cellphones, and TSMC and UMC in IC foundries. To compete with these established players requires such a massive investment that even China has been unable to mount it (Chapter 7). China's desire for indigenous innovation is thus constantly bumping into the reality that key intellectual property or crucial product components are controlled by enterprises in more advanced countries that participate in the GPNs.

Given the modularization of production, Breznitz and Murphree (2011) describe China's strategy as "the Run of Red Queen" in which Chinese enterprises shun novel product innovation, but specialize in competitive second-generation products and processes that entail incremental innovation. They criticize the techno-fetishism that underestimates the frequency and cumulative effect of the innovation that routinely occurs in Chinese enterprises. We agree with this perspective on the significance of cumulative incremental innovation in China's development thus far. But it is not sufficient to build China as an innovation nation. Both Taiwan and South Korea entered GPNs in the low-end segments in ICT industry, but have now moved to the higher value-added segments of GPNs. The success of such a transformation is contingent upon strategic control of the enterprises within a national economy. If China is to become an innovation nation, Chinese enterprises will have to work with GPNs while avoiding being captured by the networks with a lack of strategic control over the paths of China's technology development.

Several chapters in this volume examine how Chinese enterprises work within GPNs in ways that enhance their strategic control in global production. One way is through control of the system integration of technology. The rapidly growing size of the Chinese market has become an irresistible magnet for foreign companies to engage in FDI in most technology sectors, drawing technology suppliers from all over the world and creating a technological landscape that is notably more plural than those of its counterparts in East Asia. China's high-speed rail network, for example, has transferred technology from Germany, Spain, Japan, and France. Its automobile industry has a prolific array of Japanese, American, German, and Korean models, all produced through JVs with SOEs. Its 3G mobile phone network operates on three global standards: WCDMA, CDMA2000, and indigenous TDS-CDMA. A similar coexistence of a variety of technology platforms can also be found in clean-tech energy industries such as wind turbines and solar panels (Chapters 11 and 12). Chinese companies, therefore, are now well positioned to learn, compare, and select technology best suited for their markets. By moving up GPNs to become systems integrators, domestic companies can gain more in-depth understanding of different foreign technologies and, based on such understanding, can provide innovative products.

The second approach is to achieve strategic control of the technology-transfer processes. Here the ability of Chinese enterprises varies among industrial sectors. In railroads, the state monopolizes the planning, construction, operation, and regulation of the entire rail system. Because China was practically the only viable major market for high-speed technology in the 2000s, China's Ministry of Railway had tremendous leverage in negotiating favorable technology-transfer agreements with foreign technology providers, thus permitting the China Railway Corporation to integrate foreign technology and develop a distinctive Chinese high-speed rail system that it is now even attempting to export to other countries (Chapter 6).

The success of state-directed technology transfer in the case of high-speed rail is, however, more an exception than a rule. In the automobile and IC industries, the Chinese government brokered marriages between MNCs and Chinese SOEs in the 1980s and 1990s under its TMFT policy. These JVs provided China with the technological experience that subsequently made it possible to establish completely indigenous enterprises in which strategic control over all technology and market decisions resides in China. With this strategic control, indigenous Chinese companies in the automobile industry have continued to seek out foreign technology through licensing as well as through the acquisition of foreign auto companies (Chapter 5). We see similar strategies of technology acquisition in renewable energy sectors (Chapters 11 and 12).

The third way to maintain strategic control is through strengthening internal R&D. In the ICT sector where GPNs are the most well developed, successful companies such as Huawei and ZTE have relied mostly on internal R&D development while referencing foreign technology trends. National security concerns from the West have made it difficult for Chinese telecommunication companies to gain access to advanced technology (Chapter 8). Internal R&D has also been very important for machine equipment sectors where non-standardized interaction with customers is the technological norm. In the cellphone sector, the technology improvements behind smartphones have been accomplished through a combination of top-down governmental R&D on the Chinese national 3G standards and bottom-up incremental improvements among a large number of firms in the industrial ecosystem. While it continues to be necessary to collaborate with foreign companies, studies in this volume show that indigenous innovation can take place only when the Chinese companies have strong internal R&D capabilities.

Last but not least, we found that enterprises can strengthen their strategic control by targeting China's domestic market. Zhou's work on China's IT industry in the earlier 2000s suggests that the synergy between China's export production capacity and domestic market growth has underpinned the success of China's most competitive technology firms (Zhou 2008a, b). This remains the case. In particular, the Chinese market provides buoyant demand among low- to middle-class customers or price-sensitive enterprise clients. They cannot afford best-quality foreign products, but still desire reliable technology. Chapter 9 on IC design suggests that the competitive edge of China's IC designers has been in the cost of their products compared with more expensive counterparts from foreign companies, Chapter 10 describes Chinese cellphone companies as specializing in "good enough" innovation.

However, providing lower cost products cannot be a sustainable strategy without simultaneous improvements in quality. As mentioned earlier in this introduction, Japanese carmakers were able to transform Japanese-made cars from low to high quality within two decades. China's middle markets are changing rapidly with rising incomes and higher expectations for product quality. Unless Chinese firms are able to move with the shifts in demand, they are likely to lose their markets to foreign competitors that have traditionally occupied the high-end market segments. Xiaomi, popularly known as China's

"Apple," is able to produce cellphones with leading technical specifications and appealing designs, while offering prices that are less than half the mainstream global brands. This combined low-cost and good-quality strategy enabled Xiaomi to emerge as China's leading cellphone retailer in 2014, only four years after its establishment. As China's domestic market has grown in size and sophistication, only those Chinese companies that are capable of providing higher quality products will be able to survive in domestic and global competition.

In sum, whether China can become an innovation nation depends on whether it can create a governance regime that ensures the complementarity of government and business investment in productive capabilities—a collaboration that inevitably means a collaborative division of labor in the exercise of strategic control. This strategic control must be maintained and extended while Chinese companies remain integrally involved in the global economy in GPNs, which can be achieved through system integration, control of technology-transfer, and targeting distinctive domestic markets. China's emergence as an innovation nation is also contingent on employment institutions that can facilitate collective and cumulative learning within enterprises and industrial districts. And last but hardly least, China has to develop and sustain investment institutions that provide diverse but committed patient capital that supports and rewards investment in value-creating productive capabilities rather than value-extracting financial engineering.

A Brief Guide to the Chapters in This Volume

Chapter 2 traces the evolution of Chinese government policies on technological innovation. We are currently at a stage where the state is asserting its central role in promoting technological innovation following the stages of labor-intensive export-promotion and "TMFT" style import-substitution policies. The newly installed "indigenous innovation" policy employs a growing list of flexible instruments to spur innovation including direct R&D support, preferential finance and tax policies, public procurement, industrial promotion, and domestic technology standards. Foreign firms and local governments are also playing more prominent roles in bringing technological dynamism to the system.

Chapter 3 analyzes the venture capital (VC) industry in China. As a new and alternative instrument for financing innovation, distinct from state-owned or state-controlled financial institutions, the venture capital industry has supported China's new generation of ambitious entrepreneurs, and is especially active in the ICT sectors. Institutional changes in China's financial markets, especially the launch of ChiNext, the Chinese-style NASDAQ, have encouraged the explosive growth of domestic VC funds and firms. With stock markets around the world, including those in Hong Kong and China, catering to technology enterprises, VC has become a main form of financing grassroots innovation by non-state firms. However, along with the growth of VC comes a growing risk that the processes of new-firm formation and growth will become infected by a combination of American-style financialization and Chinese-style corruption.

Chapters 4 to 6 examine China's machine-based industries, as represented by mechanical engineering, automobiles, and high-speed rail. Unlike the ICT sectors, the markets for these industries are largely within China, and global integration is not as developed as in ICT. Chapter 4 focuses on the highly heterogeneous machine equipment industry. Echoing the successful German model, innovation in Chinese mechanical engineering relies on firms' internal resources, close interaction with customers, and internal learning and experimenting. Chinese firms have been engaged in indigenous innovation through either low-cost innovation by incremental improvements achieved through working closely with customers or, in contrast, deep R&D and international learning to create radical innovation.

Chapters 5 and 6 focus on automobile manufacturing and high-speed railroads in China. They highlight the advantages and limitations of the Chinese state in engineering the catch-up processes with the advanced economies. The contrasting practices of foreign technology-transfer are especially illuminating.

The automobile industry is often cited as a negative example of "TMFT." The technological capability of Chinese local carmakers developed remarkably slowly in the 1980s and 1990s, during which the government established joint ventures between Chinese and foreign firms and gave them protected market niches to encourage technology transfer. However, without the strategic control of the joint ventures, the Chinese partners in the JVs became passive players. The situation only changed in the late 1990s when indigenous companies were set up outside the Chinese governmental plans. These indigenous companies have stressed system integration and organizational learning in generating new products and processes.

Learning the lessons from the automobile industry, the high-speed rail industry took a different approach in which the China Ministry of Railway became the sole party in technology-transfer negotiations and the main systems integrator. It was able to win favorable terms for technology transfer and subsequently emphasized technology absorption and development. One cannot, however, conclude that more centralized organization works better in technology-transfer, as the success of China high-speed rail has owed greatly to the natural monopoly of this industry and the lack of foreign competition. In addition, Chinese state control of land procurement, the tradition of rail travel, China's density of population, and a long accumulation of technological expertise in the industry all contributed to the uniqueness of China's technological achievements in high-speed rail. While other sectors cannot be expected to emulate the top-down path of high-speed rail, the sector underscores the important roles of strategic control of domestic enterprises and long-term financial support from the state in the process of indigenous innovation.

Chapters 7 to 10 examine industrial dynamics in different subsectors of ICT. Chapter 7 analyzes the development of integrated-circuit (IC) foundry manufacturing. Chinese state policies in ICs parallel those in the automobile industry since both have been considered strategic industries, targeted by TMFT strategies. The state was directly involved in establishing domestic enterprises or joint ventures to jump-start these industries. The IC foundry industry however is different in that it is an integral part of the global production chain. Foundry production requires enormous financial commitment in the presence of rapid technological change that can make multibillion-dollar plants of relatively recent vintage obsolete. It has been exceptionally difficult for Chinese newcomers to break into the IC foundry industry even though China is the world's largest market for IC chips. As in automobiles, Chinese SOEs and joint ventures have been unable to become innovators. But newer firms, most notably SMIC and Grace, have been more successful as autonomous business enterprises with access to returnees, global technology, and capital raised though foreign stock markets. Yet the growth and catch-up of these foundries has been hindered by the localized, fragmented industrial financing scheme in China. The Chinese government has in 2015 started to consolidate financial support for the leading Chinese firms in this industry.

Chapter 8 examines the catch-up process of China's telecommunication equipment industry, with a focus on three Chinese firms: Huawei, ZTE, and Datang, all of which have become global leaders in the industry. Differing from the automobile and IC sectors, the Chinese government has emphasized cultivating the demand for domestically produced products as well as encouraging competition through restructuring the telecom service providers. Firms in this industry have stressed internal R&D in innovation rather than relying on technology transfer from foreign firms.

Chapter 9 analyzes the development of China's integrated-circuit design industry. IC design used to be carried out by the state IC firms. The slow development of China's IC industry has meant similar underdevelopment in the subsector of IC design. The vertically specialized model of the semiconductor industry, with the emergence of pure-play foundries, has created the possibility of growth in IC fabless design firms. Also enabling the growth of the IC design industry in China has been the increasing number of returnees who have gained education and work experience abroad. The IC design industry has also benefitted from the diverse demand created by rising Chinese ICT industries. Overall, the IC design sector demonstrates the significance of the industrial ecosystem in the catch-up process.

Chapter 10 looks into China's mobile phone industry, particularly its shift to smartphones. China's domestic cellphone makers in the 2G era had been dominated by many small *Shanzhai* phone makers based on turnkey solutions provided by Taiwan's Mediatek. But the shift to 3G has led to consolidation of the industry into a few indigenous brands. The chapter argues that Chinese domestic phone manufacturers focused on good-enough innovation because of the vast and diversified demand of price-sensitive consumers. The transition to smartphones benefitted from existing expertise and supply chains, and innovation at the platform, middleware, and application levels. The new industrial ecosystem under 3G favors larger smartphone vendors. The cellphone sector is a prime case of cluster-based bottom-up incremental innovation, similar to the IC design sectors in Chapter 9.

Chapters 11 and 12 study China's newly emerged clean-technology industries. Chapter 11 demonstrates the importance of supporting government policies and an open global market in helping the growth and technology innovation of the Chinese wind power industry. The chapter documents the technological progress of Chinese domestic wind power manufacturers in wind turbine size, patents, and costs. It provides clear evidence that governmental support in the forms of R&D in state institutions, state-funded demonstration programs, pricing policies, and industrial and trade policies have all contributed to the rapid rise of China's wind power industry. Unlike the automobile and IC sectors, Chinese local governments were not directly involved in setting up joint-venture enterprises. Instead, domestic firms have autonomously utilized licensing, merger and acquisition, or joint ventures, among other methods, to develop their technological competence.

Chapter 12 examines the rise of China's solar photovoltaic (PV) industry. The unique aspect of the industry is that its market demand initially came almost exclusively from abroad, especially European markets, in which governmental subsidies were instrumental in the initial stage of this industry. Its rapid growth in China, however, is the outcome of coordinated efforts of entrepreneurs and national and local governments. The state provided R&D funding, cheap industrial land, "patient capital" that has enabled enterprises to quickly scale up, and policies that have helped to cultivate the domestic market when the foreign markets have slackened. The state assisted the enterprises to overcome the technology, market, and competitive uncertainties in this new industry. The solar industry also benefitted from the industrial ecosystem made up of R&D in research institutes both in China and abroad, indigenous equipment manufacturers, and upstream polysilicon producers.

Acknowledgment

Partial funding for this chapter comes from the Ford Foundation under the project on Financial Institutions for Innovation and Development, directed by William Lazonick.

References

Barro, R. J., and J-W. Lee (2000), *International Data on Educational Attainment: Updates and Implications*, Harvard Center for International Development Working Paper, 42, Appendix Data Files. Cambridge, MA: Harvard University. <http://www.cid.harvard.edu/ciddata/ciddata.html>.

Bloomberg News (2011), "Huawei's $30 Billion China Credit Opens Doors in Brazil, Mexico," Apr. 24, <http://www.bloomberg.com/news/articles/2011-04-25/huawei-counts-on-30-billion-china-credit-to-open-doors-in-brazil-mexico>.

Bradsher, K. (2013), "Next Made-in-China Boom: College Graduates," *New York Times*, Jan. 16, <http://www.nytimes.com/2013/01/17/business/chinas-ambitious-goal-for-boom-in-college-graduates.html?pagewanted=all&_r=0>.

Bremmer, I. (2009), "State Capitalism Comes of Age: The End of the Free Market?" *Foreign Affairs* (May–June), <http://www.foreignaffairs.com/articles/64948/ian-bremmer/state-capitalism-comes-of-age>.

Breznitz, D., and M. Murphree (2011), *Run of the Red Queen: Government, Innovation, Globalization and Economic Growth in China*. New Haven: Yale University Press.

China Auto Web (2013), "Chinese Auto Exports Dropped Below 1 Million in 2013," <http://chinaautoweb.com/2014/01/chinese-auto-exports-dropped-below-1-million-in-2013>.

China State Council (2006), *The National Medium- and Long-Term Program for Science and Technology Development (2006–2020)*. Beijing: China State Council.

Coe, N., M. Martin, N. Hess, H. W.-C. Yeung, D. Dicken, and J. Henderson (2004), "'Globalizing' Regional Development: A Global Production Networks Perspective," *Transactions of the Institute British Geographers*, 29(4): 468–84.

Ernst, D., and L. Kim, (2002), "Global Production Networks, Knowledge Diffusion, and Local Capability Formation," *Research Policy*, 31: 1417–29.

Hopkins, M., and W. Lazonick (2014), "Who Invests in the High-Tech Knowledge Base?" *Institute for New Economic Thinking Working Group on the Political Economy of Distribution Working Paper*, 6, Oct., <ineteconomics.org/research-programs/political-economy-distribution/papers/who-invests-high-tech-knowledge-base>.

Huawei Investment & Holding Co. Ltd. (2013), 2013 Annual Report, at http://carrier.huawei.com/en/about-huawei/corporate-info/annual-report/index.htm.

Lardy, N. (2014), *Markets over Mao: The Rise of Private Business in China*. Washington, DC: Peterson Institute for International Economics, U.S.

Lazonick, W. (2007), "Varieties of Capitalism and Innovative Enterprise," *Comparative Social Research*, 24: 21–69.

Lazonick, W. (2009a), Sustainable Prosperity in the New Economy? Business Organization and High-tech Employment in the United States. Kalamazoo, MI: Upjohn Institute for Employment Research.

Lazonick, W. (2009b), "The New Economy Business Model and the Crisis of US Capitalism," *Capitalism and Society*, 4(2): article 4.

Lazonick, W. (2010a), "Innovative Business Models and Varieties of Capitalism: Financialization of the US Corporation," *Business History Review*, 84(4): 675–702.

Lazonick, W. (2010b), "The Chandlerian Corporation and the Theory of Innovative Enterprise," *Industrial and Corporate Change*, 19(2): 317–49.

Lazonick, W. (2013), "The Financialization of the U.S. Corporation: What has been Lost, and How it can be Regained," *Seattle University Law Review*, 36: 857–909.

Lazonick, W. (2014a), "Innovative Enterprise and Shareholder Value," *Law and Financial Markets Review*, 8(1): 52–64.

Lazonick, W. (2014b), "Profits without Prosperity: Stock Buybacks Manipulate the Market and Leave Most Americans Worse Off," *Harvard Business Review* (Sept.): 46–55.

Lazonick, W. (2014c), *The Theory of Innovative Enterprise: A Foundation of Economic Analysis*, AIR Working Paper, 13-02/01 Feb. (revised), <www.theairnet.org/files/research/WorkingPapers/Lazonick_InnovativeEnterprise_AIR-WP13.0501.pdf>.

Lazonick, W. (2015), *Stock Buybacks: From Retain-and-Reinvest to Downsize-and-Distribute*, Washington, DC: Brookings Institution Center for Effective Public Management, White Paper, Apr. <http://www.brookings.edu/research/papers/2015/04/17-stock-buybacks-lazonick>.

Lazonick, W. (forthcoming), "Labor in the Twenty-First Century: The Top 0.1% and the Disappearing Middle Class," in Christian E. Weller (ed.), *Inequality, Uncertainty, and Opportunity: The Varied and Growing Role of Finance in Labor Relations*. Labor and Employment Relations Association.

Lazonick, W., and Ö. Tulum (2011), "US Biopharmaceutical Finance and the Sustainability of the Biotech Business Model," *Research Policy*, 40(9): 1170–87.

Lazonick, W., P. Moss, H. Salzman, and Ö. Tulum (2014), *Skill Development and Sustainable Prosperity: Collective and Cumulative Careers versus Skill-Biased Technical Change*, Institute for New Economic Thinking Working Group on the Political Economy of Distribution Working Paper, 7: 51–4, <ineteconomics.org/research-programs/political-economy-distribution/papers/skill-development-and-sustainable-prosperity>.

Li, H. (2012), "How Will China Become World Technological Power?" *Xinhua*, Sept. 24, <http://news.xinhuanet.com/english/indepth/2012-09/24/c_131870216.htm>.

National Science Foundation (2012), chapter 2, "Higher Education in Science and Engineering," in *Science and Engineering Indicators 2012*, <http://www.nsf.gov/statistics/seind12/c2/c2h.htm>.

Organisation Internationale des Constructeurs d'Automobiles (OICA 2013), "Production Statistics" at <http://www.oica.net/category/production-statistics/2013-statistics>, accessed Sept. 2015.

Ollivier, G., R. Bullock, J. Ying, and N. Zhou (2012), *High-Speed Railways in China: A Look at Traffic*. China Transport Topics, 11. Washington, DC: World Bank Group. Dec. 1, <http://documents.worldbank.org/curated/en/2014/12/23031378/high-speed-railways-china-look-traffic>.

Pan, F., and D. Brooker (2014), "Going Global? Examining the Geography of Chinese Firms' Overseas Listings on International Stock Exchanges," *Geoforum*, 52: 1–11.

Qiu, J. (2009), "China Targets Top Talent from Overseas," *Nature*, Jan. 28, <http://www.nature.com/news/2009/090128/full/457522a.html>.

Sanderson, H., and M. Forsythe (2013), *China's Superbank: Debt, Oil, and Influence*. New York: Bloomberg Press.

World Steel Association (2014), "Crude Steel Production in Major Producing Countries and Regions in 2014 (in Million Metric Tons)," *Statista*, <http://www.statista.com/statistics/267263/world-crude-steel-production-by-region>.

Yeung, H. W-C., and N. Coe (2015), "Toward a Dynamic Theory of Global Production Networks," *Economic Geography*, 91(1): 29–58.

Ying, T. (2014), "Volvo Cars Billionaire Owner Revamps Chinese Brand Geely," *Bloomberg News*, Dec. 15, <http://www.bloomberg.com/news/articles/2014-12-16/volvo-cars-billionaire-owner-revamps-chinese-brand-geely>.

Zhao, Y. (2014), *Who's Afraid of the Big Bad Dragon: Why China has the Best (and Worst) Education System in the World*, San Francisco: Jossey-Bass.

Zhou, Y. (2008a), "Synchronizing Export Orientation with Import Substitution: Creating Competitive Indigenous High-Tech Companies in China," *World Development*, 36(11): 2353–70.

Zhou, Y. (2008b), *The Inside Story of China's High-Tech Industry: Making Silicon Valley in Beijing*. New York: Rowman & Littlefield.

Zhou, Y., and J. Hsu, (2010), "Divergent Engagements: Comparing the Roles and Strategies of Taiwan and Mainland Chinese Returnee Entrepreneurs in the IT Industry," *Global Network*, 11(3): 398–419.

2

Evolution of Chinese State Policies on Innovation

Yu Zhou and Xielin Liu

Since the founding of the People's Republic of China in 1949, the Chinese state has driven and shaped innovation in China. This function of the state is not necessarily unique to China; in fact, some governmental support of innovation has been integral to the development of many other advanced countries (Chang 2002; Lazonick 2011; Castells 1989; Block and Keller 2011). Observers of Chinese innovation often disagree on the effects of such state interventions. Some believe that its national strategic planning has given China such an advantage in innovation that their techno-nationalism will succeed at the expense of trade partners and continued technological exchange (McGregor 2010; D'Aveni 2012). Others, meanwhile, view state control as an impediment to China's long-term progress. They argue that the state monopoly over technology and finance has produced and perpetuated the rampant corruption and inefficiencies in the public sector and state-owned enterprises. From this perspective, governmental meddling in innovation has also distorted the scientific and business climate, which has led to a poorly enforced IPR regime, a stifled educational system, discrimination against non-state companies, etc. Such critics argue that state interventions have hurt national innovation (Pei 2006; Huang 2008).

In reality, the Chinese state is by no means a monolithic, static entity with a consistent vision and unchallenged control over China's economy. Since the PRC began to embrace technological change in the 1950s, numerous shifts in policy and culture, both domestic and international, have caused profound changes in China's innovation system. The evolution of these policies can only be adequately understood within specific historical contexts. International pressure and domestic criticism have both been instrumental in leading to policy paradigm shifts. The goal of this chapter is to review the evolution of China's innovation policies and practices, both in terms of its compelling legacy, striking advancements, and remaining problems. The chapter has a particular focus on two interfaces: China's integration into the global economy, and the relationship between the Chinese state sector and non-state enterprises. In particular, we will discuss the underlying forces and impacts of three highly controversial strategies, namely "techno-nationalism," "trading market for technology," and "indigenous innovation."

Historical Legacies of the Centralized Science and Technology System

The term "national innovation system" was only invented in the late 1980s in the West, and it was not used in China until the 1990s. Its earlier proxy was the "Science and Technology (S&T) System," the foundation of which was laid out in the 1950s. Although dramatic changes have occurred since the 1980s, certain objectives and organizational forms left a legacy that not only is visible today but also continues to provoke debates over the role of the state in innovation.

Mao Era: Defense-Led Techno-Nationalism, 1950–1980

Shortly after the founding of the PRC, the Korean War (1951–3) gave China a rude awakening: China's technological backwardness could seriously threaten the survival of the new regime (Feigenbaum 2003; Nie 1989).[1] In 1956, Marshal Nie Rongzhen (聂荣臻) of the People's Liberation Army was commissioned by China's State Council and Premier Zhou Enlai (周恩来) to develop a "Science and Technology (S&T) System." His institutional vision profoundly shaped China's S&T sector. In the midst of the Cold War, and against the backdrop of an impoverished nation emerging from decades of war and natural calamities, China's S&T system bore the hallmarks of militarism and central planning adopted from the Soviet Union. This system was characterized by a centralized bureaucratic hierarchy, task-led approach, defense-orientation, and technological indigenization—all products of Mao-era communist ideology and Cold War geopolitics (Baark 2010; Cao 2004; Naughton and Segal 2003; Segal 2003; Suttmeier and Cao 2004, Suttmeier 1997). The S&T system has guided China to impressive advances in selected heavy industries and the defense sector. Yet China's backwardness in civilian technology in the late 1970s also resulted from the same system, which subordinated academic and technological development to the instrumental goals of the state. By

[1] This part of the chapter draws heavily on Feigenbaum's research (2003) and the collection of memoirs by China's prominent scientists associated with the defense industry, among other sources.

prioritizing heavy industry and defense under a centralized state, China undermined the creativity and competitiveness of the larger economy.

The guiding principle of China's S&T development in the 1950s was the so-called "task-led approach." In Marshal Nie Rongzhen's (1989) memoirs, he recalls a vigorous 1956 debate among S&T policymakers to determine the future of the PRC's approach to innovation (52). One option, the conventional model of academic disciplinary growth, would allow each discipline in engineering and sciences to identify new fields of research and then pursue development as conventional independent university researchers might. The development of each discipline would then collectively push forward the development of the S&T sector. The other option, the "task-led model," gave the state a more authoritarian role in development: this way, the needs of the state would determine the research and development projects in various sectors. For instance, the state would commission researchers' civilian and defense projects, and then different research units would collaborate and cooperate to realize the state's objectives. Nie argued that since China had extremely weak science and technology infrastructure, a task-led approach would ensure the concentration and coordination of limited national resources to meet urgent national needs. However, this organizational structure—originally formulated to address the exceptional circumstances of the early days of the PRC—eventually strengthened and expanded. It turned out that the "task-led approach" was impeccably compatible with China's central planning regime, and so over the past half-century it grew into the entrenched, top-down form of S&T governance that is still operational today.

The top administrative layer of China's S&T system during the Mao era consisted of several major governmental bodies with specific sectorial or disciplinary mandates (Figure 2.1): the Central Commission of National

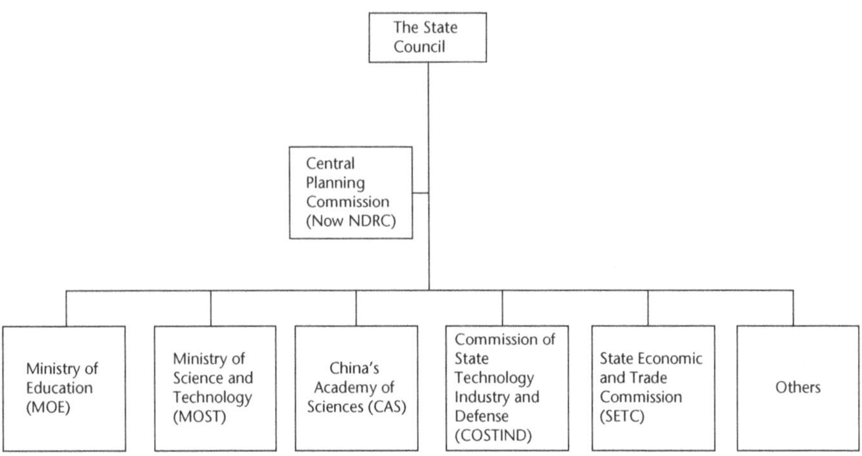

Figure 2.1. Main Administrative Bodies of China's S&T System

Planning, now named National Development and Reform Commission (NDRC); the State Economic and Trade Commission (SETC); the Ministry of Education (MOE), formerly the State Education Committee; the Ministry of Science and Technology (MOST), formerly the State Science and Technology Committee (SSTC); the Chinese Academy of Sciences (CAS); the Commission of State Technology Industry and Defense (COSTIND), which was abolished in 1998 and merged into other agencies. Major industry groups also had their own national level-ministries, including the Ministries of Light Industry, Nuclear Industry, Railways, and Industry and Information Technology (which absorbed a large portion of COSTIND). Each industrial ministry had its own research institutes and projects. Together these administrative bodies formulated and implemented national S&T programs and policies.

Beyond these policymaking bodies, the main players in China's national S&T system have been state research institutes, universities, and enterprises. The roles of these entities have changed considerably over the decades. From the 1950s to the 1980s, government research institutes (GRIs) were the main bodies that carried out research and development (R&D) work, and such institutes were established both at the national and the provincial levels. The most important of these were at the national level, notably the CAS, research institutes affiliated with national ministries and large research universities such as Peking University and Tsinghua University. There were also hundreds of industrial research institutes in different provinces, which were organized under a wide range of industrial ministries that focused on applied research and developmental tasks. For example, there were provincial branches of the Academy of Agriculture Science, and the CAS established several regional branches in Beijing, Shanghai, and other major Chinese cities. These regional GRIs conducted R&D tasks commissioned by the regional government (Liu and Lundin 2006). Most universities at that time were not involved in research, except Peking, Tsinghua, and a few other large institutions. Many specialized universities focused on industry-specific technical education, such as textile industries, railroad, telecommunication, metallurgy, and printing.

The role played by industrial enterprises in the innovation system was limited, since they functioned mainly as manufacturing and/or sales units. State-owned enterprises (SOEs) generally were not under competitive pressure, so they had little incentive to upgrade their technology. Except for a few large SOEs with in-house R&D laboratories that focused specifically on technology implementation, most enterprises had virtually no R&D functions. Instead, the government was in charge of directing new technology to enterprises, which were largely the passive recipients. Within the GRIs, research funding was allocated hierarchically and determined by the perceived needs of the state. Individual initiatives from scientists and other sectors of society were rarely encouraged. If these fell outside of the mandated needs of the state, they were often neglected entirely.

Another key characteristic of China's S&T system during this period was its emphasis on active technological indigenization, or tracking global trends and adapting other countries' technologies for its own uses. From the 1950s through the 1970s, the imported technologies from the former Soviet Union, Germany, and Japan laid the foundation for the Chinese chemical, automobile, steel, and textile industries, among others. At the time, China was internationally isolated due to the Cold War. The abrupt cutting off of foreign aid from the Soviet Union in the 1960s, and the shortage of foreign exchange in China meant that such imported technology was often outdated and extremely expensive. The GRIs were thus tasked with monitoring the technological trends of advanced countries and replicating the most important developments whenever possible. Foreign intellectual property rights protection were not a consideration, and Chinese researchers rely largely on reverse-engineering imported technology to achieve incremental innovations.

During this period, the dominant philosophy of China's S&T system was, by default, an extreme version of techno-nationalism. Keller and Samuels (2003) attribute the techno-nationalist policy to the unwillingness to open the domestic market to direct foreign investment, out of a concern that more mature foreign-based firms and technologies would snuff out nascent domestic ones. They explain, "Techno-nationalists believe that a domestic economy can be mature, and the nation secure, only if it exerts substantial control over the generation of knowledge and the standards by which design and manufacture are undertaken" (2003: 9). Since China's market was virtually closed off to the outside world, and since the state had near total control of S&T projects, S&T development served little function beyond fulfilling the obligations to the state.

While some scholars argued that China's centralized and hierarchical S&T system was more flexible and collaborative than its Soviet predecessor (Feigenbaum 2003), it was still fundamentally flawed (Suttmeier and Cao 2004). At its best, the Chinese state excelled at mobilizing national resources to execute large and complex strategic tasks such as building nuclear bombs or developing a space program. Its success was especially striking in areas without open market competition, such as the defense sector. Similarly, the state's impressive engineering and organizational capacity has been apparent in more recent mega-projects for other monopolized civilian sectors, including the development of high-speed rail, telecommunication infrastructure, and energy infrastructure. This shows that, with the exception of the disastrous period of the Cultural Revolution, Chinese elite GRIs possessed considerable technical capacity and talent.

But the system failed in internationally competitive fields where unpredictable grassroots players, both private and public, pursuing gains in a global market system, serve as sources of innovation. By severing the connection between R&D and production and restricting innovation by largely free-willed enterprises and individuals, China's S&T system mirrored the backwardness of much of China's civilian sectors by the end of 1970s. The lack of R&D autonomy and the institutional isolation from global interaction and the larger economy limited the roles GRIs could play in innovation. China clearly needed an alternative innovative engine beyond the state's centralized system. Much of the reform since the 1980s has aimed at addressing this problem.

Marketization of Research Institutes and the Emergence of Competitive Technological Enterprises in the 1980s and 1990s

China's market reform in 1978 shifted the task-led S&T system from the defense industry to the building of the domestic economy. This shift caused major upheaval for GRIs. As China's economy became open, the increasing availability of foreign products—imported or smuggled—reduced the market appeal of the inferior domestic imitations. This undercut the demand for, and rationale of, GRIs. The Chinese state thus reduced direct funding to GRIs with the stated purpose of making them more responsive to the market. Most of the industry-specific ministries were abolished, and research institutes associated with these ministries were transferred to SOEs. Toward the end of 1998, the State Council decided to transform 242 GRIs at the national level into technology-based enterprises and technology service agencies (Liu and Lundin 2006). These changes markedly reduced the dominance of the state and GRIs over the Chinese innovation system.

The remaining GRIs endured the rather traumatic initial reform period in the middle 1980s. They were forced to look for alternative income sources as the state rolled back its resource commitments. Worse, at the time, China did not have enterprises that were interested in—or capable of—commercializing R&D product technology from GRIs. SOEs had little motivation or autonomy in their pursuit of technological upgrades, and many were losing market share to imports. A new type of enterprise had to be created. Desperate for funding and outlets for their research results, GRIs and universities set up their own spin-off enterprises, and encouraged scientists to leave their research positions and engage in commercial activities, mainly in order to earn income and commercialize GRI's research. This was how China's Silicon Valley—Zhongguancun—was born (Lu 2000; Zhou 2008a).

The high-tech spin-off enterprises that populated Beijing's Zhongguancun regions in the 1980s and 1990s included some now-leading information technological companies such as Lenovo (联想), Founder (方正), and

Shuguang (曙光). Similar processes took place in other cities and sectors with strong GRIs or university presence. Most of the Chinese biotechnology companies were also spin-offs, though developed more in the 1990s or later; for example, Shenyang Sunshine Pharmaceutical Co. Ltd, Beijing Shuanglu Pharmaceutical Co. Ltd, and Anhui Anke Biotechnology Co. Ltd were all founded by former researchers from research institutes (Liu and Lundin 2006). Around 2,400 university spin-off enterprises were established before 2004, which altogether generated around US$9.7 billion in revenue (Ministry of Education 2005).[2] The marketization of GRIs and universities turned out to be short-lived, however. The government restored and increased funding to research institutes in the late 1990s, as will be discussed later. Afterwards, most GRIs stopped considering their primary function to be the development of spin-off companies. Instead, they increasingly relied on contract research and licensing for the industrial sector or sponsorship from industrial incubators (Wu and Zhou 2012).

But this spin-off period was momentous, as it produced China's first group of competitive technological enterprises that combined technological competency with market savvy (Lu 2000; Zhou 2005, 2008a, b). Following their examples, more and more private and semi-private startups were established all over China, which transformed the SOE-dominated technological landscape. Even SOEs were given more autonomy to invest and innovate based on their own business interests, and inefficient SOEs were allowed to go bankrupt in the mid-1990s. The growing market competition incentivized enterprises to pursue technological upgrades, signifying a dramatic departure from the Mao era. In the 1990s, these industrial enterprises emerged to become the core of the innovation system. Lu (2000) and Zhou (2008a) detail the development of information and technology firms during the transitional period of the mid-1980s to 1990s in Beijing. They show that the market and financial decisions of these firms were shaped both by their institutional connections with the mother GRIs and by growing access to the international technology and financial markets.

As China became integrated into the global economy, Mao's strict techno-nationalism gradually faded, although various restrictions on non-state sectors have remained in place until today. Foreign technology and foreign direct investment (FDI) became increasingly important drivers of technological change. But this shift also raised new questions and sparked debates on how to balance importation and indigenous technological change.

[2] Total revenue of 2,355 spin-offs amounted to RMB 80.7 billion, and was converted into USD using the annual average exchange rate of 8.28.

Innovation in the Globalized Age: Export Upgrading and Trading Market for Technology, 1990s to 2000s

China's economic policies since the 1980s have been based on two goals: (1) export promotion and (2) import substitution. The former has received considerably more international attention than the latter. The export promotion policies followed the classic path of the "Four Little Dragons—South Korea, Taiwan, Singapore and Hong Kong. The state provided favorable tax, regulation, and infrastructure conditions to attract foreign investment into coastal special economic zones, which in turn strengthened exports. The goals were to exploit China's comparative advantages of cheap land and labor, thereby earning foreign exchange. During the 1980s and 1990s, almost two-thirds of FDI into China came from overseas Chinese in Hong Kong, Taiwan, and Southeast Asia. They transferred labor-intensive export production to China. The synergy of their management expertise, the abundant supply of disciplined, relatively educated low-wage Chinese labor, and the willingness of local governments to do all that was necessary for export, made China one of the most rapidly growing export machines in the world (Lever-Tracy and Ip 1996).

Since the mid-1990s, FDI into China has grown, becoming more capital-intensive and technically sophisticated. The sources of capital have also diversified to include the US, EU countries, Japan, and South Korea, as well as Taiwan and Hong Kong. While export promotion certainly raised China's production capacity, the evidence of its effects on technological innovation is debatable (Crookes 2012). The technological content of China's exports has risen steadily as more multinational firms outsourced their upstream production to China. Xu (2006) calculated that, of total Chinese exports to the US, the percentage of products with technical content increased by 19 points from 1989 to 2001. Wei (2012) calculated that the share of value-added from domestic sources in China's merchandise exports was 54 percent in 1997 and 60.6 percent in 2007. However, he also found that the domestic value-added share is much lower in high-tech sectors, in the 30 to 40 percentage range. A 2011 IMF report on global trade found that the growth of China's high-tech exports is almost entirely (99.4 percent) driven by growth of value-added by foreign companies (IMF 2011: 19). This suggests that the technological upgrading in China's export industry has largely been driven by foreign companies, particularly so in the higher end of technological products. China is still largely dependent on technology from abroad.

China's promotion of exports has succeeded beyond anyone's expectation. Over the course of one generation, China emerged from near total isolation to becoming the largest global exporter by 2009, firmly coupling itself with global production networks. However, China's technological import substitution, or the more explicit policy of "trading market for technology" (TMFT), has been far less successful.

TMFT has earlier origins than export promotion, and from the beginning it was much more state-centered. It was formulated between 1979 and 1981, and it was fully articulated as state policy in 1984 (Xia and Zhao 2012; Feng 2011). The idea was to entice high-tech foreign enterprises to transfer advanced technology to China by allowing them to sell a portion of their products to the Chinese market. China had protectionist policies that strictly limited domestic sales of foreign products at that time. FDI was only allowed for exports. TMFT projects were exempted from this rule, and the preferred arrangement was to establish joint ventures between large SOEs and foreign enterprises in order to produce in China and thereby replace direct imports. Policymakers reasoned that such exposure to production processes would allow Chinese partners to learn foreign technology, and they hoped that availability of advanced products on the Chinese market would incentivize domestic firms to develop better technology independently. To make the deal more attractive, some of the joint ventures were given preferential tax rates and privileged market access. The joint venture between Shanghai Auto and Volkswagen, for example, was allowed to monopolize the Chinese passenger car market between 1985 and 2000 (see Chapter 5).

Throughout the 1980s and early 1990s, the technology China managed to attract went to labor-intensive production of exports or to the final assembly of consumer electronics that were in high demand within China. Few of these were high-tech enough in the eyes of the Chinese state, which believed that part of the reason for this disappointment was the continuous restriction of foreign-owned firms. Fully owned foreign enterprises had not been allowed until the mid-1990s, and proportions of domestic sales were also limited. In 1992, more sectors were finally opened to foreign enterprises, and many restrictions were removed. The Chinese government also made preferential industrial policies to attract investment from large multinational corporations (MNCs) into specific technology in large-scale energy, transportation, raw material, and infrastructure projects. Particular favorable conditions were granted to high-tech foreign firms (Xia and Zhao 2012).

As a result, the market share of foreign companies in China grew rapidly in the 1990s, even becoming dominant in some industries. Yet the technological developments under TMFT were still repeatedly disappointing. In the semiconductor industry, the huge state investment into joint ventures and SOEs failed to close the technological gap (Chen and Xue 2010; Chapter 10 in this book). The failures were painfully evident in the auto industry (see Chapter 5). The technology of the automobile joint ventures was entirely that of the foreign partners, while the Chinese partners were solely responsible for securing market access. As there were many potential Chinese suitors for any foreign firm, foreign partners had strong leverage to choose suitable domestic partners and to impose conditions upon them. To court the foreign firms, the promise of protected market access was made in perpetuity; it was not contingent upon continued or proven technological transfer (Xia and Zhao 2012). The market monopoly meant that foreign companies were not compelled to continue upgrading technology. Even when the initial joint ventures were set up with advanced technology, such competitive edge inevitably faded without continued upgrades (see Chapter 5 for examples). Chinese partners were not motivated to seek technology transfer as long as the monopolized company was profitable.

Ultimately, given the assured market monopoly and internal technical incompetence, the Chinese side ended up as perpetually passive players. In contrast, a number of private auto companies—such as Chery and Geely—started with far fewer resources, no protected market, and no designated foreign partners, have managed to make more rapid technological progress than the state-owned joint ventures because they had the autonomy and drive to learn and acquire technology tailored to their own commercial needs (Lu and Feng 2005; Chapter 5 of this book). Instead of imparting a contagious spirit of innovation, the arrangement of TMFT stifled the motivation and competence of the China partners in transferring technology.

Until now, Chinese companies have spent more effort on importing existing technology and products than on assimilating and improving technological processes. Assimilating means to integrate the technology into core competence of the enterprises so they are able to create improvements or adaptations of the technology. Figure 2.2 shows the changes of the ratio between spending on assimilation vs. importation based on Chinese government statistics. If the ratio is lower than 1, the enterprises spent more on purchasing the technology than integrating it into its operations. A persistently low ratio indicates little effort at learning and internal growth. From Figure 2.2, we can see that the ratios are different for different types of enterprises. SOEs include state-owned enterprises, state joint-ownership enterprises, and solely state-funded corporations. The remaining domestic enterprises include joint-stock companies, private enterprises, etc. Foreign-owned firms spent little on assimilation because they mostly use their own technology. The brief spike in the ratio for foreign-owned firms in 2009 was the result of the temporary cutback in technology purchases after the financial crisis, so it is an anomaly that can be discounted.

In 2002, barely 0.1 yuan was spent assimilating technology for every yuan spent in purchasing from abroad. The ratio has gone up since then, especially after 2008, but it is still below 1. In 2012, the ratio of SOEs was 0.57:1, non-state-owned enterprises had a ratio that, at 0.56:1, was slightly lower. In contrast, it was reported that the ratio was 2:1 in South Korea in the 1980s

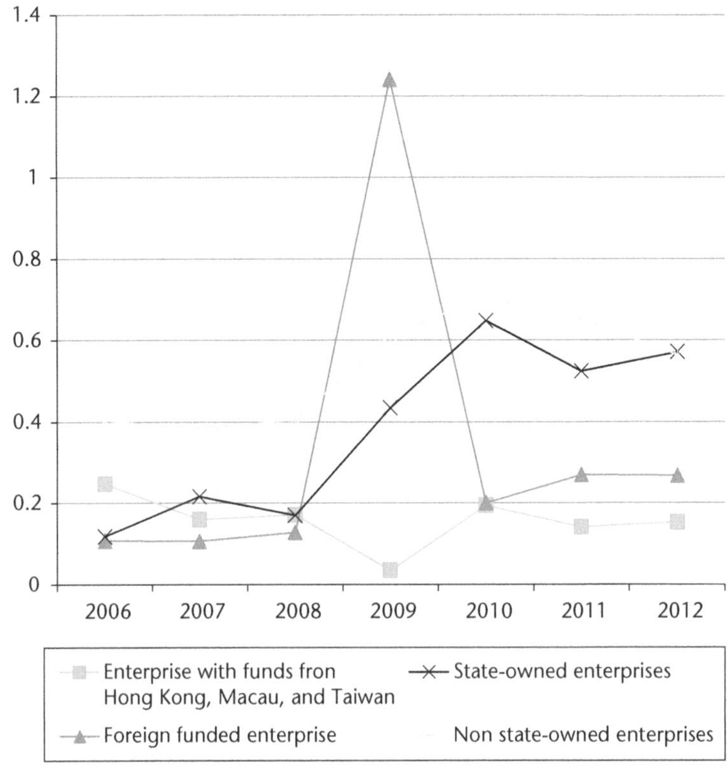

Figure 2.2. Ratio of Expenditure on Assimilation to Importation of Technology in Different Ownership Enterprises

Source: MOST and NBS, *China S&T Statistical Yearbook* 2003–11.

(KOITA various years). The low Chinese ratio suggests considerable inertia in increasing assimilation spending. For SOEs, such inertia resulted from the easy access to state funding as they are given substantial support in purchasing advanced technology from abroad, while limited market competition minimizes the incentive to assimilate technology. For non-state firms, the lack of technology assimilation is a sign of the relatively weak internal R&D capacity of these firms. Since 2008, a state emphasis on indigenous technology development has helped Chinese-owned firms to increase spending on assimilation.

Nearly two decades of export promotion and TMFT have vastly improved China's industrial capacity and global competitiveness, but far less progress has been made with China's own R&D capacity. Chinese policymakers had assumed that, because technology is embedded in the products, manufacturing such products in China would naturally lead to technology transfer. They viewed localization as indicative of successful transfers. Few realized that the global restructuring of industrial manufacturing had made such assumptions unreliable. So-called global production networks (GPNs), defined as "the globally organized nexus of interconnected functions and operations by firms and non-firm institutions through which goods, and services are produced and distributed" (Coe et al. 2004: 471), had become the main organizational form of global industry. GPNs have hierarchical structures led by the top tier of large transnational corporations (TNCs) which Ernst and Kim (2002) call "flagship corporations." As they explain, "The main purpose of these networks [GPNs] is to provide the flagship with quick and low-cost access to resources, capabilities and knowledge that are complementary to its core competencies" (Ernst and Kim 2002: 1420). The medium tier, which mediates between the flagships and lower-level players, includes enterprises in newly industrialized countries such as Taiwan and South Korea. The lowest tier consists of many small- and medium-sized companies located mainly in developing regions. China is the most important site for both the lowest and intermediate tier GPNs (Fischer 2010). Such a structure means that manufacturing products according to established blueprints no longer necessarily requires mastering the technology behind these products.

These changes in the global production system have heightened the anxiety of the Chinese state. They fear that, despite China's expanding production capacity, it could forever be trapped in low-end production. The fear was also fed by the persistently low wages and low profit margins of Chinese producers in a wide range of industries (Xing 2011; Breslin 2011).

By the early 2000s, the situation was alarming for the Chinese policymakers; they saw that globalization had undermined China's technological independence and security without benefiting China's R&D capacity. Foreign-led firms retained high profits and highly skilled parts of the operations in export-oriented industries, while most Chinese exporting companies specialized in lower-end work with slim profit margins. The R&D of multinational corporations in China was largely conducted in their respective home countries. Chinese companies and joint ventures in the domestic market continued to rely on imported core components to produce. A number of frustrated scholars and governmental policymakers even praised the technological achievements of the strategic weapons programs under Mao, and argued that a self-reliant model would be more productive for Chinese technological progress than globalization (Lu and Feng 2005; Xu 2005).

While not everyone agreed with this bleak assessment of the R&D capacity achieved under China's global integration, there was an emerging consensus by the early 2000s that a market approach and global integration could not produce satisfactory progress in technological learning and innovation. Foreign companies were eager to produce for the Chinese market, but they would vigilantly safeguard their core technology. It is up to China's own enterprises to conduct R&D and thereby learn and assimilate foreign technology. As the intermittent labor shortage hit China's coastal regions starting in 2003, it became evident that an economic model based on inexhaustible supply of cheap labor was no longer sustainable. China had to shift to a technology-driven economy. Such a transition could not be achieved by relying only on private domestic enterprises to serve as lower tiers of global GPNs, most of which operate on slim profit margins. The developmental experiences of the United States, Europe, Japan, and South Korea show that the state was an indispensable player during their respective "catch-up" processes. In all of those countries, the state invested during their early stages of development in large, highly complex, risky, and crucial public technology projects such as the internet and geographical positioning systems (Lazonick 2011; Block and Keller 2011; Mazzucato 2013). If China were to truly become an innovative nation, then the state would have to reassert itself and become a central power in technology development and capacity building programs.

Reasserting Centrality of the State in Innovation, 2000 onwards

Top-Down National Research Projects

In actuality, the Chinese state had already started to increase funding for GRIs and universities as early as the mid-1990s, reversing the marketization trend of the 1980s. This shift partly reflected China's high rate of growth in the 1990s, which increased the government's budget, but it was also influenced by the dot.com boom and the growing prominence of the knowledge economy in the West. Since Western countries were developing a knowledge-based economy, China would need to prepare to compete by investing in knowledge-intensive institutions. Though this process was starting throughout the 1990s, the increased state commitment did not become obvious until after 2000. Figure 2.3 shows the funding increase to the main S&T programs controlled by the Ministry of Science and Technology (MOST) since 2000. While all the funds have had a clear growth trend since 2001, there was a substantial jump after 2006, the year China formulated its "indigenous innovation" policy as the new national platform. Though all funds except National Natural Science stabilized after 2009, it was only a result of the proliferation of new R&D programs. The dramatic trajectory of National Natural Science Fund, a relatively new funding agency modeled after America's National Science Foundation, indicates that the state intends to consolidate various S&T funding programs with more transparent review process, a task yet to be accomplished.

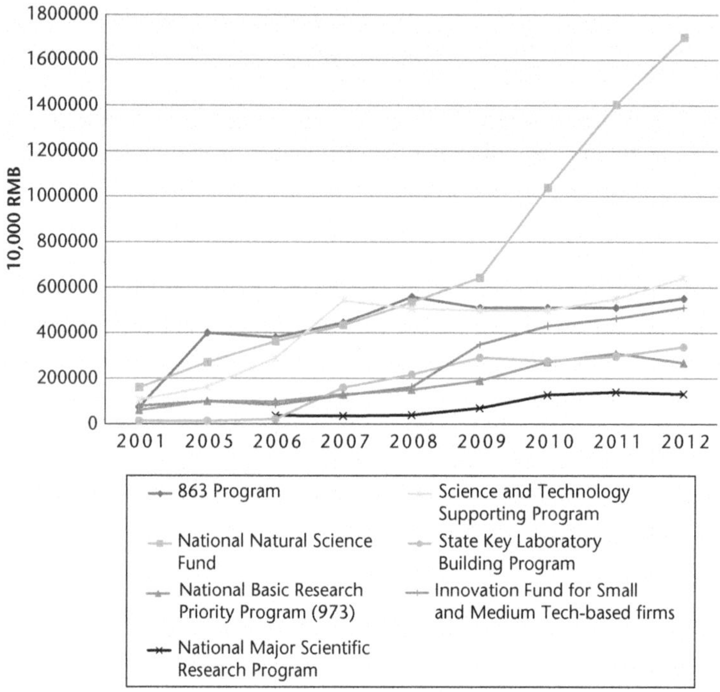

Figure 2.3. Funds Controlled by Ministry of Science and Technology
Source: MOST, *China Science and Technology Development Report*, 2008, *China S&T Literature Press*.

The new state policy was fully articulated in *National Programming 2006–2020 for the Development of Science and Technology in the Medium and Long Term* (*MLP* hereafter, see Cao et al. 2006). It is often referred to as the *zizhuchuangxin* (自主创新) policy, which has been translated as the "indigenous innovation" policy. The "indigenous" translation is somewhat misleading: *zizhu* literally means self-directing, and does not necessarily describe technology that has to originate in China or be free of foreign contribution. The policies were designed to address the perceived weaknesses of TMFT and export programs by boosting domestic R&D efforts and encouraging domestic enterprises to take strategic control over technological interactions with foreign parties. The state hopes to increase China's investment in R&D to 2.5 percent of GDP by 2020, up from the 2006 level of 1.42 percent. Since GDP is projected to grow at 7–8 percent annually, increasing R&D expenditure as a share of GDP implies a huge increase in absolute terms.

The indigenous innovation policy signals a return of the Chinese state to the center of the nation's S&T endeavor to address the perceived gap in financing long-term and risky R&D projects by commercial enterprises.

This is most evident in how highly significant national projects are organized. The policy, or *MLP*, identifies 16 mega-projects in microchip, broadband, alternative and nuclear energy, aerospace, diseases control and health, and fuel vehicle development (see the detail in Lu et al. 2012). There are striking similarities between the organizational form of these mega projects and Mao-era strategic weapon programs. First, all of the recent projects have been administrated through a top-down system. The Chinese State Council, headed by then Premier Wen Jiabao (温家宝), has coordinated and guided them. MOST, NDRC, and the Ministry of Finance form the core leadership group, and relevant industrial ministries have directed projects relevant to their fields. For example, the Ministry of Health has helped to lead a pharmaceutical mega-project (ibid. 2012).

Unlike in the Mao era, participation of enterprises is encouraged and commercialization of technology is seen as a critical step. Some projects, such as those related to advanced machine tools, even make a point of soliciting input from enterprises (Lu et al. 2012). The practice of establishing research consortia was borrowed from Japan, Korea, and the US, all of which developed such organizational forms in the 1980s and 1990s to improve critical industrial and general technology. In China now, the government is the key coordinator in the consortia. Industry, university, and research institutes form R&D teams that focus on research, development, or implementation. The industrial members of a given consortium include sector leaders and their competitors. The 3G wireless communications standard TD-SCDMA is an example of such a consortium, and will be discussed later. It is possible for large private companies such as Huawei and Lenovo to be key members of consortia, but overall the consortia tend to favor primarily GRIs and state-owned companies (Liu and Peng 2011).

Unlike the Manhattan Project in the United States, which pushed the frontier of human knowledge, these mega-projects are selected to pursue cutting-edge technology with applications in China. They are costly and challenging, and so they are beyond the organizational and financial capacities of even the largest Chinese commercial enterprises. But they are still follower projects in which the technology has been researched elsewhere and the objectives of the projects are to imitate, adapt, and further develop, rather than to innovate novel products. Research teams study technical feasibilities and set achievable goals within a given time frame (Lu et al. 2012). As was the case during the early period of the PRC, the state's objective is to master core technology pioneered outside China and make it applicable. China has given little significant attention to basic research, or research that may not have an immediate application for instrumental goals but is fundamental to the development of future knowledge.

Table 2.1. Breakdown of Investment in Research Activities after Implementation of Indigenous Innovation Strategy by Country (%)

Country	Year	Basic Research	Applied Research	Experimental Development
China	2012	4.8	11.3	83.9
USA	2009	19.0	17.8	63.2
Japan	2010	12.7	22.3	65.0
France	2010	26.3	39.5	34.2
Australia	2008	20.0	38.6	41.4
Switzerland	2008	26.8	31.9	41.3
Korea	2010	18.2	19.9	61.8
Russia	2010	19.6	18.8	61.6
Italy	2010	25.7	48.6	25.7
UK	2010	8.9	40.7	50.4

Source: *China S&T Statistical Yearbook*, 2013.

Table 2.2. Percent of Intramural Expenditure on R&D

	Basic Research (%)	Applied Research (%)	R&D Implementation (%)	% of GDP
1995	5.18	26.39	68.43	0.57
1996	5.00	24.51	70.49	0.57
1997	5.39	26.02	68.60	0.64
1998	5.25	22.61	72.13	0.65
1999	4.99	22.32	72.68	0.76
2000	5.22	16.96	77.82	0.90
2001	5.33	17.73	76.93	0.95
2002	5.73	19.16	75.12	1.07
2003	5.69	20.23	74.08	1.13
2004	5.96	20.37	73.67	1.23
2005	5.36	17.70	76.95	1.32
2006	5.19	16.28	78.53	1.39
2007	4.70	13.29	82.01	1.40
2008	4.78	12.46	82.76	1.47
2009	4.66	12.60	82.75	1.70
2010	4.59	12.66	82.75	1.76
2011	4.74	11.84	83.42	1.84
2012	4.84	11.28	83.87	1.98

Sources: NBS and MOST, *Statistical Yearbook of China's Science and Technology* (2009). National R&D Survey for 2010 (MOST et al., published 2013).

Tables 2.1 and 2.2 show that China's allocation of funds to basic research has been much lower than OECD countries, even after 2006, and that the basic research share of total R&D funding has actually declined since 1995. This bias against basic research indicates that the Chinese state remains mostly concerned with catching up with existing Western technology rather than achieving novel scientific breakthroughs. The emphasis on this strategy may also be a legacy of the bureaucratic S&T system, which is good at implementation but lacking in creativity. As China moves further forward on its S&T trajectory, it may have to increase its attention to basic research because it provides a foundation for long-term application research and represents China's contribution to human knowledge. But such research requires the state to lessen the bureaucratic control of the S&T processes and give scientists more freedom and autonomy to pursue scientific knowledge and development.

This forceful return of the state to the center of technological endeavors, and especially the central role of MOST in allocating resources, has generated considerable controversy in China, particularly in academic circles. Shi Yigong and Rao Yi—both prominent scientists and overseas returnees—published an editorial in *Science* in 2010 that criticized the funding mechanism for mega-science projects and claimed they were corrupt. They charged that the distribution of funds is based on scientists' relationships with powerful bureaucrats rather than on scientific merits. This view reflects that of many rank-and-file scientists in universities who feel stifled by increasing bureaucratic control over R&D funding. Given the powerful central planning legacy, centralization and state control over R&D run the risk of undermining openness and scientific practices in China's innovation system. In 2014, China started another round of reform aiming at the state control and funding allocation of R&D. The hope is to reduce the bureaucratic roles of MOST and increase accountability of state funding. The details of the reforms are still unclear, but what is certain is that China's effort to change the roles of state in R&D continues.

Another long-term problem of state-led development is that the Chinese state has yet to find effective ways to diffuse and transfer technology to the wider society. In the 1980s and 1990s, there had been significant technological diffusion from GRIs to the commercial sectors, as exemplified by Lenovo and Founder startups, documented by Lu (2000), and Zhou (2008a). Yet this happened during a time of great distresses for GRIs as discussed earlier. Once GRIs became amply funded by the state, the incentives of technological diffusion to the commercial sectors weakened. In addition, many commercial corporations established their own in-house R&D centers and research institute spin-offs were no longer institutionally encouraged for their mucky blends of public and private financial mechanisms. In their absence, new channels of technological diffusion are yet to be well established. Lu et al. (2012) found, after surveying China's mega-projects, that state-led mega-projects are strong in project implementation but weak in diffusing knowledge to enterprises and non-state companies. In a study on nano-technology, a group of American scholars noted that state efforts were instrumental in creating China's lead in this new industry, but that the state also has difficulties commercializing the technology due to a lack of inter-industrial flow (as compared to Taiwan) (Brahmbhatt and Hu 2009; Shih and Chang 2009; Huang and Wu 2012; Appelbaum et al. 2011).

The analysis makes it abundantly clear that China still faces considerable institutional barriers to establishing an effective and comprehensive national innovation system. Lazonick (2011) argues that to accomplish scientific projects with a high fixed cost, high risk, and a public nature, investment from the state is critical at certain stages in the development of an industrial sector. But to have a lasting and significant impact on economic growth, such investments must be complemented by the further development and utilization of productive resources at the business level. The endemic weakness in commercialization and technological diffusion of China's centralized S&T system suggests that further institutional reform is needed in order to create more linkages with industrial innovation. There is a danger, however, that the heavy reassertion of Chinese state control may cause the S&T system to revert to more centralized decision-making. If implemented in the wrong way, this could undermine the further development of the grassroots innovation that has emerged in certain areas such as the internet over the last 20 years.

Indigenous Innovation and Globalization

With the state's reassertion of its role as the central actor in the innovation system in 2006, China's S&T policy again assumed techno-nationalist hues and caused concern that China would pursue a state-controlled strategy of protectionism and discrimination against foreign companies. But contemporary China is vastly different from Mao-era China. The return of techno-nationalist practices has been repeatedly checked by the reality of a vibrant market economy and intertwining relationships with other parts of the world. Even *MLP* has explicitly acknowledged the need to strike a balance between domestic innovation and importation of technology (Ernst 2011). China's integration into the global economy also means that foreign parties can now credibly challenge Chinese state policy, rendering a complete return to techno-nationalistic practices all but impossible (ibid.). For instance, one of the most important new tactics to promote indigenous innovation in China is the use of public procurement—a practice that has generated considerable international controversy. Previously, public procurement in China was primarily determined by the quality and cost of products. The new procurement law issued in 2002 implicitly mandated that governmental purchases should favor domestically made goods. In 2006, the Chinese government went even further: public procurement should favor accredited national indigenous innovation products (NIIP), which require indigenous intellectual property rights or indigenous branding. The law was made in reference to some of the global best practices in selected sectors of the United States, Japan, and Korea (Okimoto 1989; Ahern 2010). By prioritizing goods from innovative Chinese companies, the government hoped to encourage their growth through market demand and reduce their market risk (Cao, Suttmeier, and Simon 2006).

However, since the governmental procurement market is huge—and includes central and local governmental levels—the NIIP accreditation scheme caused alarm among foreign companies who export high-tech goods to, or make such goods in, China. Foreign firms and governments protested the requirement on the grounds that it discriminated against their products, increased barriers to trade, and prevented China from fulfilling the requirements of the Government Procurement Agreement (GPA) under the WTO, an organization that China expressed a desire to join. Under pressure from the United States and EU, the Chinese government agreed to give foreign companies equal treatment and delinked government procurement with NIIP accreditation. However, this dispute is not completely settled: foreign businesses continue to worry about similar barriers at the provincial and local levels (Lubman 2010). The pushback on NIIP from China's trading partners suggests that globalization has made it increasingly difficult for China to make unilateral moves based upon techno-nationalist considerations. China encountered similar challenges when the US and EU increased tariffs against Chinese solar and wind energy products, based on the argument that state subsidies unfairly benefit Chinese corporations (Hopkins and Li, Chapter 12 in this book).

Furthermore, beyond increasing S&T fund allocation from the central government, the Chinese government has also become adept at using a diverse set of tools to encourage enterprises to seek technological development. For example, new tax policies make R&D expenditure 150 percent tax deductible and allow for accelerated depreciation in R&D equipment. There are also various fiscal incentives and subsidies for specific industries in strategic areas such as renewable energy (see Chapters 11 and 12) that encourage enterprise formation and growth. Thus, indigenous innovation policies recognize and encourage the growing R&D capacities of the domestic enterprises and their interactions with GRIs. In other words, this policy does not envision a radical reversal of market-oriented reforms that China has undergone since the 1980s, but it does redefine the central piece of the roles of the state.

As a part of the promotion of indigenous innovation, China also started to develop an IPR promotion strategy for Chinese enterprises. China's patent system was established in 1985, but it did not receive much attention until after *MLP* was issued in 2006. Here we will briefly discuss China's involvement in establishing a wireless communication standard as an example of the strategic considerations of the state (more discussion of this policy is also found in Chapters 7, 8, 9, and 10). China used to be an eager adopter of prevailing international technical standards. But in the early 2000s Chinese policymakers realized the interlocked nature of industrial standards and

IPR. This led the government to push for China's own technical standards as a way to enable innovation in the Chinese communications technology industry. At the same time, it increased Chinese bargaining power over the determination of royalties that must be paid to foreign IPR holders. The best and most documented case is China's 3G wireless standard TD-SCDMA (Suttmeier et al. 2006; Zhou 2006; Breznitz and Murphree 2011; Ernst 2011).

In May 2000, the International Telecommunication Union (ITU) certified TD-SCDMA—proposed by an SOE called Datang Telecom Technology—as one of the 3G mobile communications standards. This was seen as a milestone for China. At that time, TD-SCDMA was an infant technology compared to the dominant WCDMA in Europe and CDMA2000 in the US. The fact that Datang had created TD-SCDMA suggests that Chinese GRIs had never completely stopped developing indigenous technology. But the most significant step in this case is the reasoning behind the government's decision to adopt TD-SCDMA as a national standard: since China had already become the largest cell phone market, the idea was that this standard could boost indigenous innovation in the entire industrial chain because it reduces royalty payments to foreign firms such Qualcomm. Several key governmental agents—MOST, MII, and NRDC—were on board. It is estimated that over RMB 1.2 billion (USD 150 million) of special funds have been granted to develop TD-SCDMA since the late 1990s (Zhan and Tan 2010). Before 2006, the Chinese government created a research consortium, or TD-SCDMA alliance, to develop the industrial chain based on the standard.

The young TD-SCDMA faced many technical uncertainties and setbacks, and even Chinese domestic equipment makers and operators were reluctant to commit to this technology. Worried that TD would prove to be uncompetitive against the two existing, mature standards, the Chinese government delayed 3G licenses for many years to allow TD-SCDMA more time to mature. Even though commercial 3G networks appeared in advanced countries as early as 2001, China did not issue 3G licences until the end of 2008, after the landmark Beijing Olympics. The government also designated China Mobile, the largest mobile phone carrier in the world, as the 3G carrier for TD, in order to establish the new standard with a favorable market position (Liu 2007). This visible, aggressive support from the Chinese government eventually secured commitment to the standard from a host of foreign and domestic equipment makers (such as Datang and Siemens), chip designers (Datang, T3G Technology, Spreadtrum, and Mediatek), testing and instrument companies (Zhongyou, ZCTT), mobile phone handset makers (Da Tang, Soutec, and Do pod) and operators (China Mobile).

The lessons on innovation in this case study are mixed, however, which is reflected in several chapters in this book (Chapters 7–10). On the one hand, the forced promotion of TD delayed China's launch of 3G phones for the China

Mobile. Even as of 2012, the share of 3G users amongst China Mobile customers is significantly lower than the shares of much smaller Chinese competitors who adopted international standards (Trefis 2013). As it turns out, the wireless industry's value chain is long and complex (Chen and Tai, Chapter 10), and the government-supported standard could not be successful without the support of all players in the whole process of technology development. Given that TNCs control the core cell phone technology, their willingness to participate was necessary. The length of TD's maturation process meant that there had been considerable uncertainty, which discouraged domestic and foreign companies from investing in the equipment and further slowed the growth of the TD network (Breznitz and Murphree 2011: 74). On the other hand, China finally was able to implement an indigenously developed and internationally recognized technology standard in the industry. Since TD involves a different technology architecture from CDMA and WCDMA, the experience and expertise that various Chinese firms gained by initiating and managing the complex industrial eco-system surrounding TD should not be underestimated. Several Chinese chip designers—such as Spreadtrum, among the top five cell phone chipmakers in the world in 2012—would not have survived had there not been an indigenous platform upon which they had advantages over existing international rivals (see Chapter 9).

While TD-SCDMA is established only as a national standard, China Mobile and other major telecommunication equipment and chipmakers are now well positioned to implement 4G LTE standard, which is poised to become one of the two major international standards. The dominant market position of China Mobile in China's cell phone market (70 percent, with 700 million users) means that companies such as Apple would have to make their iPhone compatible with the technological standard (Chen and Pfanner 2013).

This contrasts with Japan's experience when adapting 3G technology. While Japan implemented 3G earlier than other countries, its telecommunication sector was decoupled from the global market, trapping the companies in the domestic market. The Japanese 3G standard was not adopted elsewhere, nor was there significant participation of international IT players in the standards, creating a phenomenon Kushida (2011) has called "leading without followers". The size and rapid growth of the Chinese cell phone market and its relative openness to foreign players suggests that a similar outcome of isolated development is unlikely to happen in China (Chen and Tai, Chapter 10).

This case is representative of two lessons. One is similar to what could be learned from China's governmental procurement policy: in a globalized era, the frontier of industrial technology will have to involve international collaboration, whether through joint ventures, strategic alliances, and technological partnership or subcontract relationships. Even leading Chinese innovative companies such as Huawei have expressed a preference for an environment of open innovation in which they can cooperate with international firms rather than develop all new technologies from scratch (news release, PR Newswire 2015). China will have to remain open internationally and engage in foreign collaboration if it intends to be a valued member of international technological alliances. The second lesson is that the capacity of China's own firms, unless buoyed by strong internal R&D aptitudes, is currently insufficient to compete with top firms in the international arena. This capacity refers not only to R&D but also to the holistic ability to have strategic control over R&D, production, marketing, and collaboration of the entire commodity chain. Such capacity will take time to develop, and setbacks are a necessary part of the process. China's 3G-standard struggle should be viewed as a way station on a long accumulative journey. The journey has already yielded the growth of a number of Chinese companies, such as Huawei, and other startups such as Spreadtrum.

The Roles of Local Government

Recently, in addition to the central government, Chinese regional governments have cultivated closer relationships with local, often private, enterprises, which represents another critical development in the landscape of China's innovation system. Today, the R&D spending of regional governments is larger than that of the central government, and it is most heavily concentrated in richer coastal regions (NBS 2009). Regional governments have specific interests in innovation that benefits localities in the country-wide competition for central government investment and subsidies, high-tech human resources, and specific strategies for industrial upgrades and growth. Some scholars (Granick 1990; Naughton 1994, 1995; Huang 1996; Shirk 1993; Oi 1999) argue that local autonomy and inter-regional competition have been the institutional driving force behind China's growth. Provinces such as Jiangsu, Shanghai, Guangdong, Shandong, and Zhejiang have regional governments that are more willing to push for indigenous innovation through regulations and spending (Liu et al. 2010). In fact, a few of China's strategic emerging industries were fostered initially by regional governments rather than by the central government. For example, the photovoltaic industry originally emerged in Jiangsu (Suntech) in 2001 with no involvement from the central government. Now, more than 17 regions list photovoltaic as a key industry. Even central governmental control has not been able to lessen the fervor for this intense regional competition, which has led to overcapacity (Chapter 12). This happened in the solar energy industry in 2008. When the NDRC established a program to restrict industry capacity to no more than 10 million kw by 2010, the real capacity had in fact already reached 12.27 million kw.

Regional governments have also played crucial roles in funding and building industrial parks and startup incubators around China (Zhang and Wu 2012; Liu et al. 2012; Zhou 2008a). These parks provide various benefits for enterprises, such as transportation and communication infrastructure, subsidized rent, tax deduction and business services for firms. Most industrial parks in China are places with interlinking industrial clusters, but some—such as Zhongguancun in Beijing, Zhangjiang in Shanghai, Shenzhen, and the Suzhou high-tech development zone—are also among the most active in innovation (Zhou et al. 2011). Local officials tend to be much more supportive of non-state firms than the central government since the non-state industry provides the crucial tax base for local governments. Such is the case in Beijing's Zhongguancun region as well as in the Yangtze River and Pearl River deltas, where non-state enterprises are highly active in diverse industries.

SOEs and Non-State Firms in Innovation Performance

The role of non-state, or private, firms in China's innovation system is an important and controversial one. There has been compelling evidence that the Chinese state has long favored SOEs in finance and industrial policy at the expense of non-state firms, and that such state involvement intensified considerably after the 2008 financial crisis (Naughton 2011). Some scholars called it "Guojinmintui" (国进民退), which can be translated as an "expansion of the state and retreat of non-state" (Breslin 2011). The reality, however, is more complicated, varies across industries and regions, and can look different according to different measurements. Here, we examine the Chinese S&T statistical yearbook to discern some patterns in the growth of R&D in non-state enterprises.

Based on data collected between 2002 and 2012, we can make the following observations about the situation of non-state enterprises in China. While such enterprises had faster growth rates than foreign firms and SOEs as a whole, they remained smaller and less capitalized than SOEs in terms of individual size, in part reflecting the sector distribution of such enterprises (Figure 2.4). But though the GDP share of SOEs is decreasing, the power of state-owned enterprise in controlling the crucial segments of the economy has not diminished, and indeed it has even expanded in several key sectors. SOEs enjoy dominant positions in resource-intensive industries, such as petrochemical and power sectors but also in national defense, finance, communication, transportation, mining, metallurgy, and machinery sectors. They control 55 percent of China's electricity supply, 48 percent of automobile output, and 70 percent of hydroelectric generation equipment (Xinhua Net 2008). They also have effective control over the core industrial and infrastructural sectors. The non-state-owned enterprises, while gaining status, are still weak compared to SOEs. They

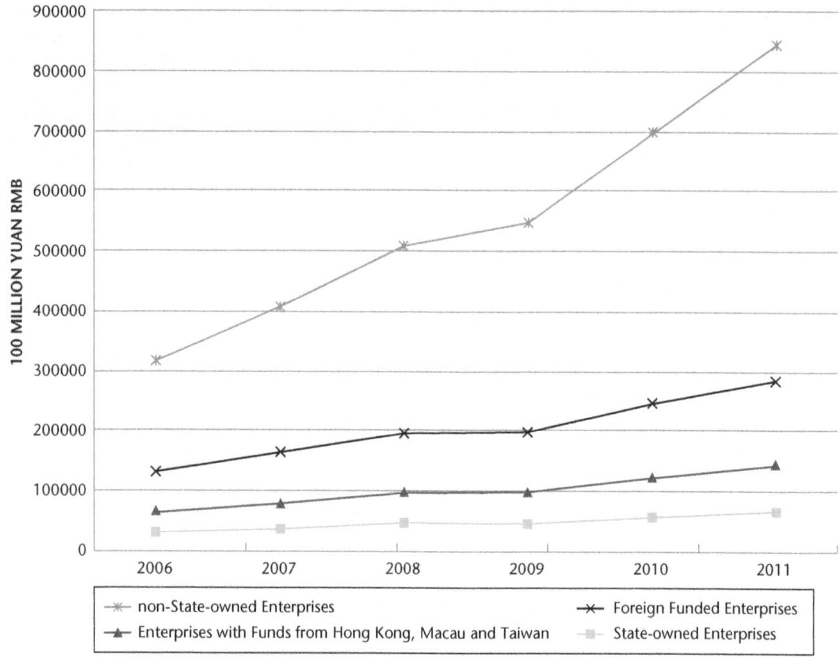

Figure 2.4. Total Value of Industrial Output of Different Ownership Enterprises (above the designated size) (2002–12)

Source: *China S&T Statistical Yearbook*, 2003–12.

are concentrated downstream, in sectors that are more consumer-oriented, with relatively low fixed costs and lower profit margins.

While effective innovation may be difficult to measure, we can use a few proxy variables to measure the R&D capacity of non-state enterprises. One is the number of scientists and engineers employed (Figure 2.5). Traditionally, only large SOEs can afford to hire a high number of scientists and engineers. But from Figure 2.5, we can see that the absolute number of scientists and engineers in SOEs declined slightly in 2008, while the share of scientists and engineers in non-SOEs increased sharply from 49.5 percent in 2002 to 62.7 percent in 2008. The absolute number of scientists and engineers hired by non-SOEs with funds from Hong Kong, Macau, and Taiwan and by foreign-funded enterprises has increased as well, although not as dramatically. This trend parallels the spending on R&D, shown in Figure 2.6.

From Figure 2.6, we can see that after 2003, all forms of enterprises have been accelerating their R&D expenditures. By 2010, SOEs accounted for 10.7 percent, declined from 17 percent in 2002, while non-SOEs accounted for 60.1 percent in 2010, declined from 64.7 percent in 2002. The share of

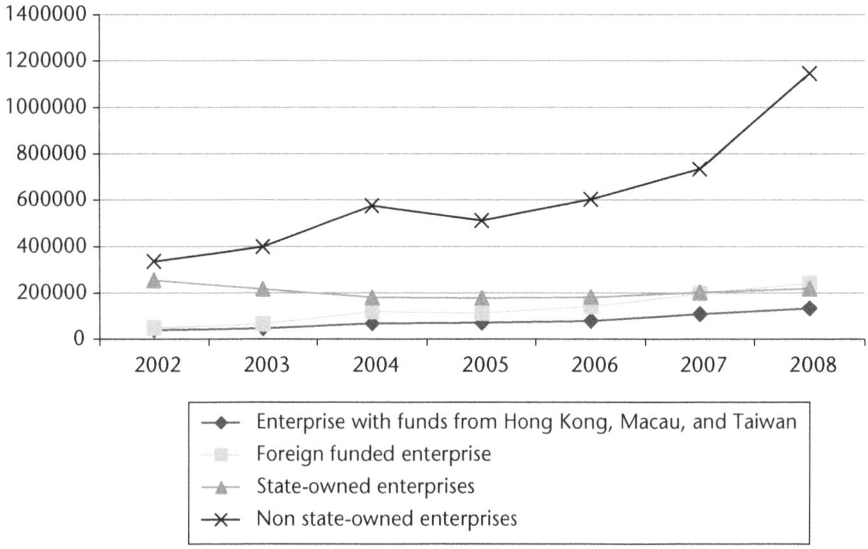

Figure 2.5. Number of Scientists and Engineers in Different Ownership Enterprises
Source: *China S&T Statistical Yearbook*, 2003–9.

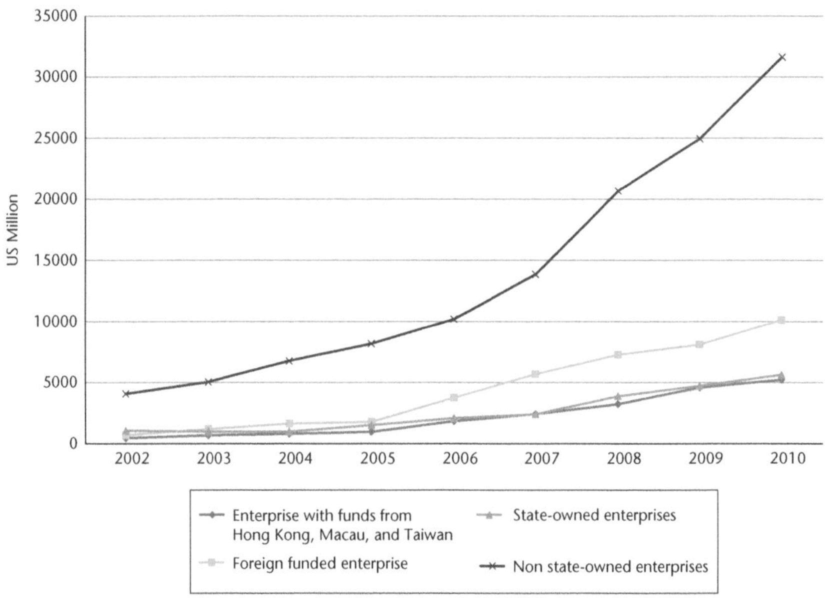

Figure 2.6. Amount of R&D in Different Ownership Enterprises
Source: NBS, *China S&T Statistical Yearbook*, 2003–11.

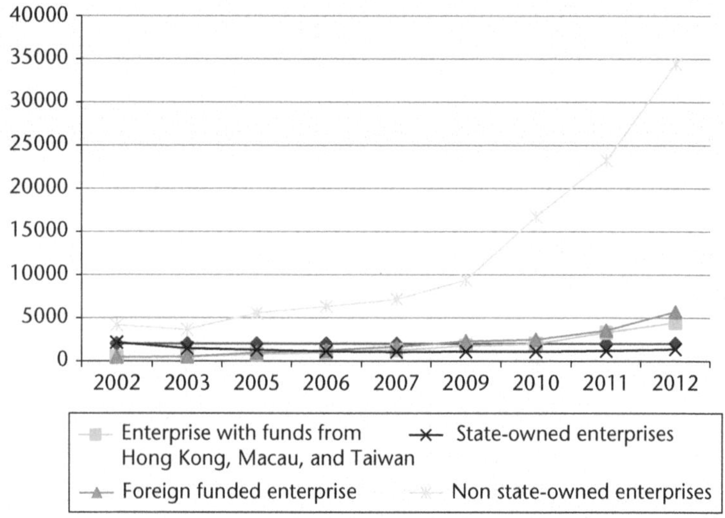

Figure 2.7. Number of Research Institutions in Different Ownership Enterprises
Source: NBS *China S&T Statistical Yearbook*, 2003–13.

Hong Kong, Macao and Taiwan firms were virtually unchanged. But foreign-owned enterprises increased their share sharply from 11% to 19% during the same period, showing a heightened foreign R&D investment in China. At the end of 2007, MNCs had set up more than 1,160 R&D labs across China (Xinhua Net 2008).

The number of R&D institutes in enterprises is another important indicator of innovation capability. From Figure 2.7, we can see that an increasing number of non-SOEs have set up their own R&D labs, especially after 2009. The ratio of firms that have R&D labs in non-SOEs in 2010 reached 74.9 percent, while the number of research institutes owned by SOEs continues to decrease, accounting for only 5.0 percent. SOEs are the only type of firms with a declining number of R&D labs, as shown in Figure 2.7.

Interestingly, although the number of labs in SOEs is in decline, Figure 2.8 shows that SOEs rank first in the average expenditure of R&D labs among domestically owned firms, especially after 2006. Non-state firms invested more than Hong Kong, Macao and Taiwan-invested enterprises after 2008, but all domestically owned were lower than foreign owned firms until 2010. This suggests that either R&D labs in SOEs are more capital-intensive and sophisticated, or that such labs have an easier time receiving state funding than other types of labs. It is likely that after China issued its indigenous innovation policy, the heightened state R&D investment went disproportionately to SOEs. Firms from Hong Kong, Macau, and Taiwan tend to concentrate on labor-intensive assembly in their operation in the mainland, so it is not surprising that they have the lowest position. Non-state firms, by this measure,

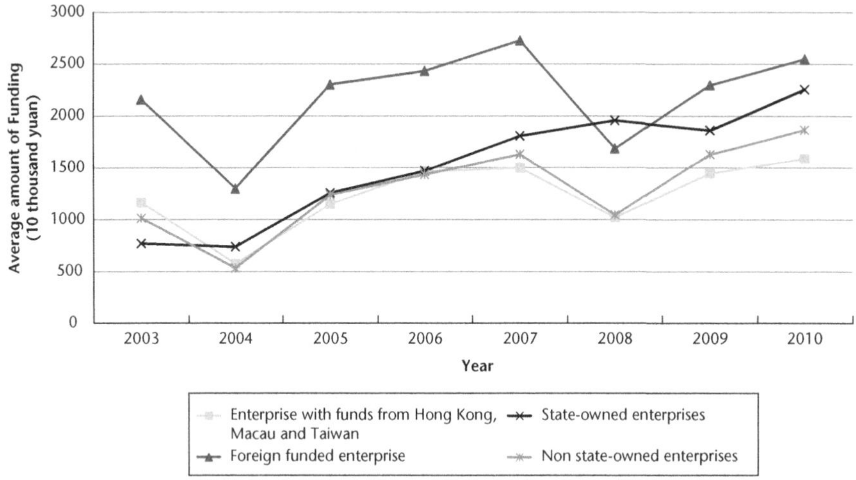

Figure 2.8. Average Amount of Funding of Every Institution of Different Ownership Enterprises

Source: *China S&T Statistical Yearbook*, 2003–11.

Note: In 2004 and 2008, the data are from "Industrial Enterprises above Designated Size," the others years are from "Large and Medium-sized Industrial Enterprises." Thus the two year dips do not reflect real decline.

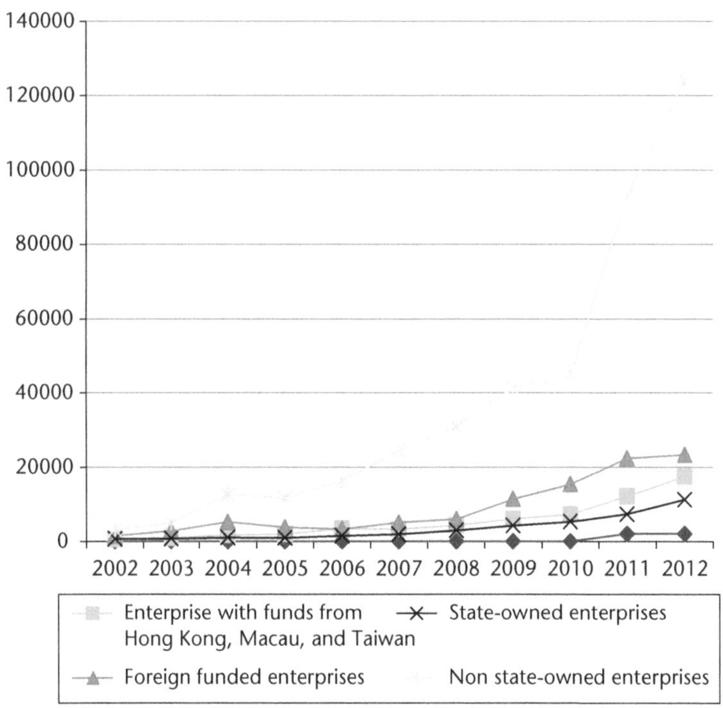

Figure 2.9. Number of Invention Patent Applications of Different Ownership Enterprises

Source: *China S&T Statistical Yearbook*, 2003–13.

have a similar trajectory to that of Hong Kong, Macau, and Taiwan firms. This shows that, despite the reform since the 1980s, non-state sectors continue to find it harder than SOEs to receive R&D funding allocations from the state.

If we examine invention patent applications statistics, another potential indicator of effective innovation (Figure 2.9), it seems that non-state firms have had the highest growth, next are foreign firms, and SOEs have the lowest patent application. Of course, patent statistics are tricky measures because they do not directly correlate to the quality of innovations, the importance of innovations, or actual novel developments. To further muddle this metric, the number of patent applications from foreign firms can also be reapplications for R&D in other countries. Nevertheless, an undeniably significant change is the growth of patent applications from non-state firms, especially after 2010.

These indicators show that Chinese non-state sectors have increased their R&D capacity tremendously in the last decade or so. This suggests that China's indigenous innovation policies do indeed have the effect of increasing R&D spending amongst all types of firms. Although the state R&D funding allocation seemed to favor SOEs, non-state companies also benefitted from the policy, although not proportionally to their shares of GDP, R&D budget, and patents. For China to have a more effective innovation system, the state not only needs to be willing to spend, which it currently does, but also has a fair and inclusive system to allow competitive non-state firms to share state R&D resources. China currently has two systems of R&D spending. The state-controlled system, with its enormous resources, has been able to make strategic advances in key sectors such as high-speed rail and IC foundry manufacturing which requires large and long-term financial commitment, but this system has not been equally effective in propagating technological assimilation and in diffusing technology to other sectors of society. The R&D spending in non-state sectors is highly responsive to the needs of the market and society, but it is not well supported by the state and faces a battle against monopolies held by SOEs and well-capitalized foreign enterprises. For example, many non-state firms could be involved in the production or service chain of China's railway system, but have largely been shut out of the state monopoly. The state monopoly has also been a fertile ground for corruption, as the arrest and conviction of the former head of Ministry of Railway, Liu Zhijun, testify. Overall, China's top-down and bottom-up R&D systems need to become better integrated and less discriminatory to non-state actors. As China's economy progresses, more innovation will undoubtedly come from the interactions of Chinese and global firms as well as from non-state companies that have to generate higher-quality, lower-cost products (i.e. to innovate).

Collectively, the graphs also show substantial changes after 2008 in all these measures. They suggest that the increasing labor costs, the growth of China's industrial capacity, and the state indigenous innovation policy together are generating forces to push different types of enterprises to increase R&D investment, thus signaling an acceleration of China's innovative capacity. While China still faces the problems of state favoritism toward SOEs, as well as inefficiencies and corruptions within the state S&T system and between the state and private business, the bleak assessment of extreme bias or the low capacity of China non-state sectors in innovation is false. All measures have shown a rapid growth and productivity of R&D investments in non-state sectors. In sum, we view that the Chinese state investment in R&D and internal R&D by business enterprises as largely complementary and mutually supportive.

Conclusion

Overall, the Chinese state has consistently been a pivotal player in China's innovation system. It not only has accumulated tremendous resources with China's economic development, but it has also initiated, organized, and funded national S&T projects. Since the 1950s, the Chinese state has shifted from a centralized, Soviet-style S&T model to a more open and enterprise-driven model in the 1980s and 1990s through export promotion and TMFT. By the 2000s, China had integrated itself into global production systems, serving both the international and domestic markets. Even so, policymakers felt that China's innovative capacity did not match the progress in its production capacity. China continued to be situated at the lower end of the global division of labor, while core R&D activities remaining largely in R&D centers of TNCs in developed countries. The joint ventures that were set up between Chinese SOEs and foreign companies in several high-tech sectors have not generated sustained technological advances, largely due to a lack of technological competency or strategic control on China's side of the joint ventures.

In 2006, recognizing that a complete embrace of globalization does not naturally bring technological innovation, the Chinese state decided to reclaim a more central role in the innovation system by issuing a newly fashioned indigenous innovation policy. This policy resumed some of the top-down management and bureaucratic control practices of the 1950s and 1960s, but it also tried to encourage participation of enterprises, including non-state and foreign enterprises. The policy utilized sophisticated new tools to encourage indigenous innovation, including tax incentives and public procurement policy as well as pushing for domestic technological standards. While it started with techno-nationalist objectives, the state has made concessions to the demands of foreign enterprises, and it has promised to maintain an international and open innovation system. China is also building on the post-1980 reforms to promise the participation of non-state actors, although critics have charged that the deeply entrenched, top-down system does not do enough in this area. In the last 30 years, regional governments have emerged as important players for financing new high-tech industries. While the state has not lessened its control of the essential sectors of the economy, and SOEs still enjoy privileged positions in state finance, the growth of non-state sectors in R&D capacity and investment is also evident. The trend of growth in R&D is especially sharp after 2010, which suggests that Chinese economy is undergoing a structural change that is primed on technological development.

The Chinese state has come a long way, but significant reform still lies ahead for China to become a truly innovation nation. In particular, the innovation system has to guard against the powerful legacy of central planning and strong bureaucratic control of the system. It also needs to develop better mechanisms to take advantage of nascent venture capital markets. In order to contribute to global scientific progress, more attention needs to be paid to basic research. Toward that end, scientists need to have more autonomy and freedom to pursue advances, unleashed from bureaucratic allocation of resources. The most recent round of S&T policy reform seems to signal such a direction (Cyranoski 2014). China's new position in the global system and its increasing trade relations and fractions with other partners suggest that interactions with foreign companies need to remain open, transparent and ensure mutual benefits. More institutional linkages will need to be built between GRIs, universities, SOEs, foreign companies, and non-state sectors to foster knowledge diffusion and creativity.

The state clearly can play a leading role in advancing innovation, but such a role can only be fair and effective if the state is more inclusive and accountable for its various constituents and trade partners. China's transition to a technology-oriented economy must be a long and cumulative process, one that makes China more, not less, integrated in the global economy. "Indigenous innovation," finally, should stress China's contribution to the world innovative system as an engaging and transforming actor. If future policies are developed with such goals in mind, China could overcome its historically limited role as a partner producer and become a truly innovative country. Given its resources and potential capability, a strong innovation system in China could not only help fortify the domestic economy but could also drive technological progress for the entire world.

References

Ahern, N. (2010), *Innovation and the Visible Hand: China, Indigenous Innovation, and the Rrole of Government Procurement,* Carnegie Papers, 114, July. Washington, DC: Carnegie Foundation.

Appelbaum, R. P., R. Parker, and C. Cao (2011), "Developmental State and Innovation: Nanotechnology in China," *Global Networks*, 11(3): 298–314.

Baark, E. (2010), "Technology and Entrepreneurship in China: Commercializing Reforms in the Science and Technology Sector," *Policy Studies Review*, 18(1): 112–29.

Block, F., and M. Keller (eds) (2011), *The State of Innovation: The U.S. Government's Role in Technology Development*. Boulder, CO: Paradigm Publishers.

Brahmbhatt, M., and A. Hu (2009), "Ideas and Innovation in East Asia," *World Bank Research Observer*, 25(2): 177–207.

Breslin, S. (2011), "China and the Crisis: Global Power, Domestic Caution and Local Initiative," *Contemporary Politics*, 17(2): 185–200.

Breznitz, D., and M. Murphree (2011), *Run of the Red Queen: Government, Innovation, Globalization and Economic Growth in China*. New Haven: Yale University Press.

Cao, C., R. Suttmeier, and D. F. Simon (2006), "China's 15-Year Science and Technology Plan," *Physics Today*, Dec. <http://www.levininstitute.org/pdf/Physics%20Today-2006.pdf>, accessed Mar. 2014.

Cao, C. (2004), *China's Scientific Elite*, New York: RoutledgeCurzon.

Castells, M. (1989), "High Technology and the New International Division of Labour," *Labour and Society*, 14 (special issue on High Tech and Labour): 237–42.

Chang, H. (2002), *Kicking Away the Ladder: Development Strategy in Historical Perspective*. London: Anthem Press.

Chen, B. X., and E. Pfanner (2013), "Cheaper iPhone Will Cost More in China," *New York Times*, Sept. 11, <http://www.nytimes.com/2013/09/12/technology/the-surprise-in-apples-cheaper-iphone-in-china-its-expensive.html>.

Chen, L., and L. Xue (2010), "China's High Tech Industry in the International Division of Labour and its Industrial Upgrading Strategies: A Case Study of IC Industry," *China Soft Science (中国软科学)*, 6: 36–46 (in Chinese).

Coe, N., M. Martin, N. Hess, H. W. C. Yeung, D. Dicken, and J. Henderson (2004), "'Globalizing' Regional Development: A Global Production Networks Perspective," *Transactions of the Institute British Geographers*, NS 29(4): 468–84.

Crookes, P. I. (2012), "China's New Development Model: Analyzing Chinese Prospects in Technology Innovation," *China Information*, 26(2): 167–84.

Cyranoski, David (2014), "Fundamental Overhaul of China's Competitive Funding," *Nature* news blog. Oct. 23. <http://blogs.nature.com/news/2014/10/fundamental-overhaul-of-chinas-competitive-funding.html>.

D'Aveni, R. (2012), *Strategic Capitalism: The New Economic Strategy for Winning the Capitalist Cold War*. New York: McGraw-Hill.

Ernst, D. (2011), *China's Innovation Policy is a Wake-up Call for America*. East-West Center, 100. <http://www.eastwestcenter.org/news-center/east-west-wire/chinas-innovation-policy-is-a-wake-up-call-for-america-not-a-threat>.

Ernst, D., and L. Kim (2002), "Global Production Networks, Knowledge Diffusion, and Local Capability Formation," *Research Policy*, 31: 1417–29.

Feigenbaum, E. A. (2003), China's Techno-Warriors: National Security and Strategic Competition from the Nuclear to the Information Age. Stanford, CA: Stanford University Press.

Feng, W. (冯伟) (2011), "How Can Market Access Trade Technology?" (市场如何才能换来技术?) *Science of Science and Management of S.&T. (科学学与科学技术管理)*, 32(10): 158–64.

Fischer, A. M. (2010), "Is China Turning Latin? China's Balancing Act between Power and Dependence in the Lead up to Global Crisis," *Journal of International Development*, 22(6): 739–57.

Granick, D. (1990), *Chinese State Enterprises: A Regional Property Rights Analysis*. Chicago: Chicago University Press.

Huang, C., and Y. Wu (2012), "State-Led Technological Development: A Case of China's Nanotechnology Development," *World Development*, 40(5): 970–82.

Huang, Y. (1996), "Central–Local Relations in China during the Reform Era: The Economic and Institutional Dimensions," *World Development*, 24(4): 655–72.

Huang, Y. (2008), *Capitalism with Chinese Characteristics: Entrepreneurship and the State*. New York: Cambridge University Press.

IMF (2011), "Changing Patterns of Global Trade," June 15. <http://www.imf.org/external/np/pp/eng/2011/061511.pdf>.

Keller, W. W., and R. J. Samuels (2003), "Innovation and the Asian Economies," in W. W. Keller and R. J. Samuels (eds), *Crisis and Innovation in Asian Technology* (1st edn), 1–22. Cambridge: Cambridge University Press.

KOITA (various years), *Major Indicators of Industrial Technology*. Seoul: KOITA.

Kushida, K. (2011), "Leading without Followers: How Politics and Market Dynamics Trapped Innovations in Japan's Domestic 'Galapagos' Telecommunications Sector," *Journal of Industry, Competition and Trade*, 11: 279–307.

Lazonick, W. (2011), "Entrepreneurship and the Developmental State," in W. Naudé (ed), *Entrepreneurship and Economic Development*, 254–70. Helsinki: World Institute for Development Economics Research, United Nations University.

Lei, X., Z. Zhao, X. Zhang, D. Chen, M. Huang, and Y. Zhao (2012), "The Inventive Activities and Collaboration Pattern of University-Industry-Government in China Based on Patent Aanalysis," *Scientometrics*, 90(1): 231–51.

Lever-Tracy, C., and D. Ip (1996), "Diaspora Capitalism and the Homeland: Australian Chinese Networks into China," *Diaspora*, 5: 239–74.

Liu, X., and C. Peng (2011), *Is China's Indigenous Innovation Strategy Compatible with Globalization? Policy Studies*, 61. Honolulu, HI: East-West Center. <http://www.eastwestcenter.org/publications/chinas-indigenous-innovation-strategy-compatible-globalization> accessed Aug. 2015.

Liu, X. (2007), "Path-Following or Leapfrogging in Catching-up: The Case of Chinese Telecommunication Equipment Industry," Paper 2007/01, presented at Centre for Innovation, Research and Competence in the Learning Economy (CIRCLE), Lund University. <wordpress.circle.lu.se/upload/CIRCLE/workingpapers/200701_Liu.pdf> accessed Aug. 2015.

Liu, X., and China Science and Technology Strategic Research Group (2010), *Chinese Report of Regional Innovation Capability* (中国区域创新能力报告). Beijing: Chinese Science Press (in Chinese).

Liu, X., and N. Lundin (2006), *Globalisation of Biomedical Industry and the System of Iinnovation in China*. Stockholm: SNS.

Liu, X., and S. White (2001), "Comparing Innovation Systems: A Framework and Application to China's Transitional Context," *Research Policy*, 30: 1091–114.

Liu, X., and H. S. Zhang (2002), "Knowledge Intensity and SOE's Performance in China," in B. Grewal and L. Xue et al. (eds), *China's Future in the Knowledge Economy*, 164–77. Melbourne and Beijing: Centre for Strategic Economic Studies, Victoria University, and Tsinghua University Press.

Liu, Y., M. Woywode, and Y. Xing (2012), "High Technology Start-Up Innovation and the Role of Guanxi: An Explorative Study in China from an Institutional Perspective," *Prometheus*, 30(2): 211–29.

Lu, F., and K. D. Feng (2005), *The Policy Choice of Developing Indigenous IPR Chinese Automobile Industry*. Beijing: Peking University Press.

Lu, H., A. Przybyla, L. Reed, M. Wang, and H. Xu (2012), "A Framework and Case Study to Evaluate China's Megaprojects and Strategic Emerging Industries," paper presented at Conference on the Structure, Process and Leadership of the Chinese Science and Technology System, UC Institute on Global Conflict and Cooperation.

Lu, Q. (2000), *China's Leap into the Information Age: Innovation and Organization in the Computer Industry*. Oxford: Oxford University Press.

Lubman, S. (2010), "China Modifies Government Procurement Policies, But Foreign Concerns Remain," *Wall Street Journal*, Apr. 19. <http://blogs.wsj.com/chinarealtime/2010/04/19/stanley-lubman-china-modifies-government-procurement-policies-but-foreign-concerns-remain>.

McGregor, J. (2010), *China's Drive for "Indigenous Innovation": A Web of Industrial Policies*, report commissioned by US Chamber of Commerce. Washington, DC: US Chamber of Commerce. <http://www.uschamber.com/reports/chinas-drive-indigenous-innovation-web-industrial-policies>.

Mazzucato, M. (2013), *The Entrepreneurial State*. New York: Anthem Press.

Ministry of Education, China (2005), *Educational Statistics Yearbook of China* (中国教育统计年鉴). Beijing, People's Education Press.

National Bureau of Statistics (NBS) (2004), *China Statistical Yearbook on High Technology Industry, 2004*. Beijing: China Statistical Press.

Naughton, B. (1994), "Chinese Institutional Innovation and Privatization from Below," *American Economic Review*, 84(2): 266–70.

Naughton, B. (1995), *Growing Out of the Plan: Chinese Economic Reform, 1978–1993*. New York: Cambridge University Press.

Naughton, B. (2011), "China's Economic Policy Today: The New State Activism," *Eurasian Geography and Economics*, 52(3), 313–29.

Naughton, B., and A. Segal (2003), "China in Search of a Workable Model: Technology Development in the New Millennium," *Crisis and Innovation in Asian Technology*, 160–86. Cambridge: Cambridge University Press.

NBS (2004–12), *China Statistical Yearbook on Science and Technology*, multiple years. Beijing: China Statistical Press.

Nie, R. (1989), "In the Frontline of Science and Technology (在科学技术战线上)," in edited by L. Nie and G. Huai (eds), *Retrospect and Prospect: 1949–1989 (回顾与展望; 1949–1989)*, 49–73. Beijing: Defense Industry Press.

Oi, J. (1999), *Rural China Takes Off: Institutional Foundations of Economic Reform*. Berkeley, CA: University of California Press.

Okimoto, D. I. (1989), *Between MITI and the Market*. Stanford, CA: Stanford University Press.

Pei, M. (2006), *China's Trapped Transition: The Limits of Developmental Autocracy*. Cambridge, MA: Harvard University Press.

PR Newswire (2015), "Huawei Promotes Open Innovation and Win-win Collaborations at CeBIT 2015." <http://www.prnewswire.com/news-releases/huawei-promotes-open-innovation-and-win-win-collaborations-at-cebit-2015-300051309.html> accessed Aug. 2015.

Segal, A. (2003), *Digital Dragon: High-Technology Enterprises in China*. Ithaca, NY: Cornell University Press.

Shih, H., and P. Chang (2009), "Industrial Innovation Networks in Taiwan and China: A Comparative Analysis," *Technology in Society*, 31(2): 176–86.

Shirk, S. L. (1993), *The Political Logic of Economic Reform in China*. Berkeley, CA: University of California Press.

Suttmeier, R. P. (1997), "Emerging Innovation Networks and Changing Strategies for Industrial Technology in China: Some Observations," *Technology in Society*, 19(3/4), 305–52.

Suttmeier, R. P., and C. Cao (2004), "China's Technical Community: Market Reforms and the Changing Policy Cultures of Science," in E. Gu and M. Goldman (eds), *Chinese Intellectuals between State and Market*, 138–57. New York: RoutledgeCurzon.

Suttmeier, R. P., X. Yao, and A. Z. Tan (2006), *Standards of Power? Technology, Institutions, and Politics in the Development of China's National Standards Strategy*, NBR Special Report, 10, June. <http://www.nbr.org/publications/element.aspx?id=1d5ffd96-9673-4d1c-85ce-bd8d8787e888> accessed Aug. 2015.

Trefis Team (2013), "As China Mobile's Earnings Flatten, More 3G Growth Needed," *Forbes*, May 6. <http://www.forbes.com/sites/greatspeculations/2013/05/06/as-china-mobiles-earnings-flatten-more-3g-growth-needed>.

Wei, S. (2012), "Testimony before the U.S.–China Economics and Security Review Commission Hearing on the Evolving U.S.–China Trade and Investment Relationship." <http://www8.gsb.columbia.edu/rtfiles/chazen/Wei-testimony-6-9-2012.pdf> accessed Aug. 2015.

Wu, W., and Y. Zhou (2012), "The Third Mission Stalled? Universities in China's Technological Progress," *Journal of Technology Transfer*, 37(6): 812–27.

Xia, L., and L. Y. Zhao (2012), "Historical Evolution of the 'Trade Market for Technology' Strategy" ('市场换技术'方针的历史演变), *Contemporary China History Studies (*当代中国史研究*)*, 19(2): 27–36.

Xing, Y. (2011), *China's High-tech Exports: Myth and Reality*. GRIPS Discussion Paper 11-05. <http://r-center.grips.ac.jp/DiscussionPapersDetails/209> accessed Jan. 2013.

Xinhua Net (2008), "About 1160 R&D Labs Were Set Up by Multinationals in China." <http://news.xinhuanet.com/newscenter/2008-02/09/content_7583212.htm> (in Chinese), accessed Aug. 2015.

Xu, B. (2006), *Measuring the Technology Content of China's Export*, Working Paper. Shanghai: China-Europe International Business School.

Xu, G. (2005), Press release of the speech as the head of the Ministry of Science and Technology, at the Symposium on China's National Medium- and Long-Term Science and Technology Strategic Plan, held in Beijing in 2005.

Zhan, A., and Z. Tan (2010), "Standardisation and Innovation in China: TD-SCDMA Standard as a Case," *International Journal of Technology Management*, 51(24): 453–68.

Zhang, F., and F. Wu (2012), "Fostering Indigenous Innovation Capacities": The Development of Biotechnology in Shanghai's Zhangjiang High-Tech Park," *Urban Geography*, 33(5): 728–55.

Zhou, Y. (2005), "The Making of an Innovative Region from a Centrally Planned Economy: Institutional Evolution in Zhongguancun Science Park in Beijing," *Environmental and Planning A*, 37: 1113–34.

Zhou, Y. (2006), "State and Commercial Enterprises in China's Technical Standard Strategies," *China Review*, 6(1): 37–65.

Zhou, Y. (2008a), *Inside Story of China's High-Tech Industry: Making Silicon Valley in Beijing*. Lanham, MD: Rowman & Littlefield.

Zhou, Y. (2008b), "Synchronizing Export Orientation with Import Substitution: Creating Competitive Indigenous High-Tech Companies in China," *World Development*, 36(11): 2353–70.

Zhou, Y., Y. Sun, Y. H. D. Wei, and G. C. S. Lin (2011), "De-centering 'Spatial Fix': Patterns of Territorialization and Regional Technological Dynamism of ICT Hhubs in China," *Journal of Economic Geography*, 11(1): 119–50.

3

Venture Capital in China

Jun Zhang

Introduction

Venture capital (VC) has increasingly been recognized by both academics and policymakers around the world as a key instrument of the development of an entrepreneurial and innovative economy. In addition to the boom of private VC funds over the past two decades, there has been a dramatic expansion of public programs around the globe that encourage the formation of VC funds in order to catalyze technological innovation and job creation (OECD 1995; Lerner 2009; Sunley et al. 2005). Venture capital may be defined as dedicated pools of capital managed by independent, professional organizations that focus on equity or equity-linked investments in young firms with expected high growth potential (cf. Gompers and Lerner 2001: 146). This definition distinguishes venture capital from three other forms of investment: (1) non-venture private equity (PE) investment in the form of buyouts, restructure and mezzanine funds, targeting reasonably mature firms with stable cash flows and limited growth potential, which sometimes has been termed "merchant venture capital"; (2) public equity investment in mature firms with high liquidity and low risk; and (3) angel capital funds operated by individuals rather than organizations. Here my focus is on "classic venture capital" rather than the PE-based "merchant venture capital," because it is the former that is especially associated with the nourishing of innovation and entrepreneurship. Classic venture capitalists are both catalyst and capitalist, providing necessary resources and contacts to facilitate the creation of new companies and supporting their growth (Florida and Kenney 1988). Merchant venture capital, by contrast, does not emphasize company-building skills, instead seeking to add value by means of financial engineering or deal-making skills which often involve corporate restructuring (Mason and Harrison 2002; Sunley et al. 2005).

Venture capitalists can provide not only necessary finance capital but also complementary "infrastructural knowledge" (Gertler and Wolfe 2006), business experience, network connections, specialized knowledge of target industries, IPO opportunities, and other incubatory advantages to enhance the venturing and learning of their investees. In other words, they potentially offer much-needed firm-building skills and resources to young firms (Bresnahan et al. 2001), even though VC's short-term orientation may not be consistent with the innovative enterprise's demand for long-term commitment (Lazonick 2007a, 2007b, 2013).

In order to understand the nature of the VC industry, it is necessary to move beyond a focus on investor–investee interactions alone and view VC as a cycle with inherently interrelated processes, including fundraising, investing, divesting, and new fundraising (Gompers and Lerner 2004: 3). Although such a cycle-based view can still be insufficient to explain the connection between VC investment and industrial innovation, it is especially helpful for understanding the origination of China's VC sector from scratch, which was simultaneously subject to many bottlenecks: short supplies of VC professionals, funding sources, appropriate investees, and exit channels, as well as no enabling institutional and policy framework. In the existing literature, the localized nature of VC has been strongly emphasized. VC firms tend to focus a significant proportion of their investment in firms in their own immediate region (Thompson 1989; Mason and Harrison 2002; Martin et al. 2005; Chen et al. 2010). However, the VC industry more recently has become a highly globalized practice, and China is the example of VC globalization par excellence. Several recent surveys have provided strong evidence that VC firms all over the world are taking the idea of global investing seriously and that the VC industry is in an indisputable process of global expansion (Deloitte 2009; Ernst & Young 2009, 2011). The economic crisis initiated in the core economies in 2007 suppressed already dwindling business and forced mainstream VC funds to search for investment opportunities in emerging markets (Mason 2009). The ensuing global integration of venture capital and powerful role of transnational players posed a significant challenge to the compatibility and adaptability of institutional arrangements, especially for a transitional economy like China. Relatedly, it also has made the interplay between foreign and domestic investors central to understanding the dynamics of VC's rise in China.

This chapter documents the rise of China's VC industry partly as a process of globalization—the initial dominance of transnational investors, the recent ascendance of domestic investors, and the associated transformation of China's institutional environment as the enabling and constraining factor. The important question of to what extent venture capital can be considered to spur innovation in the Chinese context will also be addressed. The primary data used here derive mainly from the Zero2IPO Database provided by China's leading VC consulting company Zero2IPO, which is headquartered in Beijing. Sources of the Zero2IPO Database are threefold: regular questionnaire surveys covering the most active China-focused investment institutions since 1999, periodical telephone surveys, and other published reports. The Zero2IPO Database is a comprehensive, constantly updated online database covering venture capital, private equity funds, VC-invested firms, and portfolio companies focused on mainland China. The database contains detailed itemized information of 8,000 + venture capital and private equity (VC/PE) firms, 6,000 + limited partners or investors, 20,000 + funds targeting mainland China, 40,000 + individual investors, and 20,000 + VC/PE investment deals by the end of 2013. Besides including detailed profiles of the investor and investee firms, the deal information records the closing date and investment amount of each investment deal, the round and stage of the investment, all the VC/PE investors involved and their respective amount of investment. Based on the Zero2IPO data, I have constructed an itemized census database of China's venture capital investment, which I term VentureChina database. It is the first available comprehensive VC database on China and comparable to the VentureOne database used by many students of the US VC industry (e.g. Chen et al. 2010).[1] But VentureChina data are only updated through the year 2008, which were used in Zhang (2011). Here, my discussion of the post-2008 development is primarily based on the aggregate data published by Zero2IPO.

In my classification, domestic VC firms were those whose control rights located in the hands of domestic individuals and/or organizations (including the state). For domestic VC firms, those founded or controlled by government entities, state-owned enterprises (SOEs), or state-owned universities were classified as state-owned, while others were classified as private-owned. VC firms with control rights in the hands of foreigners and with capital raised (partly) from offshore limited partners (LPs) were classified as foreign-owned.[2] Most domestic VC firms raise their funds in RMB exclusively through domestic sources, but some also have offshore funds under management. Given the institutional uniqueness of Hong Kong, VC firms originated from Hong Kong were classified as foreign.

[1] For about 15% of the deals, investment amount was undisclosed. In these cases, an estimated amount was provided based on the best available information, such as the sector, round, stage of the investment, and the stated investment criteria and track record of the investor. In some co-investment cases if only the total amount but not individual contribution was disclosed, then equal contribution from each participant was assumed.

[2] As explained later, some foreign VC firms started to raise onshore RMB funds after 2009.

The Spectacular Rise of Venture Capital in China

After a decade of explosive growth, China has become the second largest national market for global VC investment after the United States in 2013.[3] According to Zero2IPO, while there were only 10 active VC/PE firms in China in 1995, the number increased to 500 in 2005 and 5,000 in 2012. My own assessment found only 58 active VC (excluding PE) offices operated by 38 firms in 1998 but 468 offices operated by 334 VC firms in 2008. By 2008, 311 foreign VC firms from 22 countries/regions already had their investment imprint on China, and 154 firms had already launched their local offices (Figure 3.1). From 2002 to 2012, Chinese and foreign VC firms' investable capital targeting the Chinese mainland grew year by year, with an annual average compound growth rate of 17 percent. In the past two decades, over US$53 billion in venture capital have been invested into over 7,900 deals, and more than 1,000 of the invested firms have achieved IPOs (Figure 3.2).

The budding domestic VC firms had long been overwhelmed by their powerful international rivals until the last few years, as will be elaborated upon later in this chapter (White et al. 2005). Foreign VC firms, as part of their global operations, came to China with higher profiles, better resources, and greater professionalism. In contrast, local VC firms were characterized by a lack of funding, experience, and competence. In 2004, there were still no more than 10 active domestic VC organizations in China. The percentage of investments by domestic VCs in China was not increasing but in fact decreasing by 2007. By the end of 2007, active VC organizations managed $21.3 billion in funds targeting the Chinese market, and 72 percent of this was controlled by foreign or joint venture VCs. The cumulative amount of investment by VC firms with identifiable ownership by 2008 was $14.3 billion, of which only $3 billion, or 21.3 percent, was invested by domestic investors: 15.3 percent by state-owned and 6 percent private. The US has been the undisputable leader among all investors. 147 US VC firms, including 55 firms each with at least one China office, contributed 40 percent of cumulative VC investment in China by 2008. VC firms from Taiwan and other Asian neighbors such as Japan, Singapore, and Korea have also been active players in China's VC market. The 334 active VC firms in 2008 included 157 foreign, 123 domestic state-owned, and 54 domestic private ones.

[3] According to Ernst & Young (2013), China's total amount of VC investment in 2012 was US $3.7 bn and 2011 US$6.3 bn, but Zero2IPO's figure was $7.3 and $13 respectively. It is likely the Zero2IPO data on China are more accurate, and the amount of VC investment in China was more than that of the whole of Europe in 2011 (US$6.8 bn) and 2012 (US$5.8 bn).

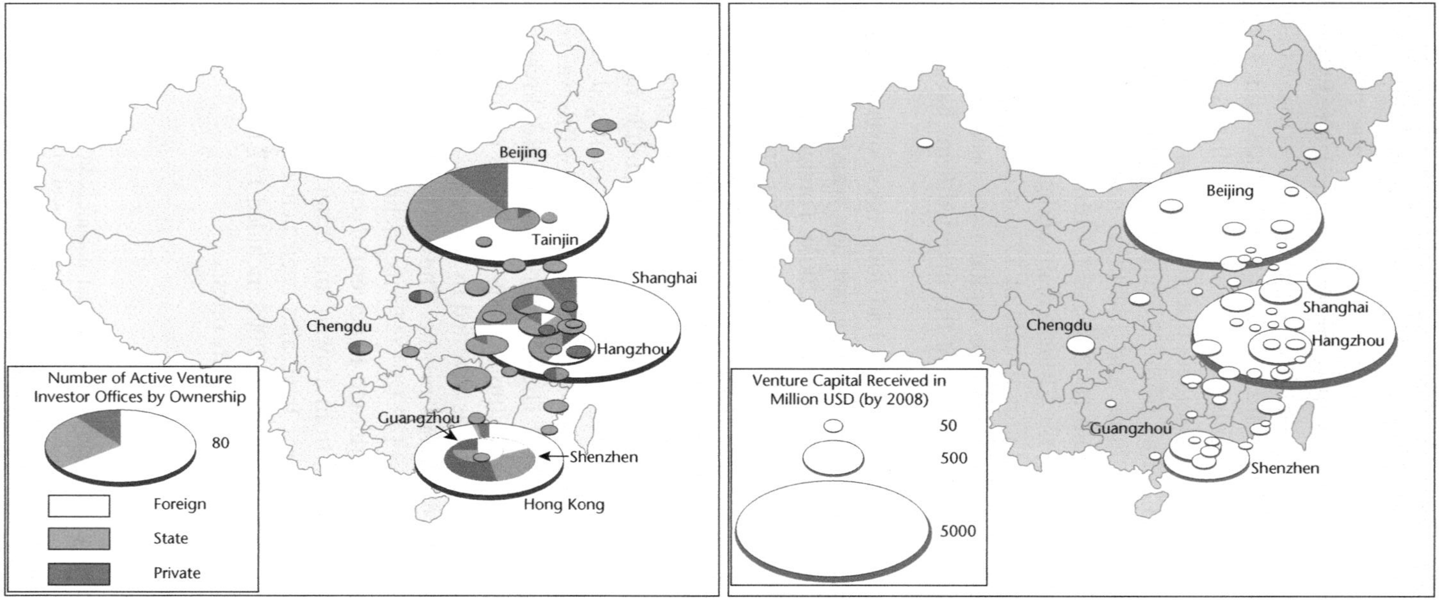

Figure 3.1. Spatial Distribution of VC Offices and Investments in 2008

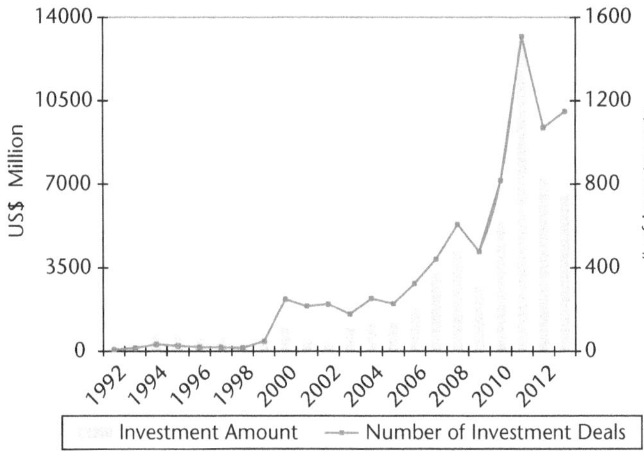

Figure 3.2. Growth of Venture Capital Investments in China: 1992–2013

However, domestic VC firms and their RMB funds have been increasing exceptionally quickly since 2008. The amount of newly raised domestic funds surpassed foreign funds in 2009, and in 2011 the invested capital stock of domestic VC exceeded that of foreign VC for the first time. In 2012, stock of domestic VC targeting the Chinese mainland reached US$26,658 m, while foreign VC held $22,181 m. In the same year, domestic VC invested US$3.8 bn in 709 deals, but foreign VC only US$2.8 bn in 295 deals.

Since 2011, China's hot VC market has encountered obstacles as a whole, and has experienced an obvious downward trajectory. The number and amount of VC investments significantly edged down in 2012 and 2013 but still surpassed the average level in 2010, with a growing momentum compared with the previous years. 2012 saw a nosedive in fundraising, leading to a new low in growth rate of investable capital in the past 11 years. Suffering from a sluggish fundraising environment, the investable amount dropped for the first time in China's VC market in 2012. But still, the general performance of domestic VC funds far exceeded that of their foreign rivals. In 2013, the number of newly raised RMB funds was 189, in the amount of US$6.4 bn; that of USD funds was 10, representing US$541 m (Figure 3.3). This major downturn since 2012 was a natural adjustment over the market overheat in 2011. It was also largely associated with the IPO suspension by the China Securities Regulatory Commission (CSRC) between November 2013 and January 2014. The market is likely to become stabilized at least in the foreseeable future with the IPO restarting after some regulatory disciplining and tightening.

In terms of the most successful VC institutions as comprehensively measured by Zero2IPO, foreign VC institutions have also been dominant

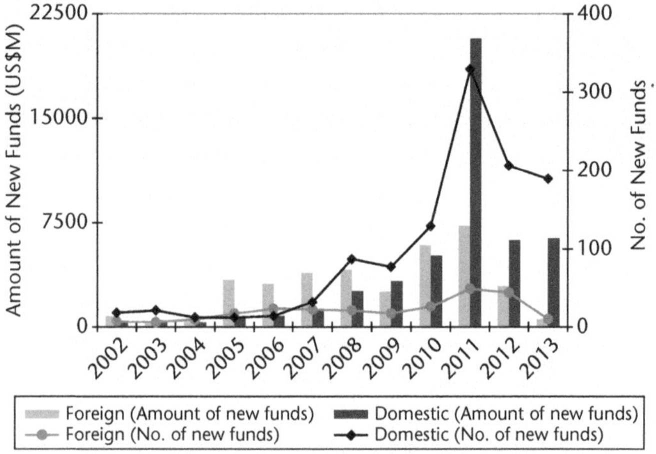

Figure 3.3. The Annual Growth of Newly Raised VC Funds: 2002–13

Table 3.1. Zero2IPO's Annual Top 20 VC Investors 2012

Rank	VC firm	Origin	HQ
1	Fortune Venture Capital	China	Shenzhen
2	Shenzhen Capital Group	China	Shenzhen
3	IDG Capital Partners	USA	Beijing
4	China Science & Merchants Capital Management	China	Beijing
5	Sequoia Capital China	USA	Beijing
6	Zhejiang Sinowisdom Capital	China	Hangzhou
7	Granite Global Ventures	Singapore	Shanghai
8	NewMargin Ventures	China	Shanghai
9	Shenzhen Co-win Venture Capital	China	Shenzhen
10	Shenzhen Costone Venture Capital	China	Shenzhen
11	Suzhou Venture Group	China	Suzhou
12	Happy-Silicon Capital Management	China	Hangzhou
13	Suzhou International Development Venture Capital	China	Suzhou
14	Shenzhen Oriental Fortune Capital	China	Shenzhen
15	Shenzhen GTJA Investment Group	China	Shenzhen
16	Qiming Weichuang Venture Capital	USA	Shanghai
17	SureLand Capital	China	Beijing
18	Cybernaut (China) Investment	USA	Hangzhou
19	Jiangsu Govtor Capital Group	China	Nanjing
20	SAIF Partners	Japan	Beijing

throughout the years. But the number of foreign institutions in the Zero2IPO annual top 20 list has been constantly declining since 2006, and in 2012 for the first time, the number of domestic firms (14) exceeded foreign ones (6) and has a large collective lead (Table 3.1). As detailed in the next section, it was the juxtaposition of a shifting international market environment and domestic institutional transition that gave domestic VC firms such a quick and complete turnaround against their powerful foreign rivals.

Venture Capital Cycle: Changing Rules, Changing Dynamics

Rather than a series of discrete investment activities, the VC firm's process is better viewed as a continuous cycle: raising a fund; investing in, monitoring, and adding value to firms; exiting deals; returning capital to investors; and finally renewing itself by raising new funds (Gompers and Lerner 2004). In order to understand the venture cycle, we need to pay close attention to China's changing institutional environment from a capital-hostile, centrally planned economy to an increasingly capital-friendly, market-oriented economy, and the interplay between foreign and domestic investors. It is well known that an iron law of the VC industry is "no exit, no entry." Exit strategies sit front and center for venture capitalists. In some cases, a perceived easy exit may lead to careless investments, as exemplified by the internet boom and bust (e.g. Zook 2002). But without foreseeing stable exit channels, fundraising for VC institutions would not be forthcoming and new firms, funds, and offices would have nothing to begin with. An active domestic stock market for growth enterprises is usually required for venture capitalists to exit through IPOs (Jeng and Wells 2000), especially for an economy as large as China's. But a market for mergers and acquisitions, the other primary exit routes for VCs, is yet to emerge. Providing an effective channel for VC divestment, especially through IPOs, is a daunting and controversial task, often involving substantial legal, regulatory and structural changes of the financial institutions of the host countries (see Lazonick 2007a for the US experience).

The PRC government has long seen science and technology as a critical part of its search for economic development and national security. It eventually legitimized the venture capital system to emerge in China in the late 1980s, primarily because venture capital was seen as a useful instrument to boost scientific and technological capabilities and outputs. In March 1985, the Central Committee of the Chinese Communist Party (CCP) promulgated the "Decision regarding the reform of science and technology system," calling for the development of a venture capital industry to support high-tech development as a national strategy (White et al. 2005). Consequently the first mainland China VC firm, the China New Technology Venture Investment Corporation, was formed in 1986. It was founded jointly by the State Science and Technology Commission (SSTC) (later reformed into the Ministry of Science and Technology, or MOST) and the Ministry of Finance. In addition to active initiatives taken by the MOST of the central government, regional governments—including local departments of finance, bureaus of science and technology, state asset management commissions, and high-tech zone administration departments—have all become active players in establishing VC funds. Domestic VC firms subsequently established were largely controlled by these local government bodies, large state-owned enterprises (SOEs) or state-owned universities.

Foreign VC firms were first allowed to register as commercial enterprises in China in the 1980s. Due to the lack of suitable investment targets, only in the early 1990s did a few foreign VC firms founded and/or operated by American-trained Chinese/Taiwanese start to test the water in China. These included such firms as the International Data Group (IDG), Walden International, and H&Q Asia Pacific. To reduce uncertainties, these international VC funds typically incorporated as joint venture funds with state entities in China. For example, IDG created the $50 million region-focused Pacific Development Venture Fund in 1993. In 1994, it established three joint ventures with the local Commission of Science and Technology in Beijing, Shanghai, and Guangzhou (Zeng 2004: 75–7).[4] However, VC investors, foreign and domestic alike, had largely stayed inactive due to the dearth of suitable investees until the global internet boom extended to China in 1998. The internet boom culminated in 2000 when five foreign VC-backed, Beijing-based, internet/telecom startups—AsiaInfo, UTStarcom, Sina, Sohu, and Netease—achieved their Nasdaq IPOs in the same year. Such high-profile VC investments induced unprecedented market entries, signaling the actual advent of the VC era in China.

China's financial sector has been and still is closely tied to the SOEs. In one cross-country survey around 2000, private firms in China were found to be among the most constrained in the world in terms of their access to capital (Haggard and Huang 2008). State banks nearly exclusively served state-owned enterprises, and most private ventures had been denied access to official sources of credit. In defiance of the national banking laws, small business owners especially in Southern coastal regions created what is termed "back-alley banking" by Kellee Tsai (2002)—a dizzying variety of informal financing mechanisms, including rotating credit associations and private banks disguised as other types of organizations.

Before 1999, venture capital was largely unknown to the Chinese. Some people had heard about the successful entrepreneurial stories of Netscape and Yahoo, but these were incredibly remote to the mainlanders. As Sohu's founder Charles Zhang, one of the best known returnee student entrepreneurs from MIT, commented, in 1996 financing opportunity for startups was next to zero. He attested that if he were to go to a bank for a loan, he would have been asked what the value of his fixed assets were and whether or not he had factories and machines as a deposit. At the time, there was no idea at all across China about how to evaluate a knowledge-intensive company and no way for an internet startup to get access to bank loans (Yang 2001: 36). Pony Ma is the

[4] By the end of the 1990s, as China's market-oriented reform deepened and legal barriers were removed, international VC funds were much less likely to have state partners.

founder of Tencent, now one of the world's largest internet companies. In a personal interview in Shenzhen in 2002, he told me that it is very likely that in 1999 he would have sold his business for less than one million Chinese yuan if he had not been lucky enough to secure the one million dollars from IDG VC.

With the quick burst of the global internet bubble, however, there was a widespread retreat of venture capitalists, especially during 2002 and 2003. The number of VC firms decreased from 296 in 2002 to 233 in 2003 (China VC Investment Research Institute 2003). During the internet and VC boom in 1999, there was already a strong expectation that a Nasdaq-like exchange for high-growth, high-tech start-ups would soon be allowed to open in Shenzhen by the authority in Beijing. The building of such an expectation resulted in the founding of numerous public and private VC firms in Shenzhen. However, the Chinese Nasdaq—ChiNext actually took a decade to open, and during those years many domestic VC firms went bankrupt.

The year 2004 marked a watershed in China's venture capitalism when a virtuous circle of fundraising–investing–exiting first occurred. In 2004, stock markets around the world started to embrace Chinese firms when domestic stock markets were still exclusively dedicated to serving SOEs. The total number of overseas (including Hong Kong) IPOs skyrocketed to 89 from 14 in 2003, according to VentureChina. This sudden explosion of lucrative exit opportunities immediately ignited global venture capitalists' enthusiasm for China. Global mainstream VC funds, including most legendary Sand Hill Road, Menlo Park VC firms in Silicon Valley, were then convinced that "every serious... venture capital firm needs a China strategy" (Maschek 2005: 41). In 2004, total VC investment in China broke the one billion USD mark for the first time.

Under China's unique institutional environment with enduring state controls of foreign exchange and investment, foreign VC firms nearly unanimously adopted an operational strategy that has been termed the "red chip model," or Sina model.[5] Foreign VC investors normally require firms incorporated in China to register a controlling shell company, or a Special Purpose Vehicle (SPV), in an offshore financial center such as the Cayman Islands. VC investments are then made in the wholly foreign-owned SPV, which, in turn, acquires control rights and profits of the China-based business entity through a complicated set of contracts (Chen 2009). With such arrangements, VC-preferred Western financial and legal structures can be created to provide a legal oasis where the VC investor can still meet the objectives of investment protection and access to liquidity in foreign stock markets while bypassing China's legal-political restrictions. The essence of the Sina model is to allow an

[5] The model was first "invented" by Sina, the first Chinese internet company listed on the Nasdaq.

onshore VC investor in China to have both fundraising and divestment stay offshore. This offshoring is necessary first because China restricts or forbids many areas to foreign investors, such as telecommunications value-added services, internet content provision, the media and publishing. It was precisely these restricted areas that have been the primary targets of foreign VC. Another major reason is that listing on domestic stock markets has been difficult, unattractive, or too uncertain until very recently. A third reason is that China's currency RMB is not yet freely convertible and foreign VCs were not allowed to set up RMB funds before 2009 (Chen 2009).

While the offshore funds are able to achieve their investment and divestment overseas via the Sina model, domestic VCs based on onshore RMB funds had no access to domestic or overseas exit channels until very recently. They faced the additional challenges of deficiencies in professional experience and brand recognition. When the "discounted" Shenzhen Small and Medium Enterprise Board (SSMEB), instead of a fully-fledged second board, finally launched in 2004, many domestic VC firms had already disappeared. SSMEB was not really a growth enterprise board; it required firms to have at least three years of consecutive profits to be listed, focusing its business on mature firms with long operation history. Many other institutional constraints such as unfavorable tax terms and untradability of institutional shares held by VC investors remained after the introduction of SSMEB. Nevertheless, the shortage of divestment channels for RMB funds has been alleviated by the arrival of SSMEB, which has since provided the most appealing option (Figure 3.4).

The few relatively successful domestic VC institutions that managed to obtain access to overseas IPOs by 2008, such as Shenzhen Capital Group, NewMargin Ventures, and Legend Capital, had invariably raised offshore funds through extensive collaboration with overseas partners. The majority of domestic VC firms, however, had no access to such networks and thus were unable to take advantage of overseas IPO opportunities.

However, since 2005 Chinese regulators started to tighten the grip aimed at closing offshore loopholes for illicit "round-tripping" of domestic capital and associated tax evasion. A series of directives issued by China's regulatory bodies such as China Administration of Foreign Exchange and China Securities Regulatory Commission established approval and registration requirements for all the experimenters in the red chip model, making offshore restructuring and overseas IPOs much more costly and challenging (Chen 2009). Meanwhile, overseas IPO opportunities started to dwindle when the global financial crisis began to have effect in 2008. The worst season for offshore funds, however, became the best season for onshore RMB funds given China's continuous economic growth and shifting institutional environment. On June 1, 2007, China's revised Partnership Enterprise Law cleared the path for the establishment of Western-style VC funds based on limited

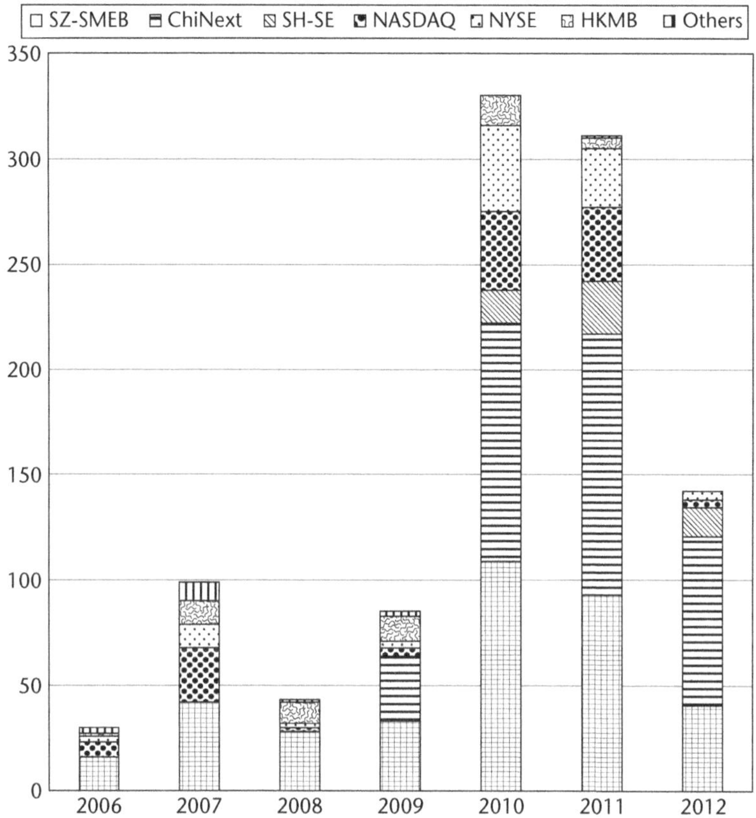

Figure 3.4. Number of IPOs of VC-Backed Firms in Different Stock Markets: 2006–12

partnership, triggering the rapid formation of private VC firms.[6] More dramatically, China's true Nasdaq-style ChiNext board, at least nominally targeting high-tech and high-growth firms, finally made its debut in Shenzhen on October 23, 2009 after a decade of postponement. ChiNext has been embraced by Chinese investors with stunning enthusiasm and thus brought enormous fortunes to the VC investors of listed firms, with an average return rate of 11.66 times by the end of 2009.

The flourishing ChiNext triggered a new boom in the venture cycle from divestment to fundraising and new fund formation. In 2009, onshore RMB funds, partly taking advantage of the global economic crisis and VC slump, for the first time surpassed foreign currency funds to become the dominant player in the Chinese VC market in both fundraising and investment deals. The

[6] The Partnership Enterprise Law was passed on February 23, 1997, was amended on August 27, 2006, and as amended became effective on June 1, 2007. Available at: http://www.gov.cn/flfg/2006-08/28/content_371399.htm.

Table 3.2. Types of Limited Partners as Sources of China's Domestic VC Funds: 2012

LP Type	No. of LPs (Total)	% of Total	No. of LPs (Disclosed)	Investable Amt. in China's VC/PE (US$m)	% of Total	Avg. Investable Amt. in China's VC/PE (US$m)
Wealthy Families & Individuals	3,773	50.2	3,597	8,673.1	1.1	2.4
Enterprises	1,289	17.2	779	27,276.9	3.4	35.0
VC/PE Firms	475	6.3	327	25,588.4	3.2	78.3
Investment Companies	441	5.9	361	32,403.5	4.0	89.8
Listed Companies	321	4.3	296	212,375.9	26.3	717.5
Government Agencies	242	3.2	149	7,221.0	0.9	48.5
Guidance Funds	207	2.8	188	16,805.1	2.1	89.4
Asset Managers	171	2.3	82	18,434.6	2.3	224.8
Fund of Funds	110	1.5	87	51,515.5	6.4	592.1
Banks	99	1.3	43	23,886.0	3.0	555.5
Public Pension Funds	96	1.3	38	166,764.5	20.7	4,388.5
Others	76	1.0	54	1,605.1	0.2	29.7
Insurance Companies	48	0.6	26	1,606.5	0.2	61.8
Trusts	45	0.6	19	8,229.9	1.0	433.1
University & Foundations	38	0.5	32	5,952.8	0.7	186.0
Family Offices	26	0.3	16	11,017.0	1.4	688.6
Endowments	24	0.3	22	962.8	0.1	43.8
Sovereign Wealth Funds	22	0.3	10	154,306.6	19.1	15,430.7
Corporate Pension Funds	8	0.1	8	32,710.6	4.1	4,088.8
Total	**7,511**	**100.0**	**6,134**	**807,335.7**	**100.0**	**131.6**

decade-long dominance of foreign VC firms over domestic firms has subsequently been reversed and sustained. The boom of VC-backed IPOs also ushered in the expansion of funding sources for domestic VC as well as China's LP market. As of the end of 2012, the Zero2IPO Database had included information concerning 7,511 LPs that disclosed a total investable amount of US$807 bn (Table 3.2). Thousands of wealthy families and individuals entered the game, although their aggregate amount of investment is still fairly limited. Institutional LPs such as public pension funds, sovereign wealth funds, and listed companies enter the market with a massive amount of capital and play an important role in investments.

The hustle and bustle of ChiNext and SSMEB to date, however, has been nearly irrelevant to offshore funds. ChiNext, with a lower threshold than SSMEB, also requires IPO applicants to be incorporated in China and to have one to two years of operational performance since incorporation. Domestic and overseas exits, therefore, represent two entirely different paths of investment with tremendous switch costs. Foreign investors who have already restructured their investee firms according to the Sina model are unlikely to undo the process and retarget ChiNext or SSMEB. Even if they can, further barriers of foreign exchange regulations and sectoral entry restrictions are still troublesome. To escape this stalemate, many foreign VCs showed that they were willing to adapt to the Chinese model by starting to set up RMB funds when they were given the green light in August 2009.[7]

The Formation of Networked Entrepreneurial Habitats

As shown in Figures 3.1 and 3.5, both VC investors and investees in China have been highly concentrated in a few urban centers, especially Beijing, Shanghai, and Shenzhen. VC-invested firms in the leading IT sector by 2008 had shown an above-average rate of urban concentration, with 77 percent in these top three centers. By 2008, 95 percent of the 279 foreign VC offices were concentrated in Shanghai, Beijing, and Hong Kong. The increase of VC offices in a given city has correlated with the increase of localized entrepreneurial investees. In the top VC centers of Shanghai, Beijing, Shenzhen, and Hangzhou, the growth in the number of VC offices has generally been proportional to the growth in the number and amount of VC recipients.

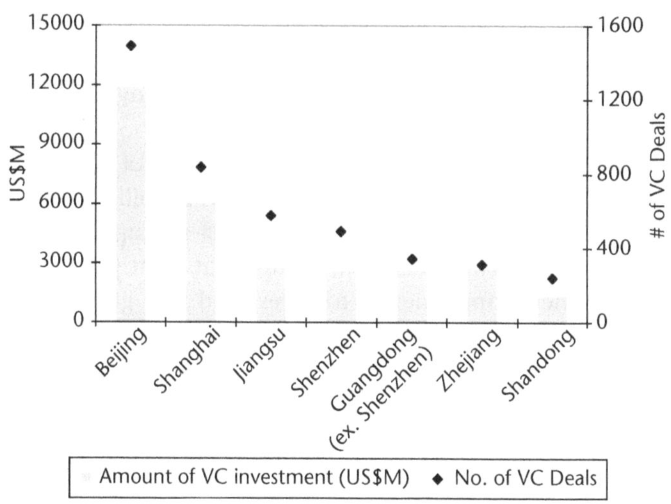

Figure 3.5. The Distribution of VC Investment in Leading Cities/Regions by 2013

[7] The Administrative Measures for Foreign Enterprises and Individuals to Establish Partnership Enterprises in China, available at: http://www.gov.cn/zwgk/2009-12/02/content_1478238.htm.

Due to the absence of proper investee firms, Hong Kong, which used to be the largest VC center in Asia, has declined significantly relative to Beijing, Shanghai, and Shenzhen (Figure 3.5), even though it has remained as a leading VC exporter (Chu 2008). Hong Kong's overwhelming strength and first-mover advantage on the supply side has not yet been, and will unlikely be, converted to its demand side. This co-location of top VC office clusters and high-tech (especially ICT) clusters indicates that the rise of top VC centers in China is contingent on the existence of a broad local base of technological entrepreneurship. The relative decline of Hong Kong in VC supply with a continuous lack of local demand also implies that it is hard to expect VC supply to create its own local demand, though positive feedback effects between the two sides can certainly exist (cf. Martin et al. 2002, 2005).

Localized concentration of VC firms and technological firms not only creates mutually reinforcing demand and supply, but also this process leads to the further specialization and sub-sectoral division of labor. A mechanism of increasing returns to scale operates through time to accumulate more knowledge, experience, expertise, networks, resources, and legitimacy within both the VC community and entrepreneur community. My study of internet development in Beijing and Shanghai demonstrates that about 50 percent of new firm formation was based on spin-offs from the established firms (Zhang 2013). Through formal and informal interactions and collaborations, these two communities became increasingly interwoven, as more and more successful entrepreneurs fed the supply of VC executives and managers. Moreover, the co-evolution of VC and technological firms leads to the localized accumulation of ancillary activities and supporting networks, e.g. VC consulting firms such as Zero2IPO, law firms, accounting and auditing firms, investment bankers, etc.

The technology sector in China is now old enough for some founders to have seen their startups grown into corporations worth billions of dollars and for other founders to have cashed out of their first startups richer and more experienced. Departing engineers from groups that went public have left to found promising new companies or to invest and give guidance to others. Among them is Lei Jun, who founded software group Kingsoft but stepped down as chief executive after its 2007 IPO. He is now better known as the man behind Xiaomi, a smartphone maker valued at $10 bn just three years after its founding. Mr Lei and Lee Kai-fu, the former Google China chief turned venture capitalist, are among the veterans helping a new generation of entrepreneurs take advantage of the rise of mobile internet in China.

In the leading VC centers, angel investors like Lei Jun have been growing very rapidly and becoming institutionalized. It is reported that 22 angel funds raised the amount of US$965 m in 2012. In 2013, US$201 m angel capital was invested in 169 deals. Over 70 percent of the angel investors were previously successful startup founders, with relevant business or technical backgrounds, who have banded together to provide capital and advice to companies in a specific industry. Beijing has dominated angel deals in China, representing 22 percent of all deal value and 30 percent of overall deal volume in 2013. In terms of the number of deals, from 2008 to 2013, 36 percent was in Beijing and 17 percent in Shanghai. On January 5, 2013, China Young Angel Investor Leader Association (中国青年天使会) was founded in Beijing by well-known investors Xu Xiaoping, Yang Ning, and other partners. In March 2014, its Shanghai office was also launched. The rise of angel investment addresses the inappropriate structure of venture investments for many small young firms. Venture funds tend to make quite substantial investments—on average, over one million USD per deal—even in young firms. Venture organizations are consequently unwilling to invest in very young firms that only require small capital infusions.

The VC hot spots in China are growing into entrepreneurial habitats, or ecosystems, and such growth is not in isolation from global VC and technology centers. The emerging VC centers in China have become increasingly interlinked through multiple ties, and at the same time they have become increasingly linked to other global VC and high-tech centers, especially Silicon Valley. Leading global VC firms, many from Silicon Valley, have become increasingly localized in China, particularly in Beijing and Shanghai, with their own local funds, offices, and management teams. On the other hand, foreign VC firms not yet localized have developed syndicated networks with local partners. By 2008, 77 percent of 334 co-investment decisions made by overseas offices had at least one local or regional partner geographically close to the target, and both their investments and partners were mainly to be found in Beijing and Shanghai (Figure 3.6). Investment syndication, or co-investment involving multiple VC firms, was most commonly found in the form of intra-urban partnerships, especially between Beijing and Shanghai. In total, 50 percent of the co-investment pairings were located in the same city as their investees, and another 10 percent in the same region. Ninety percent of all the co-investments were related to the Beijing, Shanghai, Shenzhen, and Hong Kong. Excluding overseas participants, investors in these four centers combined accounted for 98 percent of intra-urban syndication and 84 percent of inter-urban syndication. Beijing and Shanghai combined accounted for 92 percent of intra-urban syndication and 41 percent of inter-urban syndication. The thick inter-organizational co-investment networks linking Beijing and Shanghai have been co-evolving with the equally dense intra-organizational networks linking the two centers. Sixty-three active VC firms had their offices in both Shanghai and Beijing by 2008. Investment syndication helped established investors located in the core regions not only to reduce the uncertainty of localized investments but also to seize

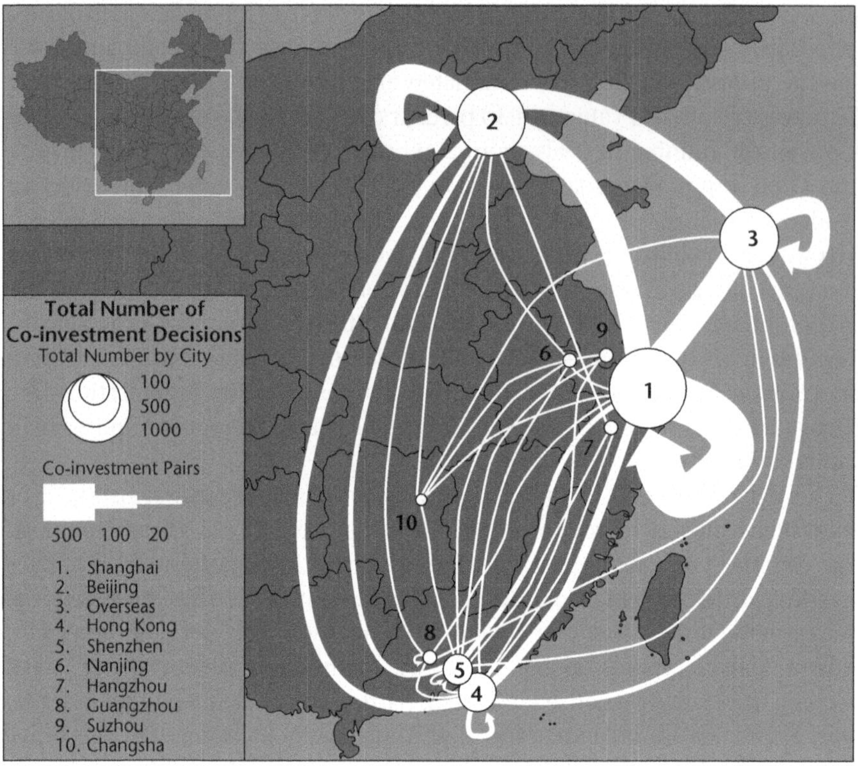

Figure 3.6. The Number of Inter-Firm Coinvestment Ties Aggregated at the Urban Scale by 2008

opportunities elsewhere. Meanwhile, most leading Chinese VC funds and incubators set up their offices in Silicon Valley, further increasing potential for cross-border knowledge spillovers and investment collaboration.

Other than the co-investment networks and multi-locational VC firms, there are other emerging organizations further weaving these leading entrepreneurial habitats together. For example, founded in 1979, Asia America MultiTechnology Association (AAMA) is Silicon Valley's largest and most established non-profit organization dedicated to the Asia American high-tech community. AAMA expanded into Asia by launching chapters in Beijing, Shanghai, the Pearl River Delta Region, Taipei, and Seoul to better serve as the bridge between Silicon Valley and Asia. Each chapter hosts events to localize the effort to promote business relationships, working closely with Silicon Valley headquarters. With a network of 10,000 participants representing over 2,000 companies in Silicon Valley and chapters in Asia, AAMA is a nexus that provides leading executives and other stakeholders with cross-fertilization of ideas and best-of-breed collaboration, resulting in successful business ventures throughout the Asia-Pacific region and global economy. AAMA's membership represents a diverse and influential spectrum of technology industries, including the internet, wireless, telecommunications, multimedia, semiconductor, software, hardware, electronics, and bio-tech industries, as well as financial and professional services industries that are integral to high-tech development. Starting in 2006, every year AAMA has invited 10 elites in business, science, investment, etc., to mentor 20 young potential entrepreneurs. They facilitate this through seminars, collective activities, and other training methods to provide one-on-one and face-to-face learning opportunities. So far 84 mentors and 174 trainees have finished the program, and many of their founded firms later achieved IPOs.

Does VC Spur Innovation in China?

Many empirical studies have found that, in the Western context, local access to venture financing and experienced venture capitalists significantly increases the number of high-tech startups in a region and the likelihood of their success (e.g. Powell et al. 2002; Florida and Kenney 1988; Cooke 2001; Zook 2002; Martin et al. 2005). In interpreting Silicon Valley's continuous ability in both incremental and radical innovation, Kenney and Patton (2005, 2006) have particularly emphasized the co-evolution of venturing firms and their "support network" consisting of venture capitalists and other service providers. But whether VC can spur innovation has also been debated. For example, the study of Kortum and Lerner (2000) shows that venture funding does have a strong positive impact on innovation: a dollar of venture capital appears to be about three times more potent in stimulating patenting than a dollar of traditional corporate R&D (see also, Popov and Roosenboom 2012). In contrast, Hirukawa and Ueda (2011) argue that the causality is reverse: it is not that VC stimulates technological innovation but technological innovation induces VC investments. Lazonick and Tulum (2011) argue that venture capitalists' desire for short-term financial gain gives them little incentive to promote learning at the organizational level, which may require long-term and sustained investments, especially in the biopharmaceutical sector, among others.

In the Chinese context, existing evidence arguably invalidates any polarized view on the relationship between VC and innovation, where the emerging economy is characterized by the scarcity of both capital and technology. Although venture capitalists are not very patient, at least they are still willing to carry their investee firms for three to five years during the most critical stage to their survival and success, when alternative sources of financing and assistance are virtually nonexistent. While the contribution of VC investors to their investees is far more than just money, the importance of cash to startups can hardly be overstated in China's primitive entrepreneurial environment characterized by the dearth of finance capital in the private sector.

Before the arrival of VC in China, due to the scarcity of financial capital and political discrimination against private ownership at the incipient stage, individual entrepreneurial startups, or de novo market entrants, especially in technological sectors, were rare and disproportionally small. This is contrary to the common prediction in the Western context where de novo entrants frequently dominate startup firm population (Helfat and Lieberman 2002). For example, among 451 leading internet startups before 1999, only 62 (13.4 percent) were founded by de novo entrants in China (Zhang 2007). But the introduction of American VC into China and subsequent growth of domestic VC organization has transformed the pattern of technological entrepreneurship in China. It has allowed young entrepreneurs outside the Southern China manufacturing zones with a strong technological background or just a good sense of opportunity to start firms on their own. Through VC, Chinese entrepreneurs' intellectual assets have been valued unprecedentedly highly, whereas previously they were obliged to beg state bureaucrats and SOEs for financial and business support. In particular, large-scale private technological startups were enabled for the first time in China with the rise of VC.

In the early 1990s, international venture capitalists invested almost exclusively in China's SOEs in low-tech industries, due to a lack of qualified private firms along with the preferential resources and domestic IPO channels enjoyed by SOEs (Zeng 2004: 64–5). This trend changed during the internet boom between 1998 and 2000, when more than 50 percent of VC investment went into internet-related firms and another 20 percent went to computer software developers. China's technological entrepreneurship became intimately tied to VC involvement. Ever since, all successful internet firms have been backed by VC, including the first-generation "big three portals," or Sina, Sohu, and Netease, and the new-generation BAT, or Baidu, Alibaba, and Tencent.

Since 2004, major foci of VC investment in China have diversified from the internet into Integrated Circuits (IC), wireless value-added, and other broadly defined ICT industries as well as clean-tech sectors. For example, Semiconductor Manufacturing International Corporation (SMIC), an IC firm founded in 2000, received US$1.46 bn VC/PE investment.[8] The VentureChina database shows that by 2008, 45 IC firms in Shanghai (over 63 percent of the national total) had received VC investment in the amount of US$2.1 bn (71 percent of national total, excluding late stage private equity investment). Among the 1,592 VC-invested firms in China by 2008, over 71 percent belongs to the

[8] More information on SMIC and the IC industry can be found in Chs 7 and 9.

three technological sectors: 931 in information technology, 137 in biotech-healthcare, and 69 in clean technology. If we consider the three broadly defined sectors as high-tech, then in combination they accounted for 71.4 percent of total investment.

Out of Zero2IPO's total record of 2,021 VC deals by the end of 2008, 1,227 (61 percent) were in the three technological sectors: ICTs (994), biotech-healthcare (138), and clean-tech (95). The amount of investment has been generally proportional to the number of deals. By the end of 2013, VC invested in 3,074 deals in ICTs (or broad IT), 632 in biotech-healthcare, and 516 in clean-tech. While these limited deals in biotech-healthcare and clean technologies are widely dispersed, 80 percent of ICT deals by 2008 are concentrated in Beijing, Shanghai, and Shenzhen (Figure 3.7). VC investments in China have been especially directed to internet and telecommunications (Figure 3.8). The proportion of VC investment in broad IT sectors has been declining, especially since 2005, due to the diversification of investment targets. But the growth of the (mobile) internet sector regained momentum recently, and so investment in broad IT sectors climbed again to reach 51 percent of the number of deals invested in 2013. In 2013, the internet accounted for 20 percent of the number of deals invested, and 16 percent of the total investment amount; telecom and value-added accounted for 17 and 10 percent respectively.

VC thus has become an increasingly important force behind China's technological entrepreneurship. It is hard to precisely assess the contribution of

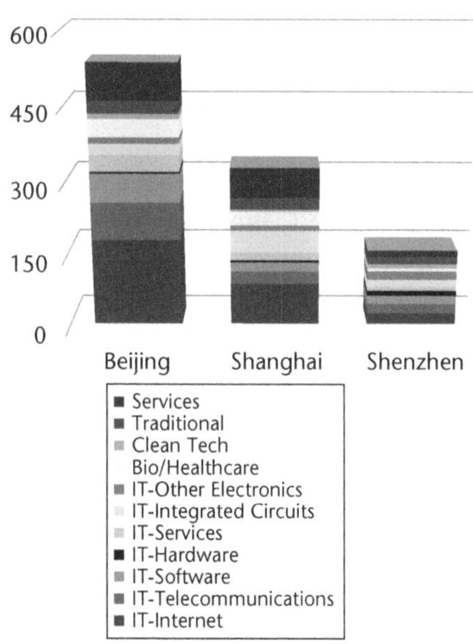

Figure 3.7. Sectoral Structure of VC-Backed Startups in Top Three Centers (2008)

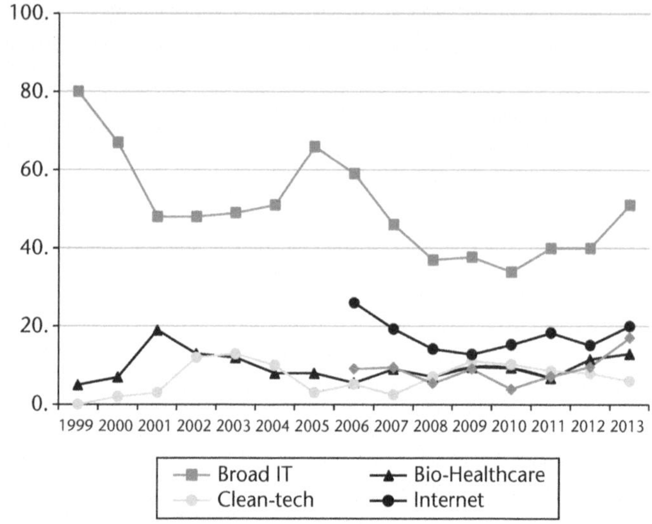

Figure 3.8. Sectoral Distribution of VC Investments: Percent of No. of Deals (1999–2013)

VC to the growth and innovation of their investees. But the fact that thousands of Chinese technological startups have voluntarily chosen to adopt VC investors is itself telling. It is safe to say that it would be a lot more difficult—if at all possible—for many entrepreneurs to get access to finance capital and the much-needed firm-building skills without VC support. Within China's emerging market economy and state-dominated financial sector, the intermediary role of VC is more salient than in advanced market economies. It is well known that the PRC is characterized by frequently arbitrary law enforcement in its party-dominated legal-political system and a long tradition of relationship-based transactions in business (cf. Peck and Zhang 2013; Piotroski and Wong 2012). The VC literature on China has highlighted the underdevelopment of formal information infrastructures, the shortage of entrepreneurial experience, and the general absence of "hard" technological competency of startup firms (Ahlstrom et al. 2007; Batjargal and Liu 2005; Bruton and Ahlstrom 2003). These factors combined render VC's resources of know-how and know-who, or *guanxi* networks (in both the business and political senses), especially valuable to their investees and their own success (Kambil et al. 2006). While domestic VC firms often have preferential access to domestic sources of information and resources, foreign firms are privileged with more professional experience in offering value-added services to their investees, and a better ability to link Chinese firms to business partners and IPO opportunities overseas. As Zhou Quan, President of IDG, once expressed, "Not only do we have frequent contact and very good *guanxi* with a large number of domestic and foreign IT firms, but also we have very good *guanxi* with many overseas (including Hong Kong, Wall Street, etc.) investment firms, security underwriter. These provide critical help to the overseas IPO for our investees."[9] In addition, familiarity with the growth experience of international companies and "foreknowledge" of business models gives some foreign VC firms such as IDG a positional advantage over their domestic counterparts. Because the business models of technological startups in China have been predominantly based on imitating their foreign role models, they usually have an imitation-lag of several months or even years.

One apparent fact is that many (once) VC-backed young firms have become increasingly innovative, although only in an incremental manner, especially in internet and telecommunications sectors. The internet has remained the leading sector of VC investment in China throughout the years. Although we have hardly seen ground-breaking technological innovations or even original business models yet, the enormity of China's user base and the cultural and political barriers have allowed VC-backed internet firms to outcompete their international rivals in the domestic market. They have collectively risen to become China's most entrepreneurial and dynamic sector, dramatically transforming China's industrial landscape. Of course this has been above all facilitated by China's huge and still-expanding domestic markets, especially in the realms of internet and telecommunications. In the last two decades, China's internet and mobile phone users have probably represented the fastest market growth in the world. By the end of 2013, the number of internet users in China had reached 600 million, and around 500 million of them are mobile internet and smartphone users. Today, far more people are online to shop, play games, search, watch videos, and use social media in China than in any other country.

The innovation of VC-backed firms in China is thus largely associated with services rather than manufacturing, and with domestic rather than international markets. Given China's huge and still-growing domestic market, any business model, even focused on a very niche market, could still be successful. A segment leader could become a powerful player no matter how narrowly the vertical market is defined. Even just emulating a well-proven and successful business model from the US is feasible for Chinese internet companies; US companies are preoccupied first with their huge local market and secondly with Europe and Japan. All these factors provide an exceptional window of time in which China startups can flourish. The market leader in a particular segment may go for an IPO, and the second and third could be potential M&A targets for MNCs who want to enter the Chinese market.

[9] Interview with Zhou Quan, Chen 2001: 66–7.

But China's huge domestic market is not pre-existing; it is created in part by startups supported by VC. For example, e-commerce in China used to be constrained by three major barriers: trust (concerning information flow), payment (cash flow), and delivery (product flow). But through persistent trial-and-error, Alibaba (along with others) has worked out a functional system that can more or less effectively cope with the problem of trust through its customer evaluation system imitating Amazon. Alibaba has approached the problem of payment through its Alipay imitating Paypal, which was once believed to be a mission impossible given the lack of credit card use in China. These "institutional breakthroughs" are remarkable achievements comparable to any technological breakthroughs, and they are now further challenging and remapping the financial, commercial, and cultural landscape of China. Taobao's sales in 2003 were a mere 22.7 m RMB; in 2007, it reached 100 bn RMB, and in 2013, 1 tr RMB. On September 18, 2014, Alibaba became a publicly listed company at New York Stock Exchange, valued at $231 bn, more than Amazon and eBay combined (Wang et al. 2014; Barboza 2014).

Of course, VC-backed firms in China even in the technological sectors are characterized by business-model innovations rather than bona fide technological innovations. Innovations are mostly about adapting existing technologies to fulfill customer requirements in a unique regulatory and social system that creates opportunities for novel solutions. It is arguably true that the enterprises funded by VC in China are largely copycats of Western technologies and business models. It is especially easy to copy internet and mobile business models or services, since foreign companies often take a long time to enter the Chinese market. More recently however, though many may still consider Chinese internet companies as knockoffs of Google, Facebook, Twitter, and eBay, these have already transformed themselves into dynamic, innovative technology companies with unique business models. For example, just three years after being introduced in China, Weixin (known as WeChat overseas) gained nearly 300 million users—a faster adoption rate than either Facebook or Twitter—giving the app a dominant position in what is now the world's biggest smartphone market. The elegantly designed Weixin is one of China's best examples of product innovation and may help the company gain business overseas. Weixin is no mere copy of an existing service but an amalgam of various social networking tools: part Facebook, part Instagram, and even part walkie-talkie. Tencent, the creator of Weixin and the Chinese internet powerhouse known for its QQ instant messenger service and its popular online games, is the most profitable internet business in China, earning more than $2 bn in 2012. The company has excelled at converting its hundreds of millions of social-media users into paying customers, mainly for virtual items in games. From its origin as a clone of AOL's ICQ, it has evolved dramatically.

Quite a few foreign firms have lost out to the hordes of Chinese entrepreneurs who copied quickly and understood their customers better. China has been one of the few countries in the world where US internet companies have failed to penetrate, partly because domestic players have quickly built up similar services that are better suited to local tastes. eBay was defeated in the Chinese online retail market by Taobao, and Google was squeezed out by Baidu. No foreign company has even come close in gaming and social media to the dominant position of Tencent. But it is important to note that the playing field has not been level. Western companies have been blocked by the Great Firewall (as happened to Twitter, Facebook, and YouTube) or slowed down and nearly impossible to use (as is still happening to Google). The Chinese government has spent a huge amount of effort on making sure that its version of the internet is distinct, not only to limit freedom of expression but also to ensure the surrounding industry serves national goals as well as commercial ones. So far the state has shown great skill in bending the technology to its own purposes, enabling it to exercise better control of its own society and setting an example for other repressive regimes. As a result, "The Chinese internet resembles a fenced-off playground with paternalistic guards" (*The Economist* 2013: 4). The government has indeed provided a roomy and attractive cage, with an army of cyber-police, hardware engineers, software developers, web monitors, and paid online propagandists deployed to watch, filter, censor, and guide Chinese internet users. Within China's borders the Communist Party has systematically put in place projects such as the Great Firewall, which keeps out "undesirable" foreign websites, and Golden Shield, which monitors online activities within China. It has also worked closely with trusted domestic internet companies such as Sina, Baidu, and Tencent. Thus what has developed so far in China is a thriving internet industry served by Chinese-led companies that it can trust to be politically reliable, rather than by unpalatable foreign ones, even though on paper many of the Chinese companies are partly foreign-owned through VC/PE investors. This is not to say, however, that the Chinese government could comfortably control the internet indefinitely. New challenges are constantly emerging to challenge its repressive control of information and society. As bureaucratic corruption and social discontent continue to fester, it is not inconceivable that the internet might ultimately turn out to be uncontrollable, and the entire authoritarian edifice may eventually founder (*The Economist* 2013).

Relatedly, the prospect for a benign and sustainable relationship between VC and innovation in China is far from sanguine. With thousands of active VC organizations in operation, tremendous expansion of the stock markets, and financial deepening of an increasingly larger proportion of Chinese firms, all under strong American influence, there is the danger of American-style financialization driven by the ideology of shareholder value maximization

(Lazonick 2013). Indeed, the nascent Chinese LPs investing in the VC funds are even less patient than their already impatient American counterparts; they are also more likely to intervene in the investment process but less likely to honor their promises (Chen 2009).

There has been a speculative pre-IPO turn of both PE and VC organizations especially after the launch of ChiNext, together with corruptive regulative practices, often involving powerful financial companies with inner connection to the top security regulators or to the offspring of former and present top party leaders (see e.g. Li 2010; Forsythe 2014). The extraordinary expansion of domestic VC funds associated with the booming ChiNext and SSMEB, therefore, might not be sustainable, even though it has permanently altered the trajectory of China's venture capitalism. Many European nations set up emerging stock exchanges, for example, the German Neuer Markt and Nasdaq Europe, in the late 1990s in the hopes of promoting increased venture capital activity (Giudici and Roosenboom 2004). However, stock prices plummeted after the ending of the stock market bubble in 2000 and these new markets suffered from poor liquidity, insider trading scandals, and accounting frauds. This finally led to the closure of the German Neuer Markt in 2003 and Nasdaq Europe in 2004 (Gompers and Lerner 2004: 345).[10] In China, similar scandals and frauds have already become widespread, resulting in the suspension of IPO approval by the CSRC in November 2013.[11] Even after rounds of IPO suspension and regulatory rectification, the accountability and viability of China's financial regime, even the entire party-state regime, is still a big question. Weakness of a poorly regulated stock market may lead to more abuse and ultimately ruin this exit vehicle. In that case, it would be questionable whether or not better alternatives could be developed.

Conclusion

The booming VC industry has played an increasingly important role in funding and nurturing new technology-based firms. The rise of the VC industry in China is a direct result of the transnationalization of leading global VC firms. The backdrop for China's integration into the global map of VC activities is the broader Chinese story of the last few decades—China's overall rise as one of the largest national economies and markets. The expansion into China of transnational VC firms was enabled by the opening up of China to the world as well as the opening up of the world to China, in particular the opening up

[10] Lazonick, based on a personal communication, believes that this is a fortunate change because since then the capital became more patient and quality of investment improved in Europe.
[11] This was the eighth IPO suspension in the history of PRC's stock markets.

of IPO opportunities for Chinese firms in major stock markets across the globe. For the VC industry governed by the iron law of "no exit, no entry," the availability of viable foreign exit channels has been especially crucial given the inadequacy and malfunction of China's domestic stock markets. The transformation of domestic institutional and policy environment, especially the launch of the ChiNext in Shenzhen as a premium domestic VC exit venue, has been instrumental in boosting domestic VC investors and enabling them to exceed their powerful foreign rivals by all measurements. However, it also led to a market overheat and escalating speculation. Within China's transitional institutional environment, the sustainability of ChiNext and China's financial system in general is yet to be demonstrated.

Of course, in China even the technological sectors are characterized by business-model innovations rather than bona fide technological innovations. Innovations are mostly about adapting existing technologies to fulfill customer requirements in a unique regulatory and social system that creates opportunities for novel solutions. Therefore, one might be suspicious about the role of VC in spurring innovation in China. However, the sheer monetary contribution of venture capital is not to be underrated. That US$53 bn in venture capital has been invested in the Chinese market in the past two decades or so is remarkable in itself. Moreover, VC has been behind almost every single successful internet company. Some VC-backed internet startups such as Tencent and Alibaba have today become global giants and transformative players in China's market and society. VC-backed internet firms have outcompeted their international rivals in the domestic market and collectively risen to become China's most entrepreneurial and dynamic sector, dramatically transforming China's industrial landscape, although this market is culturally unique and politically protected.

The supply of venture capital—whether spatially concentrated or dispersed—is unlikely to create its own, high-tech-based demand. It is more reasonable to expect active high-tech-based entrepreneurship to generate its own venture capital supply. Mutually reinforcing synergies between the supply and demand of venture capital are certainly possible and are actually in operation in a few privileged regions, but this requires in the first place that the regional economic and institutional structure is capable of generating and maintaining promising entrepreneurship at a sustainable level. In the final analysis, venture capital is just a device of capital accumulation. Venture capital and the entrepreneurs it funds will never supplant other wellsprings of innovation, such as vibrant universities, corporate research laboratories, and entrepreneurship (Lerner 2009: 9). Given its level of expected return and the nature of its operation, venture capital is unlikely to fulfill the hope of boosting innovation in less developed and economically lagging regions. With huge capital influx into the global venture capital industry, the necessary population of businesses capable of generating the returns that are sought by venture capital investors has become increasingly hard to find anywhere (Mason and Harrison 2003). The ability of venture capital in China to stimulate technological innovation and economic development is primarily limited to the leading regions and has an uncertain future, but the achievement so far has been truly remarkable.

References

Ahlstrom, D., D. G. Bruton, and S. Y. Kuang (2007), "Venture Capital in China: Past, Present, and Future," *Asia Pacific Journal of Management*, 24: 247–68.

Batjargal, B., and M. Liu (2005), "Entrepreneur's Access to Private Equity in China: The Role of Social Capital," *Organization Science*, 15: 159–72.

Barboza, D. (2014), "The Jack Ma Way: At Alibaba, the Founder is Squarely in Charge," Sept. 6, <http://www.nytimes.com/2014/09/07/business/international/at-alibaba-the-founder-is-squarely-in-charge.html>.

Bresnahan, T., A. Gambardella, and A. Saxenian (2001), "'Old Economy' Inputs for 'New Economy' Outcomes: Cluster Formation in the New Silicon Valleys," *Industrial and Corporate Change*, 10: 835–60.

Bruton, G. D., and D. Ahlstrom (2003), "An Institutional View of China's Venture Capital Industry: Explaining the Differences between China and the West," *Journal of Business Venturing*, 18: 233–59.

Chen H., P. Gompers, A. Kovner, and J. Lerner (2010), "Buy Local? The Geography of Venture Capital," *Journal of Urban Economics*, 67: 90–102.

Chen, L. (2001), *Expedition to NASDAQ: The Legends of Venture Capital in China* (Yuanzheng NASDAQ-Fengxian Touzi de Zhongguo Chuanqi). Shenyang: Liaoning People Press.

Chen, Y. (2009), "Entrepreneurship and VC in China," INSEAD Building Business in China Seminar, Apr. 26, Beijing.

China VC Investment Research Institute (2003), *China VC Investment Industry Report*, <http://www.southcn.com/tech/invest/200403090423.htm>.

Chu, Y. W. (2008), "Deconstructing the Global City: Unravelling the Linkages that Underlie Hong Kong's World City Status," *Urban Studies*, 45: 1625–46.

Cooke, P. (2001), "Regional Innovation Systems, Clusters, and the Knowledge Economy," *Industrial and Corporate Change*, 10: 945–74.

Deloitte (2009), *Global Trends in Venture Capital 2009 Global Report*. <http://www.evca.eu/uploadedFiles/News1/News_Items/2009_Global_Trends_in_VC_Report_final_web.pdf>.

Ernst & Young (2009), *From Survival to Growth: Global Venture Capital Insights and Trends Report*. <http://www.amazemind.com/vcInvestingConference/docs/2009_Global_VC_Report.pdf>.

Ernst & Young (2011), *Globalizing Venture Capital: Global Venture Capital Insights and Trends Report*. <http://www.ey.com/Publication/vwLUAssets/Globalizing_venture_

capital_VC_insights_and_trends_report_CY0227/$FILE/Globalizing%20venture%20capital_VC%20insights%20and%20trends%20report_CY0227.pdf>.

Ernst & Young (2013), *Turning the Corner: Globalizing Venture Capital: Global Venture Capital Insights and Trends 2013*. <http://share.endeavor.org/pdf/Turning%20the%20corner_VC%20insights%202013.pdf>.

Florida, R., and M. Kenney (1988), "Venture Capital, High Technology and Regional Development," *Regional Studies*, 22: 33–48.

Forsythe, M. (2014), "Alibaba's I.P.O. Could Be a Bonanza for the Scions of Chinese Leaders," July 20. <http://dealbook.nytimes.com/2014/07/20/alibabas-i-p-o-could-be-a-bonanza-for-the-scions-of-chinese-leaders>.

Gompers, P., and J. Lerner (2001), "The Venture Capital Revolution," *Journal of Economic Perspectives*, 15: 145–68.

Gompers, P., and J. Lerner (2004), *The Venture Capital Cycle*, 2nd edn. Cambridge, MA: MIT Press.

Giudici, G., and P. Roosenboom (2004), *The Rise and Fall of Europe's New Stock Markets*. Bingley: Emerald Group Publishing, Elsevier.

Gertler, M. S., and D. A. Wolfe (2006), "Spaces of Knowledge Flows: Clusters in a Global Context," in B. Asheim, P. Cooke, and R. Martin (eds), *Clusters and Regional Development: Critical Reflections and Explorations*, 218–35. London: Routledge.

Haggard, S., and Y. Huang (2008), "The Political Economy of Private Sector Development in China," in L. Brandt and T. Rawski (eds), *China's Great Economic Transformation*, 337–74. Cambridge: Cambridge University Press.

Hirukawa, M., and M. Ueda (2011), "Venture Capital and Innovation: Which is First?" *Pacific Economic Review*, 16: 421–65.

Helfat, C. E., and M. B. Lieberman (2002), "The Birth of Capabilities: Market Entry and the Importance of Prehistory," *Industrial and Corporate Change*, 11: 725–60.

Jeng, L. A., and P. C. Wells (2000), "The Determinants of Venture Capital Funding: Evidence across Countries," *Journal of Corporate Finance*, 6: 241–89.

Kambil, A., V. W. Long, and C. Kwan (2006), "The Seven Disciplines for Venturing in China," *MIT Sloan Management Review*, 47: 85–9.

Kortum, S., and J. Lerner (2000), "Assessing the Contribution of Venture Capital to Innovation," *RAND Journal of Economics*, 31: 674–92.

Kenney, M., and D. Patton (2005), "Entrepreneurial Geographies: Support Networks in Three High-Tech Industries," *Economic Geography*, 81: 201–28.

Kenney, M., and D. Patton. (2006), "The Coevolution of Technologies and Institutions: Silicon Valley as the Iconic High-Technology Cluster," in P. Braunerhjelm and M. Feldman (eds), *Cluster Genesis: Technology-Based Industrial Development*, 38–60. Oxford: Oxford University Press.

Lazonick, W. (2007a), "The U.S. Stock Market and the Governance of Innovative Enterprise," *Industrial and Corporate Change*, 16: 983–1035.

Lazonick, W. (2007b), "Evolution of the New Economy Business Model," in E. Brousseau and N. Curien (eds), *Internet and Digital Economics: Principles, Methods and Applications*, 59–113. Cambridge: Cambridge University Press.

Lazonick, W. (2013), "From Innovation to Financialization: How Shareholder Value Ideology is Destroying the US Economy," in M. H. Wolfson and G. A. Epstein (eds),

The Handbook of the Political Economy of Financial Crises, 459–511. New York: Oxford University Press.

Lazonick, W., and Ö. Tulum (2011), "US Biopharmaceutical Finance and the Sustainability of the US Biotech Business Model," *Research Policy*, 40: 1170–87.

Lerner, J. (2009), *Boulevard of Broken Dreams: Why Public Efforts to Boost Entrepreneurship and Venture Capital have Failed—and What to Do about it*. Princeton: Princeton University Press.

Li, X. (2010), "The Unspoken Rules of IPO behind the 'No. 1 Case of PE Corruption'" (PE 腐败第一案"背后的上市潜规则), *First Financial Daily* (第一财经日报). <http://finance.jrj.com.cn/2010/08/0601287884961.shtml>.

Martin, R., C. Bernd, B. Klagge, and P. Sunley (2005), "Spatial Proximity Effects and Regional Equity Gaps in the Venture Capital Market: Evidence from Germany and the United Kingdom," *Environment and Planning A*, 37: 1207–31.

Martin, R., P. Sunley, and D. Turner (2002), "Taking Risks in Regions: The Geographic Anatomy of Europe's Emerging Venture Capital Market," *Journal of Economic Geography*, 2: 121–50.

Maschek, M. (2005), "China is Ripe for VCs with the Right Approach," *Venture Capital Journal*, 45: 41–3.

Mason, C. (2009), "Venture Capital in Crisis?" *Venture Capital*, 11: 279–85.

Mason, C., and R. Harrison (2002), "The Geography of Venture Capital Investment in the UK," *Transactions of the Institute of British Geographers*, NS, 27: 427–51.

Mason, C., and R. Harrison (2003), "Closing the Regional Equity Gap? A Critique of the Department of Trade and Industry's Regional Venture Capital Funds Initiative," *Regional Studies*, 37: 855–68.

OECD (1995), *Government Programmes for Venture Capital*. Paris: OECD.

Peck, J., and J. Zhang (2013), "A variety of capitalism . . . with Chinese characteristics?" *Journal of Economic Geography*, 13: 357–96.

Piotroski, J. D., and T. J. Wong (2012), "Institutions and Information Environment of Chinese Listed Firms," in F. Joseph and R. Morck (eds), *Capitalizing China*, 201–42. Chicago: University of Chicago Press.

Powell, W., W. Koput, I. Bowie, and L. Smith-Doerr (2002), "The Spatial Clustering of Science and Capital: Accounting for Biotech Firm-Venture Capital Relationships," *Regional Studies*, 36: 291–306.

Popov, A., and P. Roosenboom (2012), "Venture Capital and Patented Innovation: Evidence from Europe," *Economic Policy*, 27: 447–82.

Sunley, P., B. Klagge, C. Berndt, and R. Martin (2005), "Venture Capital Programmes in the UK and Germany: In What Sense Regional Policies?" *Regional Studies*, 39: 255–74.

The Economist (2013), "Special Report: China and the Internet," Apr. 6.

Thompson, C. (1989), "The Geography of Venture Capital," *Progress in Human Geography*, 13: 62–98.

Tsai, K. S. (2002), *Back Alley Banking: Private Entrepreneurs in China*. Ithaca, NY: Cornell University Press.

Wang, L., E. Lam, and C. Bost (2014), "Alibaba IPO Pours Shares into Shrinking Pool of Stock," Sept. 25, <http://www.businessweek.com/printer/articles/227610-alibaba-ipo-pours-shares-into-shrinking-pool-of-stock>.

White, S., J. Gao, and W. Zhang (2005), "Financing New Ventures in China: System Antecedents and Institutionalization," *Research Policy*, 34: 894–913.

Yang, G. (2001), *Biography of Internet Heros (网络英雄传)*. Beijing: The Writers Publishing House.

Zeng, F. (2004), *Venture Capital Investments in China*. Pardee: RAND Graduate School dissertation series, the RAND Corporation.

Zhang, J. (2007), "Market Transition and the Spatial Dynamics of Internet Development in China." Ph.D. dissertation. University of Minnesota.

Zhang, J. (2011), "The Spatial Dynamics of Globalizing Venture Capital in China," *Environment and Planning A*, 43: 1562–80.

Zhang, J. (2013), "Related Variety, Global Connectivity and Institutional Embeddedness: Internet Development in Beijing and Shanghai Compared," *Regional Studies*, 47: 1065–81.

Zook, M. A. (2002), "Grounded Capital: Venture Financing and the Geography of the Internet Industry," *Journal of Economic Geography*, 2: 151–77.

4

China's Mechanical Engineering Industry

Offering the Potential for Indigenous Innovation?

Ingo Liefner and Gang Zeng

Introduction

The mechanical engineering industry provides investment goods for the whole economy (Kalkowksi and Manske 1993). It produces machines, tools, equipment, robots, and production facilities, and thus supplies the process technology for virtually all other industries. Unlike most other industries, the mechanical engineering industry is neither centered on a specific product innovation, such as the automobile or the semiconductor, nor on a distinct technological trajectory, such as electricity or digitalization. The mechanical engineering industry is old and has its roots in related crafts in many areas. It reacts to the demands of other industries and integrates process innovations developed elsewhere, such as computer-aided design, to fulfill such specific demands.

Due to the diversity of the mechanical engineering industry and the fact that it continually responds to technical changes and new demands from other industries, singling out the technical development path of the mechanical engineering industry, as can be done with the semiconductor industry, is not possible. Hence, this chapter does not tell one story of technological catch-up and the development of innovative capabilities from the angle of milestone products or processes. It instead focuses on the industry-specific conditions for development and innovation that are rooted in the diverse organization of the mechanical engineering business.

Analysis of this field is further complicated by the fact that mechanical engineering usually involves all kinds of companies, ranging from very small and highly specialized producers to large firms supplying standardized machines or producing large facilities. Within a given product range, firms differ with regard to their emphases on scale, precision, and reliability versus price, and thus mirror different demands. Some sub-fields of the mechanical engineering industry use standardized technologies and enjoy stable demand while others are shaped by rapid changes in production technologies and/or demand. In the case of China, different forms of ownership influence company strategies and industry development as well. Given the heterogeneity and complexity of the mechanical engineering industry, there is an insufficient number of industry leaders that might function as role models for other firms, unlike in the information technology (IT) industry. Thus some large companies are not at all representative of the industry's development. The importance of small firms limits the use of aggregate statistical data as well, as these firms are usually not fully covered.

This chapter aims to identify important structural features of China's mechanical engineering industry, including its historical development, its spatial structure, and its average technological capabilities based on available secondary data. Secondly, it compares the structures found in China with the structures that characterize the German mechanical engineering industry, which is often considered the international leader in innovative and high-precision machinery. Thirdly, this chapter discusses how the general and the China-specific features of the mechanical engineering industry offer the conditions for successful indigenous innovation. These arguments are based on a detailed discussion of two selected cases of the most innovative mechanical engineering firms to be found in China.

In this chapter, we argue that mechanical engineering offers unique conditions that are more favorable for creating "indigenous innovation" than many other globalized industries. Innovation processes typical of mechanical engineering, emphasizing localization and customization, ideally match with the current needs of China's rapidly growing and diversifying economy. We will take a close look at the way innovation processes are organized in mechanical engineering, using the well-documented organization of innovation in the German mechanical engineering industry as a benchmark model. We will then examine two companies in depth in order to explore how firms in the mechanical engineering industry may become innovative. The two cases represent two distinct ways of combining the industry characteristics and the characteristics of the local business environment to develop innovation strategies. Both pathways of cultivating innovativeness have great potential to succeed.

Growth and Spatial Structures of China's Mechanical Engineering Industry

The mechanical engineering industry is defined as providing investment goods to all parts of the economy (Kalkowksi and Manske 1993). It produces machines and mechanical equipment. However, mechanical engineering is not singled out as a distinct category in the statistics provided by China's statistical offices; instead, mechanical engineering firms are treated as part of the larger group of equipment manufacturing industries. The equipment manufacturing industries include the manufacturing of metal products, general purpose machinery, special purpose machinery, transport equipment, electrical machinery and equipment, communication equipment, computers and other electronic equipment, and the manufacturing of measuring instruments and machinery for cultural activity and office work (Peng and Huang 2011). Within this broad field, the mechanical engineering industry mainly comprises the manufacturing of general purpose machinery and special purpose machinery.

While the equipment manufacturing industry as a whole reached a gross production value of 27,659 bn yuan RMB in 2011, the share of manufacturing of general purpose and special purpose machinery was about one quarter of that figure (NBSC 2013a, 2013b).

Figure 4.1 illustrates the development of the production value of China's mechanical engineering industry. Due to data constraints, the first part of the figure, covering the years 1952–77, shows data for the whole manufacturing industry prior to Chinese economic reform, whereas the second part, from 1978 to 2011, shows aggregate data for general purpose machinery and special purpose machinery. In 1949, the manufacturing industries produced only 5 percent of China's overall industrial output. Between 1949 and 1998, a period of rapid industrialization and sharply rising need for production equipment, the production value of the equipment industries grew by a factor of 2,800, and the average annual growth rate stood at 17.2 percent, much higher than average industrial output growth (NBSC 1999: 164).

According to a report in the journal *China Equipment* (Anonymous 2007), the development of China's mechanical engineering industry follows a pattern similar to that of manufacturing industries, which can be divided into three phases.

1949–1978: Development of a National System

Following the foundation of the People's Republic of China in 1949, the Soviet Union assisted China in completing 156 key development projects throughout the 1950s, of which 68 involved the mechanical engineering

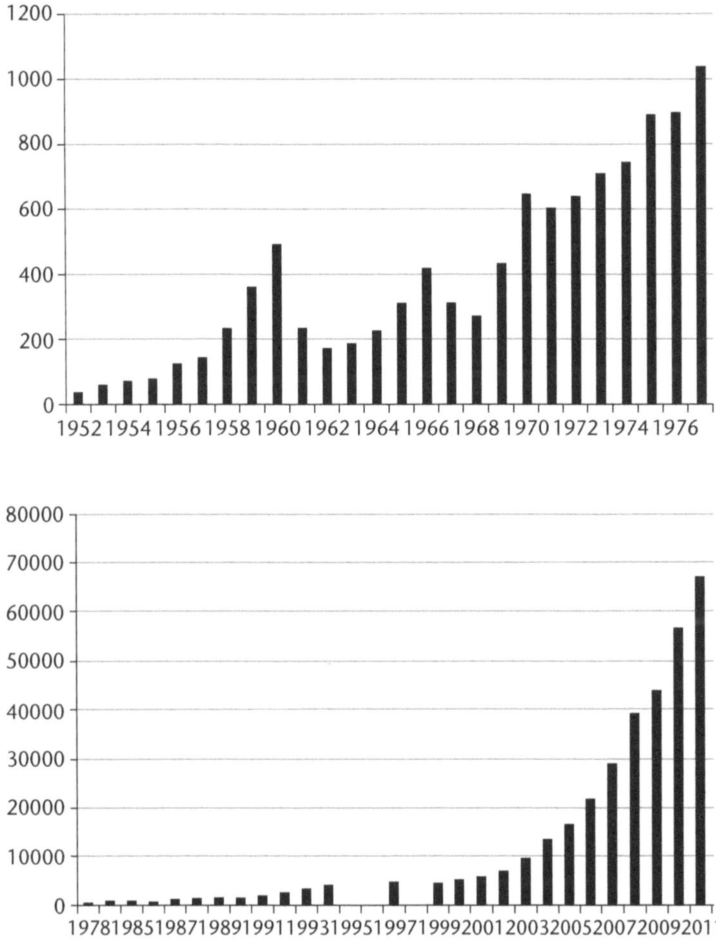

Figure 4.1. The Growth of China's Mechanical Engineering Industry (1952–77 and 1978–2011)

Gross production value, in 100 million yuan RMB.
Source: NBSC 2013b.

industry. Industrial assistance also came from East Germany. In the course of these projects, engineers were trained and research institutes were established, and China started to build a systemic research and production competence in the field of mechanical engineering.

Following the political split from the Soviet Union in 1958, China become internationally isolated, which forced it to focus on growing national capabilities independently. Given the geopolitical situation, the central government's industrial policy was dominated by a drive to increase China's ability to withstand potential military attacks by the Soviet Union or the US. As part of this strategy, China's industry was to be organized in the form of seven independent regional systems: the Northeast, North, Northwest, Central, Southwest, South, and East. This strategy fostered growth of the mechanical engineering industry in China's interior, and it helped to build up a limited ability to innovate. The emphasis on heavy industry also meant that, besides the traditional light industrial center of Shanghai, other regional centers such as Shenyang, Beijing, Xian, Wuhan, Chongqing, and Niuzhou (Guangxi) also developed their industrial complexes (cp. Murakami et al. 1996). Overall though, the isolation from international development and general backward state of China's economy meant that China's machinery sector was several generations behind the global frontier (Anonymous 2007).

1979–2005: The Primacy of Foreign Equipment and Technology

With the opening of China, the government's priority shifted toward the import of machines and equipment from the West. In July 1983, the state council announced a set of measures to stimulate the use of foreign equipment in the fields like coal mining, electricity, and petrochemical processes. Similar trends are also documented in the automobile (Chapter 5 of this book), telecommunications (Chapter 8), and semiconductor industries (Chapter 7). As a consequence of these measures, China went into a substantial deficit in foreign trade with machines and equipment.

Starting in 1978, the regional economic development priority was placed on the coastal zone, given its historical advantage of industrial infrastructure and human resources. With the emphasis on efficiency and opening up, the coastal regions were positioned to make the first moves in industrial modernization. With China's promotion of export-led growth, foreign direct investment (FDI) and related technology imports mainly benefitted the coastal areas and led to a massive increase in industrial capacity in regions such as the Yangtze River Delta, the Pearl River Delta, and the Shandong peninsula.

Since 2006: Focus on Systemic Capabilities for Producing Indigenous Innovation

As other chapters in this book (e.g. Chapter 2) make clear, the growing reliance on foreign sources for advanced equipment generated considerable concern for the Chinese leadership after 2000. With the 11th five-year plan (2005–10), the state council announced a move toward prioritizing indigenous innovation. The labor-intensive manufacturing industries located in coastal areas were encouraged to move to sites in inland China for the lower wages and improved infrastructure in the interior. Meanwhile, the coastal areas were encouraged to place more importance on developing the mechanical engineering industry. This was intended to increase Chinese producers' competitiveness in general and their ability to upgrade their products and move up product value chains in particular.

In the near future, the regional distribution of China's mechanical engineering industry is likely to be affected mainly by existing production, research, and education capacities (cp. Figures 4.2 and 4.3). The Chinese mechanical engineering industry's spatial pattern will thus show concentrations in the coastal Yangtze River Delta, the Pearl River Delta, and Northeast China with centers in Shenyang and Dalian, but also in the inland cities of Wuhan, Xian, and Chongqing (Anonymous 2007: 31). This spatial pattern is confirmed by Figure 4.2, which represents data for the mechanical engineering industry, including the manufacturing of general purpose machinery and special purpose machinery (NBSC 2013b).[1]

As the provider of machines and tools for manufacturing, the mechanical engineering industry mainly sells its products to other industrial companies. The capabilities, the quality, and the precision of machines and mechanical equipment influences innovation, product development, and production costs of the customer firms, and is thus a crucial force behind domestic productivity growth (Kinkel and Som 2007). Conversely, the increasing and diversifying demands for machinery also attract investment and drive innovation in the machinery sector. With the fast expansion of China's production capacities in many industries (e.g. electronics and ICT, automotive, and production materials) and the massive investment in infrastructure, China's market for machines has witnessed sustained high growth rates, ranging from around 20 percent p.a. to more than 40 percent p.a. in the last decade (Bfai 2008). Today, China's mechanical engineering industry benefits greatly from industrial growth, public procurement, and government-led initiatives to expand infrastructure and stimulate the growth of certain industries. This has been particularly true in industries such as wind and solar energy (cp. Liu and Lundin 2009, also Chapters 11 and 12 of this book).

Technological Capabilities

Chinese firms have had a much shorter time period to build internal innovation-related resources than German or Japanese producers. *China*

[1] Besides differentiating between general purpose and special purpose machinery, the mechanical engineering industry can also be differentiated according to product groups and industrial uses. An important section within the mechanical engineering industry is the tool manufacturing industry, and China's machine tool and tool builders' association (CMTTBA 2012) provides one of the rare sources of regionally differentiated data. The spatial distribution of production in the tool manufacturing industry, however, is quite similar to that of mechanical engineering in total.

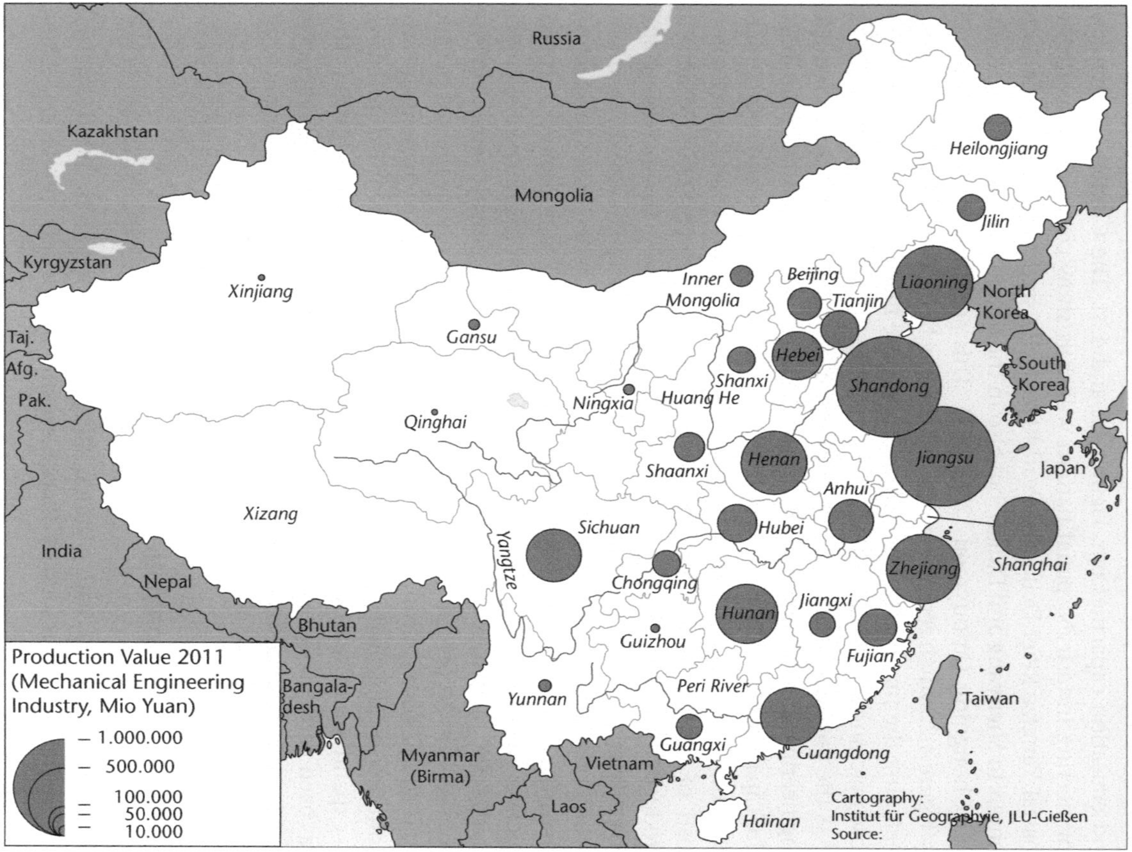

Figure 4.2. The Spatial Distribution of China's Mechanical Engineering Industry

Equipment (Anonymous 2007: 31) acknowledges that the Chinese mechanical engineering industry is much less sophisticated than their counterparts in Germany or Japan. It leans heavily toward producing standardized machines (cp. NBSC 2013c) and using imported specialized components (Bfai 2008).

The technological capabilities and innovation successes of such a heterogeneous industry are very hard to assess. Technological capabilities are usually measured by input indicators, throughput indicators, and output indicators (Kroll and Frietsch 2014). The inputs that go into mechanical engineering innovation processes, such as the qualification of engineers, knowledge created in public research laboratories, or R&D expenses, are hard to identify for the numerous small firms and related data restrictions. Innovation throughput indicators, such as patent applications and the number of patents held or used, are available but may characterize only a fraction of innovation activities. In an industry oriented toward small companies and customer-specific solutions that lack a general application, patents can be a tricky metric. The innovation output, in the sense of new products, is hard to observe as well for the same reasons. It is obvious that many customer-specific products must at least be "new to the producing firm" and to the customer (OECD and Eurostat 2005). However, radical innovations that develop new, superior solutions to change the quality, precision, and effectiveness of standard processes, taking product performance beyond the levels previously reached, are much more interesting than minor adaptations or incremental changes. They unfortunately are not covered by any statistic. The following brief assessment must therefore focus on the available information that can only partly disclose the technological level of mechanical engineering in China. The statistics we will use concern public R&D laboratories (input), patents (throughput), and export markets (output).

Public research organizations can have a positive influence on the potential for innovation and technical progress in mechanical engineering firms (e.g. Anselin et al. 1997). Knowledge is shared between firms and public organizations in many ways, including interchanges of skilled workers, joint development projects, and sponsored research. This input from public laboratories correlates with increased production in mechanical engineering; universities and research organizations can potentially provide well-trained engineers and new ideas to local producers. Hence, the spatial pattern of innovation in China's mechanical engineering industry can be expected to show concentrations of innovative activities in places where universities and research organizations are located. Figure 4.3 shows the locations of the most prominent national research laboratories in mechanical engineering. The map reveals that the distribution of these laboratories is consistent with the pattern of production output value in Figure 4.2.

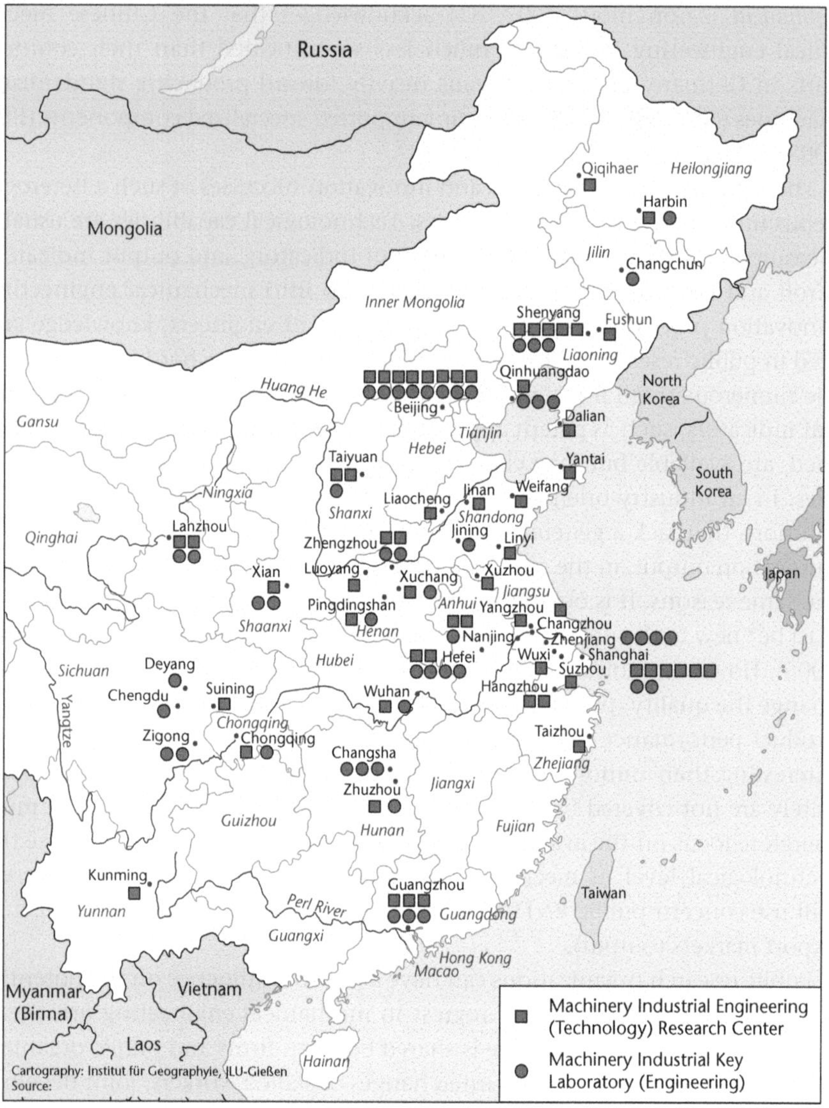

Figure 4.3. The Spatial Distribution of China's National Mechanical Engineering Research Laboratories

One step further in innovation and technical progress is the development of product or process inventions that may qualify for patent protection, one metric we can use to gauge the throughput. Although many inventions in mechanical engineering may never end up with patents, patents still reveal structures of an innovative industry (Lyu et al. 2014). Lyu et al. (2014)

examine patents in the equipment manufacturing industry and co-patents involving more than one applicant in particular, taking the cities of the Yangtze River Delta as an example. They find that the number of universities and companies holding patents in this technology field is rising sharply.

According to Lyu et al. (2014), the most active individual patenting players are universities. Companies generally have a much lower number, frequently holding only one or two patents. The fact that few companies have more than one or two co-patents may indicate that most firms in this industry are small and highly specialized. In contrast, some universities are very large, encompassing many laboratories and institutes, which may explain their higher patent numbers. We also have to consider the possibility that Chinese universities, leaning heavily toward applied research and development instead of basic research, may be relatively more involved in product development than their peers in other countries (cp. Kroll and Frietsch 2014). Moreover, universities are subject to incentives offered at the provincial level, public funding of patent application fees in particular (cp. Li 2012).

By connecting the co-patent applications, it is found that the main applicants do not form one consistent network but rather fall into many sub-networks. The largest fully interconnected sub-network covers only about 15 percent of all 2009–10 co-applicants, a small fraction when compared to other innovative networks (cp. Liefner and Hennemann 2011). This result underlines the importance of customer-centered innovation processes that evolve around customer–producer interaction and only loosely involve other players.

Output figures deliver the most accurate picture of technical capabilities as they reflect past progress manifested in products and prices. The following brief analysis of sales markets of mechanical engineering products and the values of products sold gives at least partial answers to the questions of innovation and technical progress. First, export markets indicate the overall quality of products sold, as it is reasonable to assume that companies located in advanced economies demand high quality whereas companies located in middle-income countries demand lower prices. Second, the unit value of exports and imports can be used as a technology-related output indicator. A high unit value of exports indicates technological advancement of the exporting firms whereas high unit value of imports indicates importers' reliance on technology developed elsewhere. Of course, unit value can only be meaningfully interpreted for rather homogeneous product sets and under conditions of market pricing. Other restrictions concerning the use of unit values as technology indicators are discussed by Aiginger (1997).

Commodity trade figures supplied by the United Nations Statistics Division (UN comtrade 2014) show that Chinese commodity exports in 2012 added up to more than US$2,000 bn. The main export destinations were the US,

Table 4.1. China's Trade and Unit Values for 8428: Lifting, Handling, Loading Machinery (2012)

Trade Partner: exports to... imports from...	Export Value (US$)	Rank	Import Value (US$)	Rank	Unit Value of Exports (US$)	Unit Value of Imports (US$)
(all)	3,505,061,204		2,577,283,689		1,881	7,030
USA	129,230,172	7	143,142,706	5	224	23,841
Japan	96,313,216	13	590,014,792	2	908	9,661
Germany	28,724,724	28	735,328,359	1	227	44,008
Republic of Korea	90,850,620	14	335,022,779	3	6,147	20,965
Singapore	123,432,529	8	28,752,561	12	4,543	16,619
Thailand	107,320,919	11	8,024,396	19	7,781	18,362

Source: Data from UN comtrade; authors' calculation.

Hong Kong, and Japan, followed by the Republic of Korea, Germany, and the Netherlands. The picture changes, however, when looking at exports of the mechanical engineering industry. For many mechanical engineering product groups, India is the major export destination, and other middle-income countries are important export markets as well. Table 4.1 shows the figures for a commodity group that can be regarded as typical of mechanical engineering, which is 8428 (HS1992): lifting, handling, loading machinery. For this product group, figures for trade quantity are available that allow for the calculation of unit values.

Table 4.1 shows that imports in that product category mainly come from the US and from advanced economies in Europe and Asia. Middle-income and low-income countries are not among the top import countries. This goes hand-in-hand with high unit value of imports. Exports go to many newly industrializing countries and other middle-income countries. The unit value of exports is much lower. Some countries are at the same time top export and import partners. In these cases, China's unit values of exports are always lower than the unit value of imports.

Other top export destinations from China include India (1), Australia (2), Indonesia (3), Russian Federation (4), Malaysia (5), Brazil (6), Viet Nam (9), Turkey (10), Hong Kong (12), Saudi Arabia (15), Iran (16), Philippines (17), Colombia (18), and Mexico (19).

Other top import countries for China: Italy (6), Netherlands (7), Switzerland (8), France (9), Austria (10), United Kingdom (11), Sweden (13), Finland (15), Spain (16), Canada (17), Czech Republic (18).

Examining historical trade data reveals the long-term development of China's production capabilities and technological upgrading. Although unit values reflect levels of technological sophistication as well as general shifts in production technologies and changes in demand, the available data for this commodity group (8428) for the time period from 1992 until 2012 still displays a coherent pattern. The following features stand out (UN comtrade, own calculation):

- China was a net importer of machines until 2005. Since then, China's exports in this category have exceeded its imports. From 1992 to 2012, imports have been growing by a factor of 13, while exports have been growing by a factor of 100.
- Since 1992, Germany, Japan, and the United States have consistently ranked among the top five countries for imports.
- More changes occurred in the ranking of export destinations but newly industrializing countries always occupy the top ranks.
- The ratio of the unit value of exports to the unit value of imports of trade with Germany, Japan, and the US has narrowed from the range of 1:200–1:400 in 1992 to the range of 1:10–1:200 in 2012.

The difference between unit values of exports and unit values of imports has decreased more clearly for the total trade in this commodity group. Including all imports and exports, the ratio of the unit value of exports to the unit value of imports used to stand consistently at about 1:40 during the years 1997 to 2002, but by 2012 it had narrowed to 1:3.7. Hence, China's mechanical engineering companies have developed their export capabilities with regards to volume and quality. The diminishing differences between unit value of export and import suggest that China has been upgrading its exports to higher value products. This is especially the case regarding export to middle- to low-income countries. The substantial gaps in trade with advanced countries remain, although in some areas they are getting narrower. The disproportionate expansion of exports also suggests that the content of imports has probably shifted. As China expands its localization capacity, what has to be imported may become more concentrated to relatively fewer components of higher value. For example, if China once imported an entire car, after some development they might have only needed to import the engine, eventually importing just parts of the engine.

Despite this trend, China's technological capabilities in mechanical engineering—more precisely, the capabilities reflected in the commodities exported—must be regarded as weak compared to the capabilities of advanced economies. Chinese producers specialize in relatively low-cost machines and tools, while China imports expensive machines and tools. This is a clear indication that mechanical engineering firms in China are capable of producing machines and tools for the world market, but mainly in the low-cost segment and with a focus on serving middle-income countries and emerging markets that can be assumed to be price-sensitive. In this way, mechanical engineering does not much differ from most other Chinese industries.

Consistently, Chinese firms have managed to upgrade their capabilities to levels that allow for successful exports but not to enter the top-quality market segments (Zeng et al. 2011). From an innovation systems perspective, mechanical engineering seems to be similar to other industries as well. Input into the innovation process has risen markedly, as indicated by the presence of public research organizations and evidence of product and process inventions. Chinese producers have learnt to serve export markets with new and more sophisticated products, but the fundamental pattern of reliance on foreign sources for core technology remains (cp. Liefner and Wei 2014; Kroll and Frietsch 2014).

Indigenous Innovation and the German Model of Mechanical Engineering

Given the fact that the Chinese mechanical engineering industry seems to concentrate on low-technology products and markets, why do we maintain that mechanical engineering offers better opportunities for indigenous innovation than many other industries? On the one hand, trade figures only tell us about aggregates and averages, and among the myriad of mechanical engineering firms, some will always stand out. But more importantly, it is the characteristics of innovation processes in the mechanical engineering industry that may favor indigenous innovation. Specifically, it is particularly worth examining a certain type of innovation often viewed as the strength of the German mechanical engineering industry. This model of innovation in German mechanical engineering overlaps well with the notion of indigenous innovation in China.

A German Model of Innovation

The term "German model" has received scholarly attention over the last two and a half decades, but it was paid most attention in the late 1990s, as distinguished from the Anglo-American model. Important ingredients of the German model are close cooperation between management and workforce organizations in strategic decision-making, structured negotiation processes to embed company strategies in local and national policy, and business organization that pushes companies toward specializing in customer-oriented high-quality production (e.g. Kern and Schumann 1998; Lane 2000; Dörre 2001). With the rising interest in the role of institutions for economic structure and development, economic geographers also became interested in the German model (e.g. Gertler 2010; Gertler and Vinodrai 2005). One of the important questions raised in this literature concerns the distinctiveness or convergence of business practices in a globalizing economy. This strand of literature covers a very broad spectrum of institutions and other characteristics that are said to be linked to the German model. However, this understanding of the German model is not well suited to typecasting innovation processes in the German mechanical engineering industry.

The success of Germany's mechanical engineering industry has stimulated much conceptual work on the broader German production model, but there is a second strand of literature that concentrates specifically on understanding the organization of innovation processes in German mechanical engineering. A short review of this "German model of innovation in mechanical engineering" helps to identify important success factors for innovation in mechanical engineering. While many success factors attributed to the German model are attributes of the particular constellation of actors and functions specific to German mechanical engineering, it should be noted that they might be found elsewhere and might be applicable to other countries.

The factors, processes, and outcomes of innovation in the German mechanical engineering industry differ in many regards from innovation in other industries. Innovation in the mechanical engineering industry is first and foremost focused on the needs of a particular customer. Kalkowski and Manske (1993) use the expression "instructed innovation" to highlight the importance of the customer in defining exactly the purpose that a particular machine should be developed to fulfill. The innovation process is evolutionary in the sense that the specification of the purpose by the customer and the development of alternative technical solutions by the producer can co-evolve during close producer–customer interaction (Kalkowski and Manske 1993; Hirsch-Kreinsen and Seitz 1999). This characterization, however, mainly holds true for the producers of single custom machines. Producers specializing in lines or programs of similar machines or standardized machines sold to a variety of customers need less intense producer–customer interaction (Kinkel and Som 2007). However, the fundamental features of innovation processes apply to them as well.

Innovation in the German mechanical engineering industry is mainly generated through learning by producing. The development of alternative technical solutions, tests of pilot machines, changes in machine components, their architecture and interaction, etc., are carried out in-house during the work on the machine under development (Kalkowski and Manske 1993). The main resource required for this is the knowledge, experience, and creativity of the technicians and engineers working for the producer. Therefore, the employees who constitute a company's mechanical engineering core determine the direction and outcome of product innovation (Kalkowski and Manske 1993).

This underlines the importance of a workforce that is not only formally qualified but also experienced and familiar with the particular line of technical solutions in which their company specializes. At the same time, they must be creative when it comes to improving the technology that they already master. Here, an essential feature of the German production model comes into play, which is the importance attached to apprenticeship, the importance of highly skilled technicians who find practical solutions for the concepts developed by engineers, and the long-term stability of core groups of technicians and engineers in most companies (Finegold and Wagner 1998).

Another feature of many German mechanical engineering companies is family ownership and the involvement of family members as chief engineers and managers (Schiersch 2013). This goes hand-in-hand with investment decisions and strategies meant to achieve long-term success at the expense of foregoing short-lived business opportunities and risky growth paths. The personal relationship between owners and workers frequently extends over more than one or two generations and ensures that a particular sense of quality and technology can be cultivated continuously.

It is obvious that the latter two features of the German mechanical engineering industry—qualification and stability of a core team of technicians and engineers and active family ownership—are not usually present in China. In China, deficits in the formal education of engineers, limited availability of technicians, a high rate of labor fluctuation, growth barriers for private companies, and short-term decision-making may significantly reduce the prospects of the whole industry when it comes to replicating the German development path. However, these impediments can be controlled and deliberately changed by company owners and managers who decide to regard quality improvements and innovation as the main resources of their companies. Thus innovation remains rare in today's mechanical engineering companies in China, but it can still be considered possible for all firms with sufficiently strong internal capabilities.

This is particularly true since external inputs to the innovation process through commodity chains are less significant in mechanical engineering than in other industries. Theoretical concepts such as the innovation systems approach (Cooke et al. 1997) highlight the importance of external actors and their knowledge inputs to innovation processes in general. They argue that innovation is usually the outcome of interactive learning processes and thus depends on the innovative capacity of a company's external environment as well as its internal capabilities. Despite the rise in spending for science and technology in China, as well as the visible successes in education and research (Kroll and Frietsch 2014), most studies of innovation systems in China find that interaction and knowledge sharing are underdeveloped (e.g. Wei et al. 2011). However, innovation in mechanical engineering depends less on interaction with partners than with customers in the innovation systems. Thus, it is less likely than other industries to be affected by this particular weakness of the Chinese innovation system.

In the German mechanical engineering industry, cooperation with universities or other research organizations has a limited function in the innovation process. Input from basic research is irrelevant, and cooperation with universities is generally used to conduct joint tests of new components or technical solutions, often carried out at the site of the producer. Moreover, universities often analyze new technical solutions after they have been developed in order to understand their function in a systematic way, which is useful for teaching and future research (Kalkowski and Manske 1993). Small companies in the German mechanical engineering industry, however, have reportedly tried to use universities' R&D capacities to substitute for a lack of internal resources (Kinkel and Som 2007). Universities and R&D organizations, as well as business associations, can give support in both standardizing mechanical components and exploring the general suitability of fundamentally new techniques (Hirsch-Kreinsen and Seitz 1999). With the ongoing concentration on rather narrow core competencies and related production activities, innovation in mechanical engineering is thus embedded into a wider network of partners (Hirsch-Kreisen and Seitz 1999). For the product innovation itself, however, the producer's internal resources and producer–customer interaction remain the key success factors.

In China, private small and medium-sized enterprises (SMEs) may wish to cooperate with universities and research organizations because of limited internal resources. Many of the SMEs, however, may be too weak in technological terms to engage in university–industry interaction. For reasons of institutional proximity (Knoben and Oerlemans 2006), state-owned enterprises may have even better access to universities and research organizations and therefore their knowledge bases and facilities. Many state-owned enterprises, however, feel little competitive pressure and thus lack an incentive to seek scientific knowledge for innovation. Despite these China-specific barriers to university–industry interaction, companies with sufficiently strong internal resources should not face major constraints regarding external interaction and knowledge sharing.

Indigenous Innovation

If we define indigenous innovation as self-directed as in Chapter 2 of this book, this refers to innovation processes under the control of domestic firms that can combine knowledge from different sources. Indigenous innovation should not be dependent on or dictated by foreign firms. The concept of indigenous innovation, established with the "National Guidelines for Medium and Long-term Plans for Science and Technology Development (2006–2020) of China," outlines the direction in which innovation in China should develop within the coming years. China intends to increase spending on R&D, raise the contribution of technological progress to economic growth, reduce dependence on foreign technology, and become a top producer of patents and publications. The concept calls for China to leapfrog into global competitiveness in new science-based industries (Cao et al. 2009). According to Liu and Lundin (2009), at least three factors have contributed to the formation of these guidelines. One is disappointment about the limited effects that FDI has had on the development of China's domestic knowledge base and innovation outcomes. The second is the aim to replace imitation and copying with true innovation. A third is the need for technological progress as a way of making high growth rates sustainable. Cao et al. (2009) also highlight two more factors: the aim to escape from a Western-dominated global innovation environment and the wish to progress in the development of military technology.

The concept of indigenous innovation includes at least two desirable characteristics. One is leapfrogging, or being at the forefront of the development of radically new fields of technology (cp. Soete 1985). The other is national ownership of intellectual property rights, or IPR. This chapter takes the perspective of individual companies in mechanical engineering and their innovation processes. Thus, the macro-level concepts of leapfrogging and national IPR ownership need to be translated into firm-level concepts.

The most popular conceptual characterization of innovation in business administration has recently been that of disruptive innovation versus incremental innovation. As Danneels (2004) points out, disruptive innovation refers to the use of a radically new technology in a company's products and usually goes hand-in-hand with finding new markets and customers. Incremental innovation, in contrast, does not affect the core technology applied and does not require looking for new markets. Between these two characterizations, disruptive innovation comes close to matching the term "leapfrogging" at the firm level.

However, in China, disruptive innovation can have two different expressions and starting points. One is a radical shift in product technologies that leads to the development of new products, while the other is the rise of new markets and types of customers who express demand for products with new features. The first type of technology-driven innovation comes close to the "radical innovation" referred to by Henderson and Clark (1990). The latter type, market-driven innovation, is gaining importance with the increasing purchasing power of Chinese (and other Asian) middle-class consumers and companies, who demand well-functioning products at affordable prices. The innovation needed to satisfy their demand can be called customer-oriented innovation, cost-saving innovation, "low-cost/high-tech" innovation (Schanz et al. 2011), or "frugal innovation" (Sehgal et al. 2010; Zeschky et al. 2011). This kind of innovation usually requires new, more efficient ways of combining existing technologies, coming close to the notion of architectural innovation (Henderson and Clark 1990). Asian firms, both in China and elsewhere, seem to be successful in developing this kind of low-cost/high-tech innovation (Ernst 2008).

For the purpose of this chapter, we translate the macro-level notions mentioned in the "National Guidelines" into the company-level concept of disruptive innovation, distinguishing between radical and low-cost/high-tech within disruptive innovation.

The Association of Mechanical Engineering Companies of China has developed operational definitions that characterize indigenous innovation in this industry (2006). According to these definitions, indigenous innovation means own inventions, own integration of existing technologies, and own improvements or modification of imported technologies. This definition comprises radical and architectural innovation, and innovations can include imported components. National IPR ownership means that firms should hold patents or other IPR, develop commercial applications based on these, and earn profits with these applications. Therefore, indigenous innovation should result in new products, new markets, and new brands (Association of Mechanical Engineering Companies of China 2006).

Hence, the concept of indigenous innovation in mechanical engineering in China and the German model of innovation in mechanical engineering overlap in some important aspects. The most salient are reliance on a companies' own resources, the importance of in-house R&D, and limited dependence on the knowledge of external partners.

Innovation-Related Structures in China's Mechanical Engineering Industry

In order to assess whether mechanical engineering companies in China might have the chance to act like German mechanical engineering companies, we also have to take a closer look at the state of the structural features that support the German model, i.e. family ownership and active involvement of owners as well as qualification and stability of the core workforce.

The forms of ownership and company sizes in the Chinese mechanical engineering industry are very diverse. Parts of China's mechanical engineering industry are dominated by SMEs, just like the mechanical engineering industries of many developed economies (Murakami et al. 1996; Hirsch-Kreinsen and Seitz 1999). For example, private ownership of companies is strong in the machine tool industry, representing 78 percent of all firms and 69 percent of production value. In this field, state-owned firms are fewer in number but larger in size, accounting for 4 percent of the firm population but 13 percent of output of the machine tool industry (CMTTBA 2012). Following these are firms invested in Hong Kong, Taiwan, Macao, and other foreign countries (CMTTBA 2012). State ownership is more important in other branches of the mechanical engineering industry, as can be confirmed by a look at the industry's leading firms (China Machinery Industry Federation 2013). Among the top 10 domestic producers,[2] eight are state-owned enterprises and two are private firms. Domestic ownership—private and state-owned—is thus much more important in China's mechanical engineering industry than it is in many other industries. A high number of firms are privately owned and small, constituting a similarity with the German situation.

The qualification of the core workforce of Chinese mechanical engineering companies is difficult to assess for the aggregate of firms. Human capital development is clearly affected by China's efforts to enhance its domestic innovation potential. Two mechanisms are used in this effort: one is massive public and private investment in education and S&T, and the other is the regional concentration of the available resources. Much has been written assessing the aims of China's "National Guidelines for Medium and Long-Term Plans for Science and Technology Development (2006–2020)" (Schwaag-Serger and Breidne 2007; Cao et al. 2009), which sets out how China intends to become a truly innovative nation. The growth of input into education, S&T, and R&D is impressive, and China is entering the world stage of technology-oriented countries (Kroll and Frietsch 2014). While the impressive growth of national input and output indicators mainly indicates public and private investment in innovation capacities, the concentration of available resources to the leading regions (and organizations and firms) is also remarkable (cp. Marginson 2011). In the case of China, national university entrance exams and competitive university access lead to a stark concentration of human potential at the leading universities, many of which are located in Beijing, and the Yangtze River Delta. Public funding programs such as 211 (in which selected universities receive favorable allocation of state funding) have a similar effect. This concentration of state resources at selected organizations and places boosts conditions for innovation there, while many other organizations and regions fall behind. It thus seems justified to assume that at least firms located in prosperous regions should have the chance to attract engineers with suitable skills. However, China's market for skilled labor is notoriously unstable and has high attrition rates. This may negatively affect most companies and hamper the creation of company-specific technological profiles even in the regions that are endowed with a well-educated workforce.

[2] The companies are: (1) 中国机械工业集团有限公司 China National Machinery Industry Corporation，state-owned. (2) 徐州工程机械集团有限公司 Xuzhou Construction Machinery Group, state-owned. (3) 上海电气(集团)总公司 Shanghai Electric (Group) Corporation, state-owned. (4) 中联重科股份有限公司 Zoomlion Co., Ltd, state-owned. (5) 三一集团有限公司 SANY Group Co., Ltd, private. (6) 潍柴控股集团有限公司 Weichai holding group Co., Ltd, state-owned. (7) 中国东方电气集团有限公司 Dongfang Electric Corporation, state-owned. (8) 盾安控股集团有限公司 Duan Group Co., Ltd, private. (9) 广西玉柴机器集团有限公司 Yuchai Machinery Group Co., Ltd., state-owned. (10) 哈尔滨电气集团公司 Harbin Electric Corporation, state-owned.

Several other and China-specific factors are also important in creating favorable conditions for upgrading and innovation, at least in selected industries. State procurement, in particular in the fields of infrastructure development, transportation, telecommunication, and energy, is one significant factor. In these fields and others, the government uses its control over demand and financial power to give direction for industry development and to build national industries (Liu and Lundin 2009). To some degree, public procurement protects domestic firms from international competition and helps nurture Chinese firms to grow and gain competitiveness based on sales in the domestic market. During the period of the 12th Five-Year Plan from 2011 to 2015, for example, all 16 cities of the Yangtze River Delta region thus started prioritizing the mechanical engineering industry (Lyu et al. 2014). Public banks, which are to a large degree influenced by policy, also use their funding to support favorable conditions for domestic firms in certain industries (Ohm 2011). The sheer growth of the Chinese domestic market, government-induced or not, is another factor supporting technological advances in certain fields. The Chinese mechanical engineering industry has benefitted greatly from the growing and increasingly diverse customer base (Bfai 2008).

China has tried to use policy to create effective local clusters and certain growth dynamics. The Chinese version of cluster development (loosely based on arguments discussed by Porter 1998 and others) includes the establishment of high-tech parks offering specialized infrastructure and investment incentives. These parks are often equipped with research and education organizations, incubators, and business service providers in order to compose a set of actors that will stimulate innovation dynamics. The availability of these individual components in such clusters, however, is no guarantee for local cooperation and innovation (cp. Zeng et al. 2011). There are also many other individual and local initiatives and related entrepreneurialism that receive no academic attention unless they prove to be successful (cp. Heilmann 2008).

Some factors generally tend to hamper successful innovation in China. For instance, most industry leaders still come from advanced economies. As a result, Chinese companies continue to depend on foreign knowledge to progress technologically. Because of the attractiveness of MNC affiliates as employers and places of innovative activity, it is these firms that employ the most qualified personnel, both in terms of technological and language skills. This creates a significant barrier for knowledge inflow into the domestic economy (cp. Liefner 2013, 2014). But many of these factors are less relevant in mechanical engineering, in which companies have less need to coordinate international cooperation and knowledge sharing. Since the majority of the customers are Chinese, language and business culture are relatively homogeneous. Insufficient IPR protection and limited regional availability of innovative partners, generally a problem for Chinese innovation, have less influence on mechanical engineering than on other industries. Because custom machines are developed to serve a particular need expressed by one customer, they usually need no IP protection. If customer and producer share the same goals, protecting technical knowledge is not necessarily relevant. The limited regional availability of suitable partners for innovation does not represent a barrier to mechanical engineering either. This is the case as long as contacts to universities, research organizations, associations, or other firms have only a general support function, not a direct impact on the development of a new machine. This issue may of course affect the local availability of skilled personnel, but it does not preclude carrying out indigenous innovation activities.

Hence we assume that at least some of the factors usually considered to hamper innovation in China are less important for mechanical engineering. Meanwhile, some factors that stimulate innovation, such as market growth, public procurement, and support for domestic firms, may be particularly important in mechanical engineering. Other factors that are usually regarded as essential, like the active involvement of owners and the quality of a firm's human capital, can be influenced by management decisions.

We hypothesize that mechanical engineering exemplifies an industry with favorable conditions for producing truly indigenous innovations—products that are not only new to China but new to the world. We thus identify company cases of successful indigenous innovation that meets the definitions of the Association of Mechanical Engineering Companies. We expect successful indigenous innovation to be developed mainly for Chinese customers, in fields supported by procurement policies and/or growing domestic market size, and based on internal technical capabilities. We expect to witness indigenous innovation with medium-sized private companies who deliberately choose an innovation-oriented business approach. Indigenous innovation will have characteristics of disruptive innovation; whether it is radical or low-cost/high-tech cannot be determined a priori.

Given the limited availability of statistics or survey data, the following analysis has to rely on company case studies. The cases can only indicate larger trends and illustrate them with a concrete story, not give proof in the sense of a statistical test. However, identifying a single success case is already an indication that indigenous innovation may indeed be possible.

Sketching Firm-Level Approaches to Innovation: The Cases of Yizumi and Propower

The third section of this chapter showed that China's mechanical engineering firms do not usually possess strong technical capabilities or produce new-to-the-world products. But the last section outlined why, theoretically, at least some firms should have the opportunity to become innovative in the line of the German model, depending on company-specific resources and strategies. This section of this chapter discusses findings from two case studies in the mechanical engineering industry. These highlight which success factors make indigenous innovation possible and shape innovation processes and outcomes. They allow us to generate more fine-grained hypotheses, but they do not allow us to establish an empirically based model for innovation behavior. The case studies were carried out at two medium-sized companies, both technological leaders in their respective market niches. The first company analyzed is Yizumi (广东伊之密精密机械有限公司, Guangdong Yizhimi Jingmi Jixie Youxian Gongsi); the second is Propower (上海普羅新能源有限公司, Shanghai Puluo Xinnengyuan Youxian Gongsi).

The two companies are not prominent in terms of size, age, financial capabilities, or public support—their most notable feature is innovativeness. In this respect, they are neither representative nor typical of China's mechanical engineering companies as a whole. They are rare examples of innovative medium-sized firms that illustrate what is possible within the given framework and what routes could be followed by China's other mechanical engineering firms.

Both companies were visited in 2011 and 2012. The visits included long and intensive interviews with CEOs and CTOs as well as visits to production facilities. The firms provided very detailed information on their internal resources, their external partners, their products and product innovations, and their long-term business strategy. Information was also gathered in the forms of reports and written documents and discussed during interviews with the company owners and managers. Information available from independent websites, documents, and reports was used for the purpose of triangulation.

Yizumi

Yizumi is a medium-sized company with 1,100 employees. It has three main products: die casting machines, injection molding machines, and rubber injection machines. About 70 percent of the machines sold are custom machines while the rest are standardized. Yizumi was established in 2002 in Shunde, Guangdong province, a region with a high concentration of household appliance manufacturing industries. Additional production sites in China (Wusha and Suzhou) are in operation. The company receives decent attention from mechanical engineering business organizations and individual customers who cite Yizumi as their high-quality suppliers.[3] Yizumi

[3] For example, Metalplast, Czech Republic.

has been recognized as a "Chinese well-known Trademark" (Diecasting-China.com 2012).

In 2011, Yizumi acquired the small US company HPM (Hydraulic Press Manufacturing) and its IPR, and it still runs a team of American and Chinese researchers at the original firm site in Ohio. They plan to set up additional production sites outside China in other emerging markets (including Brazil, India, and Eastern Europe). Eighty percent of the machines are sold in China. The major export destinations are emerging markets in Eastern Europe, the Middle East, and Central Asia. The company is privately owned, and all owners are Chinese, originating from Guangdong, Hong Kong, and Taiwan. The role of the owner from Hong Kong is primarily one of strengthening the capital base, while the Taiwanese owner concentrates on technology. The owner from Guangdong runs the company as its CEO and CTO.

The central goal of Yizumi is to be regarded as a reliable supplier of high-precision and high-performance machines. The company wants to be in a leading position in the development of customer-oriented technology. Among China's domestic producers, Yizumi sells at comparatively high prices domestically compared to their competitors, covering a considerable share of the top-quality market segment. Internationally, Yizumi's prices fall into a more middling range, taking advantage of export opportunities into markets that seek reasonable product performance at an affordable price. With exports to customers in Eastern Europe, the Middle East, and Central Asia, Yizumi's export profile is comparable to the overall export structure of its particular product group (8454, HS1992) (UN comtrade 2014).

According to the CEO, Yizumi's long-term business strategy is based on high technology and continuous innovation. This statement was echoed in a report on Yizumi published in *Plastic News*, April 2013. For Yizumi, innovation specifically means changes in the product architecture that create higher value at lower costs, following demands expressed by Chinese customers in particular (Toloken 2013). As of today, the company's machines still include high-tech components imported from German and Japanese suppliers, and some core technology components are bought from other Chinese firms. However, the machine architecture is developed internally. Moreover, the recent acquisition of the US affiliate has opened access to top-level technical expertise.

On average, 4 percent of turnover is invested in R&D. With this share, Yizumi falls into a range typical for mechanical engineering firms in developed economies (cp. Meyer-Krahmer and Schmoch 1998). The company employs a core team of engineers from all over China, and personnel fluctuation is low, guaranteeing the continuity of development routines and growth of the knowledge stock. The chief engineer is one of the company owners, who previously worked at a competing company in the Pearl River Delta

(PRD). The technology acquired in the US has so far not had an impact on the products developed in China. After successful absorption, however, it will allow major technological advances to be used in China and for promoting sales in the American market.

While the development of new custom machines is primarily the result of internal resources, Yizumi has also received some public support in the form of the Chinese state Torch program which aims at helping SMEs. It has also had joint projects with various universities and colleges as well as with firms in the PRD. These projects were mainly in the fields of technical training, not product development.

Hence Yizumi uses mainly internal resources for developing customer-oriented innovation. In the active involvement of owners and the core engineering team, Yizumi comes close to typical mechanical engineering firms in Germany. The company has a clear system for continuously advancing its knowledge base. It uses a gradual approach to innovation and improves its machines to meet a balance between performance and price demanded by the customers. Improvements in the products' architecture are the most important ingredient of its innovations, while some key components are bought and integrated. Generally, Yizumi's innovation can be labeled low-cost/high-tech. If this path is continued, the company can expect growing sales as well as a move towards increasingly sophisticated machines and more demanding markets and customers. The home region of the PRD provides the company with a diverse and evolving demand profile, but otherwise it has no direct effect on the development of the resources on which innovation depends. However, the company ownership structure is typical of the PRD, with the chief engineer moving between similar companies nearby on his career path. Some activities that indirectly support the technical expertise of Yizumi, including university–industry interaction, also benefit clearly from its location.

Propower

Propower is a total solution provider for the photovoltaic (PV) industry. Besides arranging the construction of complete PV power stations, the medium-sized company develops and produces poly-silicon purification furnaces based on proprietary technology to produce solar-grade poly-silicon. Propower sells its products mainly to China, but it has customers all over the world including other Asian countries (e.g. Taiwan) and Europe (e.g. Germany). The company was started in 2007 in the city of Lingang, Shanghai, and it employs 260 people. It should be noted that this location is at the heart of China's most concentrated photovoltaic industry in the Yangtze River Delta. It is privately owned by individual shareholders and involves Chinese venture capital (VC). In 2013, Propower won a "SNEC Terawatt Diamont

Award" for creating the world's biggest poly-silicon casting furnace (snec.org 2013). The founder, Mr Shi Jun, obtained a Ph.D. from the Shanghai Institute of Technical Physics of the Chinese Academy of Science in 1991, and he has been involved in technology research and the machine automation industry for two decades.

Propower benefitted greatly from the explosive growth of China's solar industry supported by the Chinese government. The rapid increase of China's manufacturing capacity of solar panels drove the market for poly-silicon purification equipment. Initially, almost all equipment had to be imported, but domestic production picked up in the mid-2000s from the simpler operation to more sophisticated ones (Solarstar PV News 2013). Propower aimed to provide reliable and cost-saving technologies and develop the leading technologies and machines for poly-silicon purification. According to its CEO, it values technology and innovation over price, and it regards its proprietary technology as superior to alternatives provided by world market leaders from developed economies, such as Germany and Switzerland. The company's key patent "Polysilicon ingot casting furnace and polysilicon ingot casting method" (WO/2012/139362) was published by WIPO October 18, 2012. The patent is held by Propower and the inventing engineers Jun Shi, Weifeng Zong, Chuan Shui (Propower's CTO), Chen Tong, and Suling Chen, all engineers trained in China. Propower claims that furnaces based on this technology reduce the overall costs of producing solar-grade poly-silicon by 40 percent (as compared to the established process method from Siemens) and the silicon produced has a better performance when used at high temperatures. The cost reduction has been discussed in related forums and media, such as Baitron and the PV magazine (baitron.com; Ali-Oettinger 2011). The patent is both used in machines produced internally and also franchised to other producers.

Propower devoted the first three years of the company's existence solely to conducting R&D, funded by the firm's VC. The founder was previously involved in the machinery business for a long time, and this experience has helped him to develop extensive business and technological networks in the industry. Two types of resources were crucial for the innovation in poly-silicon purification. One is a team of researchers that develops, modifies, and tests machine prototypes internally. In their work, they benefit from available public and private funding and their own professional knowledge concerning the processes of furnace construction and purification. The other key resource in this process is the Research Alliance for the Purification of Poly-silicon, founded in 2009 and headed by the president of Propower. This alliance brings together companies, university research institutes, for example from Shanghai Jiaotong University and Northwest University in Xi'an, and institutes from the CAS. Dalian University of Technology was another key partner of Propower in setting up demo plants in Ningxia. Besides Propower, eight companies have joined the alliance, seven of them Chinese and one from Spain. Companies focus on the industrial application of the knowledge circulating in the alliance. They pay membership fees, while research institutes participate for free. The alliance has received public support from MOST for constructing demonstration plants and test facilities, through program 973 for university–industry collaboration, and through program 863 (another longstanding state-funded research program). The technology developed by the alliance is used by alliance members and also licensed to other entities, who pay royalties that benefit the alliance.

The technologies invented for poly-silicon purification, as well as the furnaces that implement and apply this technology, are developed at the site of the Propower company. Engineers from Propower construct furnaces and test their function, ability, and performance, and they modify the machines to improve the results. Other researchers from the alliance join the development and test work, conducting their performance analyses on site with the pilot machines. While this innovation process is very typical of the mechanical engineering industry, it involves collaboration as an integral piece.

Propower's invention involves the process technology of poly-silicon purification and the layout size of the casting furnace. It is a complex innovation that to some degree can be labeled "radical." It proposes a method different from today's standard technology and this results in a substantial increase in efficiency and reduction in cost. Propower's products have been sold to leading markets in the photovoltaic industry such as Wacker Chemie (Germany), an unusual export destination for Chinese producers of this group (8417, HS1992) (UN comtrade 2014). The company's major competitors are located in Europe and include ALD Vacuum Technologies (Germany) and Centrotherm (Germany).

The innovation activities of Propower mostly depend on internal technical capabilities, as the company was set up to explore the new technology route. The local availability of well-trained scientists and engineers is a supporting factor. Local alliance members have important roles in the innovation process, too, and Propower thus benefits from the innovative potential of the elite R&D resources in Shanghai and the YRD. However, the alliance has members from all over China, and the extent to which local networking opportunities support the innovation process should therefore not be overstated. Propower's CEO also suggests that they are still actively learning from Germany companies for their quality systems, technician training, and work ethics. Thus it is clear that Propower is an internationally open company. Unlike most other Chinese companies having a business approach centered on copying technology from others, Propower has an innovation-focused company strategy. The company chairman feels that copying may not fully enable these firms to catch up, and that inferior quality and professionalism may make them less competitive with foreign rivals. Instead, the Propower strategy is oriented toward achieving technological world leadership in the company's core field. This strategy certainly has its risks. If Propower is not able to show satisfying commercial results and justify the huge R&D investment already made, VC may be withdrawn and bankruptcy would be the unavoidable outcome.

Cross-Case Analysis

From all accessible sources of information, both companies examined here can be considered innovative. However, the types of innovation that Yizumi and Propower achieve are quite different. Similarities and differences can also be found in resources used, markets and strategies used, and the degree to which company location influences innovation. A graphic comparison, which admittedly runs the risk of overstating differences for reasons of clarification, is shown in Figure 4.4.

Both companies have a strong engineering core that form the firms' core competence. In both companies, the top management positions are filled by engineers who own company shares and participate in technology development. In both companies, the top management also has a clear vision and strategy that centers on a well-informed assessment of the companies' technological capabilities as well as targeted markets and competitors. The market

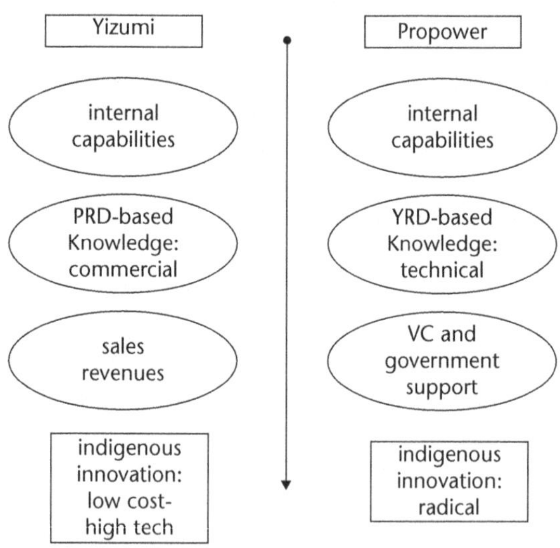

Figure 4.4. Yizumi's and Propower's Different Pathways towards Indigenous Innovation
Source: By the authors.

environments of both firms are comparable as well; Yizumi is in a field with growing demand for machines with a well-established basic technology, and Propower enjoys growing demand that is more or less guaranteed by the state's involvement in the development of the solar energy industry. The local environment influences both firms to some degree in their innovation processes. Both the YRD and the PRD are rich in skilled labor, have an abundant customer base due to the regional specialization, and generally offer healthy business conditions. These features are all quite similar to those discussed as pillars of the German model.

The differences between the two companies deserve more attention, however: they put very different emphases on commercial success vs. technological originality. Yizumi is much more oriented toward the former while Propower gives priority to the latter, using VC to overcome an initial period without any revenues. Correspondingly, Yizumi targets market segments that seek low-cost/high-tech innovation, and Propower aims to dominate the technological development path that corresponds to their core activity, although they also based their competitiveness on more efficiency and cost saving for their customers. Customer–producer interaction seems to be the dominant factor in innovation dynamics for Yizumi, as 70 percent of the machines they sell are custom. Propower's innovation is comparably more technology-driven, based on developing genuine ideas rather than specializing for customer needs.

So Yizumi's innovation clearly falls into the category of low-cost/high-tech innovation. Yizumi uses imported core components when necessary to achieve top performance and its innovation is centered on providing excellent value for money. The company tries to upgrade its innovative capability using revenue from sales. In contrast, Propower's innovation can be considered radical. Propower uses a newly invented production technology to offer machines that are different from existing solutions. Propower has invested significant resources in its innovation, largely funded by venture capital, and has not prioritized the needs of individual customers or minimizing costs. To their benefit, they are in a field with an assured customer base, given the rapid ascent of China's solar industry.

The different innovation outcomes of both companies may be the result of differences in internal resources, management, markets, and strategies. Thus, they fall within the paradigm of the resource-based view (Wernerfelt 1984). However, they also seem to be affected by the conditions in the two companies' home regions. Yizumi is located in the PRD, where a business climate has been shaped by decades of successful export sales to very different markets. Companies in the PRD have had investment from truly international sales companies based in Hong Kong and technologically leading Taiwanese firms, and benefit from functioning business interrelations (Meyer et al. 2012).

However, the PRD is not rich in technologically strong universities and CAS institutes that might create a better climate of technological leadership. That climate is exactly what Shanghai and the YRD, where Propower resides, can offer, in addition to the country's largest cluster of solar firms (Liefner and Wei 2014). There is no shortage of high-quality partners in research, such as those that form the Propower research alliance. Public support for innovation is available in both locations; however, Propower has better opportunities for making use of support because of the proximity of research partners. Scholars have found that, within the information and communications (ICT) industry, YRD companies tend to specialize in more capital-intensive and knowledge-intensive production while the PRD is known for more quickly responsive, labor-intensive firm networks (Zhou et al. 2011). The findings here do show similar characteristics. Overall, however, the home regions' major influences on both firms are through human resources and local training opportunities for employees, as well as locally available capital and managers. Interactive learning or other forms of cooperation and "open innovation" may be present, but they cannot conclusively be held responsible for the differences in innovation processes and outcomes based on these two case studies.

Conclusion: Toward Indigenous Innovation in China's Mechanical Engineering Industry?

The mechanical engineering industry in China has been witnessing high and sustained growth. Chinese firms in this industry are successfully serving the demand for low-cost machinery in domestic and export markets. The majority of firms have to be considered non-innovative and technologically lagging when compared to mechanical engineering companies in advanced economies. China's huge investment in education and science and technology has clearly helped provide the input of engineers' qualifications and new ideas, but it has not fundamentally changed the inferior position of Chinese products as compared to the Western countries. So far, China's mechanical engineering industry as a whole does not exemplify the paradigm of indigenous innovation. Nevertheless, individual innovative companies are found employing different strategies based on their perception of the opportunities and their internal capabilities.

This chapter has also argued that the particular characteristics of innovation processes in mechanical engineering should create opportunities for indigenous innovation. Innovation in mechanical engineering heavily relies on firms' internal resources, on close interaction with customers, and on internal learning and experimenting. External knowledge sources is not essential.

Extraordinarily high market growth rates and state support create favorable conditions for mechanical engineering, as does the intense industrial competition.

In order to clarify which conditions are most likely to support innovation, we examined the German model of innovation in mechanical engineering as a well-documented success case. German mechanical engineering companies benefit greatly from having a stable core of highly qualified technicians and engineers and from having company owners that actively participate in production or management. It seems reasonable to expect that some of the many mechanical engineering firms in China can create similar conditions for innovation. The two company cases examined in this chapter illustrate that Chinese mechanical engineering firms are closely tied to market dynamics. Chinese companies can produce new and innovative products if they consistently invest in internal R&D growth and utilize available regional capital and technical resources. Our analysis could not conclusively answer the questions of whether this behavior is likely to become more widespread within the industry, or whether innovative companies will be more successful than their low-tech peers. However, the apparent success of innovative firms like the two analyzed here will most likely attract followers.

Of course, there are other limitations of our analysis that should be acknowledged. Future research should extend towards analyzing survey data. Moreover, it seems necessary to examine the role of state-owned companies in mechanical engineering. Some of them are very large, close to the government, and institutionally similar to universities and large R&D laboratories. They should be able to gain access to knowledge generated in universities and research laboratories, and they should have the ability to develop long-term strategies without paying attention to short-term market fluctuations. Whether they in fact use this potential is a very different question. State-owned firms face less competitive pressure and lack an important incentive to innovate. The government's influence on firms in the mechanical engineering industry extends beyond state-owned firms, however, in the form of organizing cooperation and the development of industry clusters. But private SMEs in the mechanical engineering industry can be generally regarded as quite distant from the state and independent regarding their innovative activities.

The conclusions of these case studies can be easily paired with the related literature. The Yizumi approach, with a focus on stepwise improvements and upgrading, architectural innovation, and commercial success, fits with thoughts discussed by Hobday (1995), Ernst (2009), or Peighambari et al. (2014). Propower's approach to innovation, which is clearly more risky, complements empirical contributions on interactive learning and innovation in China (e.g. Liefner et al. 2012).

The chapter has explained the conditions under which indigenous innovation is possible in China's mechanical engineering industry and how the innovation processes and their outcomes may look. Important conditions that positively affect indigenous innovation are the following:

- sufficient internal resources, including a stable core group of engineers and technicians, and sufficient revenue from sales or VC to be invested in R&D;
- management that sets out and becomes part of a long-term, innovation-centered business strategy;
- a market environment that is characterized by growth and competition;
- sufficient stimulus and direction for innovation, stemming either from customer interaction or genuine technological ideas;
- a dynamic regional environment that supplies ideas, people, and relations that match and support the company's business strategy.

Indigenous innovation can be truly disruptive and can take at least two forms, low-cost/high-tech innovation and radical innovation. The former requires more market knowledge; the latter requires deeper technological knowledge and a rapidly growing market. There is some opportunity in both types for interaction between firms and research organizations, and regional factors can influence pathways to innovation.

Despite this chapter's hypothesis-generating approach, an important underlying assumption has been confirmed: there are cases of medium-sized firms that develop products new to the world, mainly relying on domestic knowledge, and these firms' behavior seems to be partly motivated by their surrounding regional business environment. Within the PRD, a firm's innovative ability can be enhanced by different forms of contacts with foreign-invested local firms and their employees. Shanghai offers regionally confined advantages as the core of the YRD, an industrial cluster with a differentiated regional network. The two paths to indigenous innovation that have been identified may thus exemplify two different regional pathways. Indigenous innovation in the PRD can be thought of as driven by application and stimulated by management and technological knowledge originating outside of China. Shanghai is the center of an older mechanical engineering region, so indigenous innovation there will be based more on knowledge originating within the cluster. Both pathways toward indigenous innovation can be expected to yield success.

However, today the majority of Chinese firms in mechanical engineering are still placing emphasis on exploiting comparatively low production costs. Despite the technological catch-up evident from trade statistics, China's mechanical engineering industry in general will stay behind the latest technological developments for the near future. It is hard to predict how much time will be necessary to developing a truly innovative mechanical engineering industry because it depends on the development of the success factors mentioned. As long as exploiting cost advantages and short-term opportunities is a suitable business strategy, investment in production capacities and generating short-term profits will usually dominate over long-term strategic capacity building. Thus, the long-term success of Chinese indigenous innovation in the mechanical engineering industry will depend on whether China can produce and sustain the factors discussed and encourage more firms to innovate.

Acknowledgments

We gratefully acknowledge funding from the German Research Foundation (DFG Li981/5-2, DFG Li981/8-1). We wish to thank Professor Xun Li, Sun Yatsen University, Guangzhou, for his support for the Yizumi case study. We are also thankful for reviews and comments on previous versions of this chapter provided by Yu Zhou, Bill Lazonick, and Yifei Sun.

References

Aiginger, K. (1997), "The Use of Unit Values to Discriminate between Price and Quality Competition," *Cambridge Journal of Economics*, 21: 571–92.

Ali-Oettinger, Shamsiah (2011), "Chinese Polysilicon a Key Focus at SNEC PV," *PV Magazine*, Feb. 22.

Anonymous (2007), "China's Five Development Stages of Equipment Manufacturing Industry," *China Equipment*, 11(1) (in Chinese).

Anselin, L., A. Varga, and Z. Acs (1997), "Local Geographic Spillovers between University Research and High Technology Innovations," *Journal of Urban Economics*, 42, 422–48.

Association of Mechanical Engineering Companies of China (2006), "Definition of Indigenous Innovation." <http://wenku.baidu.com/view/2fbf57bef121dd36a32d8298.html> (in Chinese).

Baitron.com, *"Makers Achieve Breakthrough in Polysilicon Production."* <http://www.baitron.com/article-197.html 06.01.2014>.

Bfai (2008), Branche kompakt, VR China Maschinenbau und Anlagenbau, Köln.*

Cao, C., R. P. Suttmeier, and D. F. Simon (2009), "Success in State-Directed Innovation? Perspectives on China's Plan for the Development of Science and Technology," in G. Parayil and A. P. D'Costa (eds), *The New Asian Innovation Dynamics: China and India in Perspective*, 247–64. Basingstoke: Palgrave Macmillan.

China Machinery Industry Federation (2013), *Top 100 China Machinery Industry Enterprise*, Apr. 9, <http://top100.mei.net.cn/news.asp?id=151&f=3> (in Chinese).

CMTTBA (China Machinery Tool & Tool Builders' Association (2012), *China Machine Tool Industry Yearbook 2011*. Beijing: China Machine Press.

Cooke, P., M. G. Uranga, and G. Etxebarria (1997), "Regional Innovation Systems: Institutional and Organisational Dimensions," *Research Policy*, 26: 475–91.

Danneels, E. (2004), "Disruptive Technology Reconsidered: A Critique and Research Agenda," *Journal of Product Innovation Management*, 21: 246–58.

Diecasting-China.com (2012), "*Yizumi is recognized as 'Chinese Well-Known Trademark,'*" June 1, <http://www.diecasting-china.com/news/show.php?itemid=22>.

Dörre, K. (2001), "Das deutsche Produktionsmodell unter dem Druck des Shareholder Value," *Kölner Zeitschrift für Soziologie und Sozialpsychologie*, 53(4): 675–704.

Ernst, D. (2008), "Innovation Offshoring and Asia's Electronics Industry: The New Dynamics of Global Networks," *International Journal of Technological Learning, Innovation and Development*, 1: 551–76.

Ernst, D. (2009), *A New Geography of Knowledge in the Electronics Industry? Asia's Role in Global Innovation Networks*. Policy Studies, 54. Honolulu: East West Center.

Finegold, D., and K. Wagner (1998), "The Search for Flexibility: Skills and Workplace Innovation in the German Pump Industry," *British Journal of Industrial Relations*, 36(3): 469–87.

Gertler, M. (2010), "Rules of the Game: The Place of Institutions in Regional Economic Change," *Regional Studies*, 44(1): 1–15.

Gertler, M., and T. Vinodrai (2005), "Learning from America? Knowledge Flows and Industrial Practices of German Firms in North America," *Economic Geography*, 81(1): 31–52.

Heilmann, S. (2008), "Policy Experimentation in China's Economic Rise," *Studies in Comparative International Development*, 43: 1–26.

Henderson, R. M., and K. B. Clark (1990), "Architectural Innovation: The Reconfiguration of Existing Product Technologies and the Failure of Established Firms," *Administrative Science Quarterly*, 35: 9–30.

Hirsch-Kreinsen, H., and B. Seitz (1999), *Innovationsprozesse im Maschinenbau*. Arbeitspapier des Lehrstuhls Technik und Gesellschaft, 4/1999. Dortmund.*

Hobday, M. (1995), "East Asian Latecomer Firms: Learning the Technology of Electronics," *World Development*, 23(7): 1171–93.

Kalkowski, P., and F. Manske (1993), *Innovation im Maschinenbau: Ein Beitrag zur Technikgeneseforschung*. SOFI-Mitteilungen, 29. Göttingen.

Kern, H., and M. Schumann (1998), *Kontinuität oder Pfadwechsel? Das deutsche Produktionsmodell am Scheideweg*. SOFI-Mitteilungen, 26. Göttingen.

Kinkel, S., and O. Som (2007), *Strukturen und Treiber des Innovationserfolgs im deutschen Maschinenbau: Verbreitung und Effekte von innovationsunterstützenden Technik-, Organisations- und Kooperationskonzepten*. Mitteilungen aus der ISI-Erhebung zur Modernisierung der Produktion, 41. Karlsruhe.*

Knoben, J., and L. A. G. Oerlemans (2006), "Proximity and Inter-Organizational Collaboration: A Literature Review," *International Journal of Management Reviews*, 8(2): 71–89.

Kroll, H., and R. Frietsch (2014), "Regional Structures and Trends in China's Innovation System: An Indicator-Based Account of the Last Decade's Developments," in

I. Liefner and Y. H. D. Wei (eds), *Innovation and Regional Development in China*, 41–72. London: Routledge.

Lane, C. (2000), "Globalization and the German Model of Capitalism: Erosion or Survival?" *British Journal of Sociology*, 51(2): 207–34.

Li, X. B. (2012), "Behind the Recent Surge of Chinese Patenting: an Institutional View," *Research Policy*, 41: 236–49.

Liefner, I. (2014), "Explaining Innovation and Regional Development in China: How Much Can We Learn from Applying Established Western Theories?" in I. Liefner and Y. H. D. Wei (eds), *Innovation and Regional Development in China*, 21–40. London: Routledge.

Liefner, I., and S. Hennemann (2011), "Structural Holes and New Dimensions of Distance: The Spatial Configuration of the Scientific Knowledge Network of China's Optical Technology Sector," *Environment and Planning A*, 43(4): 810–29.

Liefner, I., and Y. H. D. Wei (2014), "Introduction: Innovation and Regional Development in China," in I. Liefner, Y. H. D. Wei (eds), *Innovation and Regional Development in China*, 1–18. London: Routledge.

Liefner, I., C. Brömer, and G. Zeng (2012), "Knowledge Absorption of Optical Technology Companies in Shanghai, Pudong: Successes, Barriers and Structural Impediments," *Applied Geography*, 32(1): 171–84.

Liefner, I., Y. H. D. Wei, and G. Zeng (2013), "The Innovativeness and Heterogeneity of Foreign-Invested High-Tech Companies in Shanghai," *Growth and Change*, 44(3): 522–49.

Liu, X. L., and N. N. Lundin (2009), "The National Innovation System of China in Transition: From Plan-Based to Market-Driven System," in G. Parayil and A. P. D'Costa (eds), *The New Asian Innovation Dynamics: China and India in Perspective*, 27–49. Basingstoke: Palgrave Macmillan.

Lyu, G.-Q., G. Zeng, and J. L. Guo (2014), "Innovation Network System of Industry-University-Research Institute of Equipment Manufacturing Industry in the Changjiang River Delta," *Scientia Geographica Sinica*, 34, 1051–9 (in Chinese).

Marginson, S. (2011), "Higher Education in East Asia and Singapore: Rise of the Confucian Model," *Higher Education*, 61: 587–611.

Meyer, S., D. Schiller, and J. R. Diez (2012), "The Localization of Electronics Manufacturing in the Greater Pearl River Delta, China: Do Global Implants Put Down Local Roots?" *Applied Geography*, 32(1): 119–29.

Meyer-Krahmer, F., and U. Schmoch (1998), "Science-Based Technologies: University–Industry Interactions in Four Fields," *Research Policy*, 27: 835–51.

Murakami, N., D. Q. Liu, and K. Otsuka (1996), "Market Reform, Division of Labor, and Increasing Advantage of Small-Scale Enterprises: The Case of the Machine Tool Industry in China," *Journal of Comparative Economics*, 23: 256–77.

NBSC (National Bureau of Statistics of China) (1999), *New China's Fifty Years (1949–1999)*. Beijing: China Statistics Press.

NBSC (2013a), *2012 China Industry Economy Statistical Yearbook*. Beijing: China Statistics Press.

NBSC (2013b), *1949–2012 China Industry Economy Statistical Yearbook*. Beijing: China Statistics Press.

NBSC (2013c), *Industrial Science and Technology Yearbook*. Beijing: China Statistics Press.

OECD and Eurostat (eds) (2005), *Oslo Manual: Guidelines for Collecting and Interpreting Innovation Data*, 3rd edn. Paris: OECD.

Ohm, S. (2011), *Der Einfluss des Staates auf wirtschaftliche Aufwertungsprozesse, untersucht am Beispiel der Elektronikindustrie im Perlflussdelta (China)*. Dissertation, JLU Gießen.

Peighambari, A., S. Hennemann, and I. Liefner (2014), "Success Factors for Upgrading and Innovation in the Electronics Industry: An Analysis of Private Small and Medium-Sized Enterprises in the Pearl River Delta," *International Journal of Technology Management*, 65(1–4): 49–69.

Peng, Z. W., and Y. Huang (2011), "An Empirical Study on Spatial Agglomeration and Affecting Factors of Chinese Equipment Manufacturing Industry," *Soft Science*, 25(5): 57–60 (in Chinese).

Porter, M. E. (1998), "Clusters and the New Economics of Competition," *Harvard Business Review*, 76(6), 77–90.

Schanz, C., S. Hüsig, M. Dowling, and A. Gerybadze (2011), "Low Cost—High Tech Innovations for China: Why Setting up a Separate R&D Unit is Not Always the Best Approach," *R&D Management*, 41(3): 307–17.

Schiersch, A. (2013), "Firm Size and Efficiency in the German Mechanical Engineering Industry," *Small Business Economics*, 40: 335–50.

Schwaag-Serger, S., and M. Breidne (2007), "China's Fifteen-Year Plan for Science and Technology: An Assessment," *Asia Policy*, 4 (June), 135–64.

Sehgal, V., K. Dehoff, and G. Panneer (2010), "The Importance of Frugal Engineering," *Strategy + Business*, 59 (Summer), 20–5.

Snec.org (2013), *SNEC 2013 Top Highlights Award Winner*, June 1, <http://www.snec.org.cn/website/Common.aspx?lang=en&menuId=3&htmlId=962>.

Soete, L. (1985), "International Diffusion of Technology, Industrial Development and Technological Leapfrogging," *World Development*, 13: 409–22.

Solarstar PV News (2013), <http://guangfu.bjx.com.cn/news/20131107/470895.shtml>.

Toloken, Steve (2013), "China's Yizumi Looks at Injection Press Assembly Factory in India," *Plastic News*, Apr. 15.

UN comtrade (2014), United Nations Statistics Division Commodity Trade Statistics Database, Jan. <http://comtrade.un.org>.

Wei, Y. H. D., I. Liefner, and C. H. Miao (2011), "Network Configurations and R&D Activities of the ICT Industry in Suzhou Municipality, China," *Geoforum*, 42(4), 484–95.

Wernerfelt, B. (1984), "A Resource-Based View of the Firm," *Strategic Management Journal*, 5: 171–80.

Zeng, G., I. Liefner, and Y. F. Si (2011), "The Role of High-Tech Parks in China's Regional Economy: Empirical Evidence from the IC Industry in the Zhangjiang High-Tech Park, Shanghai," *Erdkunde*, 65: 41–51.

Zeschky, M., B. Widenmayer, and O. Gassmann (2011), "Frugal Innovation in Emerging Markets," *Research Technology Management*, 54(4): 38–45.

Zhou, Y., Y. F. Sun, Y. H. D. Wei, and G. C. S. Lin (2011), "De-centering 'Spatial Fix': Patterns of Territorialization and Regional Technological Dynamism of ICT Hubs in China," *Journal of Economic Geography*, 11(1): 119–50.

5

Chinese Indigenous Innovation in the Car Sector

Being Integrated or Being the Integrator

Kaidong Feng

Introduction

In the past three decades, the Chinese automobile sector has experienced dramatic expansion. Not only has the market grown to an unprecedented size, there is also emergence of diverse types of manufacturing companies, making the sector's development more complex. Analytical frameworks such as GPN (global production network), focusing on the globally integrated production system (Liu and Dicken 2006), can explain only part of the story, as the new Chinese indigenous car-makers are not an integrated part of GPNs. The developmental state thesis argues that the national and regional Chinese governments have been the driving force behind the automobile industry (Thun 2006; Naughton 2007, quoted from Chu 2011). While government policy has had a strong influence on the industrial structure of the automobile industry, especially prior to 2001, this chapter will argue that the rise of indigenous innovation in the car sector was not primarily a result of Chinese governmental design. Rather, it was initially caused by forces outside of governmental control.[1]

[1] Chery, as a leading firm of indigenous innovation, was indeed built up by the regional government of WuHu City. But the development of Chery, including the scale and the developmental pattern, was beyond the initial expectations of its founders. In the initial plan, Chery would be built into a small or medium-sized firm focusing on one or two product models only; and the purpose was just to enlarge the local GDP (based on interviews with some founders of Chery).

China's automobile industry follows the East Asian "technology borrowing model" of technological imitation and reverse engineering (Amsden 1989; Kim 1997, 1998). Most leading Chinese firms depend on studying technologies from abroad. However, the success of "imitation to innovation" is not assured and it depends on organizational systems. This chapter will show that it is exactly the differences in organizational systems that determine the variations in learning performances and outcomes.

The automobile industry is regarded as a strategic industry in China and many other countries. Drucker terms it the "industry of industries" (1946: 149), given its significance in both stimulating and sustaining the growth of other relevant industries. In East Asia, the rise of automobile industry signified the success of industrial catch-up. For example, in Japan during the 1980s–90s, the auto-industry accounted for 10–15 percent of Japanese manufacturing output, 10 percent of Japan's total industrial expenditures in equipment and R&D, and 20 percent of its exports (Fujimoto 2007). The Japanese automobile industry produced about 30 percent of the global automobile industry's output. Similarly, the Korean government enacted an "Automotive Promotion Act" in its First Five-Year Economic Development Plan during the rapid industrialization period and assigned the auto-industry a central strategic role in its economic system (Kim 1997).[2]

Like many other large economies, China emphasized the automobile industry early on in its industrialization. Before the economic reforms in 1978, the automobile industry was already a strategic area, although central planning only emphasized the commercial vehicles sector (trucks, buses, etc.). From 1978, policymakers began to focus on developing the automobile industry after they witnessed the effects of the modernized automobile industry in the West. A successful automobile industry potentially offers such benefits as greatly value-added manufacturing, forward and backward linkages in stimulating relevant industries, stable tax income, vast employment opportunities, increased export opportunities, and transportation and industrial optimization, among other benefits (Zhang and Gao 2001).[3] As early as 1978, a delegation of General Motors was invited by the Chinese government to visit Beijing, and both sides started to discuss the possibility of forming a joint venture to produce cars (Ming 2006; Chu 2011). In "the 7th five year plan (FYP) outline," issued in 1986, the central government declared the automobile industry a pillar industry of the Chinese economy. This was followed by the BeiDaiHe conference in 1987, where the central government initiated a set of strategies for promoting the industry. In 1993, as part of the national

[2] It should be noted that 1962 was the first year of the "Miracle on the Han River."

[3] The authors of this book, Zhang and Gao, were among the draftsmen of the "Policy on Development of Automotive Industry" in 1994.

industrial strategy, the automobile industry (including the car sector) was again officially designated as a pillar industry.[4] As a result, the Chinese automobile industry has gradually expanded. In 2006, its output accounted for 3.7 percent of Chinese GDP.[5]

The automobile industry is a prime example of adopting the import substitution strategies known as "Trading Market for Technology," or TMFT (see Chapter 2 of this book). Although hardly alone, TMFT has lasted the longest—more than two decades—and has the most pervasive influence in this sector. The main idea behind TMFT is to attract foreign investment in the technology sector by opening China's market to multinational corporations (MNCs). The automobile industry was among the first to be opened to foreign direct investment (FDI). It took the institutional form of joint ventures between selected foreign MNCs and China's state-owned enterprises, or SOEs. From the 1980s to the 1990s, despite the strong support from the government, the Chinese car sector did not achieve its expected level of success. By the end of 1990s, the so-called "Old Three,"[6] namely three outdated car models imported originally by the Sino-foreign joint ventures (JVs), still accounted for around 50 percent of the domestic market, a market structure unfit for any healthy car sector.

Since the late 1990s, new indigenous firms outside the original central plans entered the market and changed the landscape of the industry. They had little support from policymakers in the central government and were not generally attractive to FDI. Thus, they were forced to pursue independent product development early on through reverse-engineering and imitation of foreign products, as well as through cooperation with foreign technical firms. Over time, these new entrants gradually built up their in-house technological capability to emerge as a new force in the Chinese car sector.

Partly in response to the lessons learned from the automobile industry, a policy paradigm of encouraging "indigenous innovation" was enacted in 2006. In conjunction with relevant policy initiatives, such as the "National Medium and Long-term S&T Plan (2006–2020)," the Chinese government

[4] It was back in 1985 that the automobile industry was defined as a pillar industry of the Chinese economy. However, there were swings in policy between 1985 and 1993. In 1993, the 14th CPC National Congress formally relaunched a strategy, to once again develop the automobile industry as a national pillar. This decision was supported by a follow-up "Policy on Development of Automotive Industry" (1994 version) in the next year.

[5] The industries relevant to automobile counted for one sixth of the total employment, referring to the interview with Guobao, Zhang (current deputy director of the State Development and Reform Commission) by XinhuaNet <http://news.xinhuanet.com/auto/2007-09/23/content_6775367.htm>.

[6] The "Old Three" were "Jetta" (the "Jetta-A2" of Volkswagen, launched by Volkswagen in 1979), "Santana" (the "Passat-B2" of Volkswagen, in 1981), and "FuKang" (the "Citroën-ZX" by Citroën, in 1991). Note that in the headquarters of corresponding MNCs, the production of Passat-B2, Jetta-A2, and Citroën-ZX were stopped respectively in 1988, 1992, and 1998.

strengthened its support of indigenous innovation in products and technologies. This promoted the growth of new indigenous firms as well as indigenous car development among all domestic firms. The Chinese automobile market has maintained high-speed growth since 2001, overtaking Japan, United States, and Germany to become the largest in the world (in output, in 2008; in sales, 2010). New indigenous firms played a critical role in stimulating market growth and competition based on new products.

Numerous scholars such as Womack et al. (1990), Clark and Fujimoto (1991), and Fujimoto (2007) have extensively analyzed innovations in car industry. This chapter adds to the literature by analyzing the development patterns of Chinese firms including SOEs and new indigenous firms. The differences between the large SOEs involved in TMFT practices and the new indigenous firms were not necessarily manifested in the performance of their products or technologies. Rather, as this chapter will demonstrate, the differences among these firms lay in the very foundations of their organizational systems and learning patterns. Accordingly, this chapter will focus on the origins of these companies, organizational systems, and learning patterns. In the second section of the chapter, the origins and practices of the TMFT policy are discussed in order to analyze the stifled innovative capability in the Chinese car sector during most of the 1980s–90s. The third section focuses on the rise of the new entrants around the late 1990s and explores their influence on policymaking and incumbent firms. In the fourth section, the recent changes in the car industry are discussed.

TMFT Practices in the Automobile Industry

The TMFT literally means to exchange domestic market share for advanced technologies with foreign partners (in Chinese, 市场换技术). The policy was developed to realize two goals: promoting import substitution and technological catching-up, and forcing reform of SOEs. Implicit in these goals was the view that foreign equity cooperators are models of modern corporate governance. As a national policy, it set off intense competition among the regions and the automobile industry emerged as a focal point of such competition.

The Origin of TMFT Policy and Automobile Industry

The Chinese automobile industry was the first to adopt the TMFT policy. In the first half of the 1980s, China's sudden opening to the international world led to a significant increase in automobile imports and caused a huge trade deficit. In 1984, the amount of imports exceeded the entire output of the domestic car sector. In 1985, the amount of imports was 20 times that of the

Table 5.1. Imported Cars in Early Half of 1980s (unit: set)

Year	1981	1982	1983	1984	1985
Cars imported	1401	1101	5806	21651	105775
Cars imported/all vehicles imported	3.37%	6.85%	23.08%	24.40%	29.88%
Cars imported/Cars produced domestically	40.87%	27.32%	96.03%	360.25%	2031.40%

Source: *China Automotive Industry Yearbook*, 2003: 26.

domestic production capacity—not counting smuggling, which was estimated to yield as much as five times that of legal imports[7] (see Table 5.1).

The rapid growth of imports subjected the foreign exchange (forex) reserve to severe pressure. In 1985, due to the increased expenditure on importing cars, only US$140 million forex reserves were left for the whole country (Xu and Ouyang 2011). Since the forex was a strategically important asset for importing foreign capital goods, the shortage became an urgent problem. The astonishing level of smuggling was also a challenge to the central planners. In the organized car smuggling of Hainan government, the capital involved reached 4.21 billion RMB in 1984, which was 1 billion more than the GDP of the Hainan district (Xu and Ouyang 2011). Ministers of the industries realized that the root cause of the rising imports and smuggling of automobiles was the significant technological gap between Chinese-made cars and cars made in advanced countries. The central planners decided to develop a policy that would simultaneously fix the two problems: the forex shortage with overheated imports and the lack of technological capability of domestic car makers. Import substitution was a natural answer.

In a report to the state council in 1983, Bin Rao,[8] the director of CNAJC (China National Automobile Joint Company[9]) and the principal officer of China's automobile industry at the ministerial level, advocated cooperative production in the automobile and other industries in the hopes of accelerating indigenous technological capability building and import substitution. This was the foundation of the TMFT policy.

[7] The prevalence of smuggling was due to the high tariff on cars and the transitions of the market reform. In the 1980s, even though the tariff rate varied with the different type of cars, the general rate was as high as 200%. The changing social system generated a rising rich or privileged population that was eager for cars. In the 1980s, there were even cases that regional government or individual officials were involved in the smuggling of cars. Among them, the smuggling in 1983–5 supported by the regional government of Hainan was the worst scandal.

[8] Rao was the leading founder of FAW, SAW (now DongFeng), and the Shanghai–Volkswagen project. He acted as the director of Bureau of Automobile Industry and the Minister of No. 1 MMI (Ministry of Mechanical Industry) from 1979. Therefore, Rao was an influential person in the relevant industries in China.

[9] The CNAJC was transferred from the automobile division of No. 1 MMI, and acted as the practical administrative agent of the automobile industry at the ministerial level at that time.

The TMFT policy was thus designed with two goals in mind: to expedite bilateral cooperation and to improve learning from foreign partners with advanced technologies. For these purposes, the Chinese government granted access to the domestic car market to potential foreign partners. Even though there were still disputes over the details of the strategic plan,[10] policymakers generally agreed that establishing Sino-foreign JVs would be the most effective way to accomplish both goals.

Several Sino-foreign cooperative projects had already been in negotiation before this point. In 1983, Beijing Auto and the American Motors Corporation signed a contract to establish the first JV in the Chinese automobile industry, referred to as Beijing-AMC. Back in 1978, Rao himself had already proposed to importing a production line and cooperating with Volkswagen, which finally led to the SAIC (Shanghai Automobile Industrial Corporation)-Volkswagen JV in 1985.[11]

At the time of the proposed policy, China had trouble attracting high-tech foreign enterprises given its low income and inadequate infrastructure. To increase the attraction, the Chinese government implemented a series of favorable policies to ensure more profitable business for both the domestic and foreign participants in JVs. According to two taxation laws issued in 1991 and 1994 respectively, the income tax rate for foreign-funded enterprises (FFEs) was only 11 percent–15 percent, while it was 23 percent for domestic firms and above 30 percent for SOEs. Moreover, income tax of FFEs was exempted for the first two years of profit-making and reduced by 50 percent for the following three years. The import duties and value-added tax for machinery and equipment were also exempted for FFEs. General import tariffs were also adjusted to support Sino-foreign JVs in the automobile industry. The tariff on imports of completed products was raised while the tariff on imports of components was lowered. This led to the popularity of the Complete-Knocked Down (CKD) or Semi-Knocked Down (SKD) production by Sino-foreign cooperation (Xia et al 2002).

There were also common and not so codified policies to stimulate TMFT practices. For example, to ensure success of the joint ventures, the Chinese central or regional governments encouraged those SOEs with excellent performance to become the partners of MNCs. These SOEs were allowed to discard their non-performing assets. Through such practices, quality human

[10] The leading SOEs had different opinions. For example, the FAW insisted on "fairly advanced technologies" for car development, being keen to cooperate with foreign companies on medium-size or even luxury cars; meanwhile, the SAW stressed the scale of production. It insisted that the production capacity of 300,000 set/year was required for effective learning (of processing technologies) from the foreign sides. The opinion of the SAW received more support among the policymakers. However, these different opinions all emphasized equity-based cooperation with foreign sides.

[11] The contract was signed in 1984, and the JV was built in 1985.

resources, factory facilities, valued equipment, and the best business lines were usually allocated to JVs. For example, in the case of Shanghai-Volkswagen, Shanghai Auto was appointed as the Chinese partner because it had the largest domestic car production line at the time.[12] Additionally, the entire automobile industrial community in Shanghai—comprising more than 200 firms—was mobilized to support the requirements for establishing this JV (Thun 2006).[13] To increase the chances of profitability for JVs, the Chinese government also limited new entrants to the automobile market. Such limitation reduced the competitive market pressure and contributed to the stagnation of technological improvement in the JVs until the early 2000s.

The favorable treatment for JVs was also a result of GDP competition among regional governments. Automobile production can increase local GDP significantly, and so more resources were allocated by local governments to facilitate TMFT, including land allocation, low-interest loans, and even the training of personnel.

In short, these policies under TMFT framework built packages attractive both to MNCs and SOEs. In the automobile industry, 71 JV projects and five technical cooperation projects were signed between 1983 and 2000. Among these projects, 58 projects introduced foreign technologies in the form of equipment, blueprints, licenses, and training. Including projects concentrating on parts and components, 557 JVs had been established in Chinese automobile industry by the end of 1998.[14] The goal of import substitution was achieved with the growth of local automobile production. The market share of imports dropped from their peak—95.3 percent in 1985—to just 37 percent in 1995. In the 2000s, they have further decreased under 10 percent.[15] However, in spite of the strength in domestic production, technological learning on the part of the local partners has been far less successful.

[12] The "Shanghai" (named "Phoenix" before 1964) model was produced by Shanghai Auto at that time with an output of 7,000 sets per year, which was a local design developed in the 1950s that targeted Benz-170. Its production was stopped in 1991, because the factory facilities and human resources were asked to transfer to the building of Shanghai-Volkswagen.

[13] Thun (2006) emphasizes that the production localization of Santana was limited mainly to Shanghai, and the amount of local suppliers grew from 18 to 248 during 1987–97. The localized rate of production in Shanghai reached about 90% in 1997. Differing from FAW and DongFeng, the Shanghai Auto originated from an SOE invested in by regional government rather than the central government. However, since the Shanghai-Volkswagen cooperation was a leading national project, the scope of mobilization for the production localization went beyond Shanghai geographically. SOEs in Yangzhou, Suzhou, Wuxi, and Guizhou province were even mobilized. While Yangzhou, Suzhou, and Wuxi are located in the lower-middle reaches of the Yangtze River like Shanghai, Guizhou province is in the southwest of China, far away from Shanghai.

[14] Referring to *China Automotive Industry Yearbook* (1999).

[15] *China Automobile Industry Yearbook*, respective years.

The Learning Pattern under TMFT

At the beginning, JVs seemed to promise everything. China's policymakers sought "efficient" and "low cost" technological transfer methods, which were largely represented as technological trade of highly packaged production solution or "turn-key" projects. Policymakers recognized the problems of unequal capabilities between the partners, but they expected such problems to be solved by introducing the international supply chain of MNCs or by establishing complementary JVs locally. The joint ventures were expected to enable domestic firms to learn from foreign collaborators by working with them closely, so as to replace the Chinese corporate governance with advanced foreign ones. What they did not anticipate was how the operations of JVs would be shaped by the different motivations of the two sides. In fact, the particular organizational form created a learning pattern that was consistently dominated by the foreign sides. Effective technological transfer and capability building did not materialize to the degree expected.

DIVERGENT MOTIVATIONS OF DIFFERENT PLAYERS

In theory, TMFT should help build the capability of Chinese firms. For Chinese planners, the goal was to increase local production of imported models, based on two assumptions. First, they associated advanced technology and success with economies of the scale. They believed bigger was better and that the production capacity of 300,000 set/year was a basic line for the survival of a single automobile maker. Planners used this standard in the development of domestic investment plans[16] to justify the restrictions on new entrants to the industry.

Secondly, policymakers assumed that manufacturing products inherently led to technological learning (Lu and Feng 2005). The physical foreign product models were expected to embody the relevant technologies, thus production localization would undoubtedly improve local mastery of related knowledge and therefore increase indigenous technological capability. In the "Outline of National Industrial Policy in 1990s" announced in 1994, this model was optimistically summarized as "import–assimilation–absorption" (see Figure 5.1). In reality, since the 1980s, the global production system has

[16] This policy was formulated by central planners in the 1994 version of "industrial Policy on Automobiles." Nominally it was set to prevent overheated investment, but in fact it served as another method to enhance the central planning for industries. None of the TMFT JVs had met this requirement at their initial stage. For example, the production capacity of Shanghai-Volkswagen in its first phase was only 50,000 set/year. In 2001 (right before the policy change), the production capacity of the entire FAW, DongFeng, and SAIC groups was 419,800, 262,900, and 440,400 set/year respectively (Xia et al. 2002: 224), but it must be noted at that time these large SOE groups had already established multiple factories, therefore, this policy was mainly employed to exclude entrants.

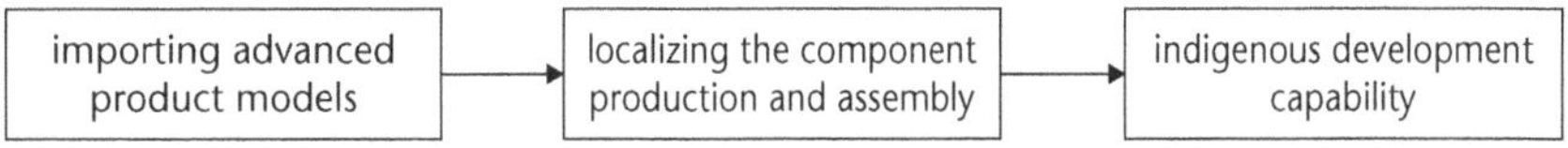

Figure 5.1. Linear Bottom-Up Model of Technological Learning
Source: Lu and Feng 2005: 24.

Table 5.2. Production Localization Rate of Components of Santana (Shanghai-Volkswagen)

Year	1984	1986	1987	1988	1989	1990	2000
Production localization rate, %	NA	2.7	12.6	30.6	60	70	92

Source: Collected by author.

evolved into global production networks (GPN) that separate the R&D process from the manufacturing process. Manufacturing could be divorced from the core intellectual properties behind the design, making learning by doing increasingly difficult.

Production localization was viewed as the magic bullet to reduce costs and key measures of successful technological transfer. In 1988, Zhu RongJi[17] then the Mayor of Shanghai, required the SAIC to promise that the rate of production localization of the "Santana" model[18] would reach 25 percent by 1988 and 50 percent by 1989. Based on the original JV contract, the Chinese government expected it to be 80 percent in 1991, only seven years after its establishment (see Table 5.2).

The SOEs involved in JVs were eager to speed up the rate of production localization because it increased profits by reducing supply costs. JVs had to compete both with other domestic firms and with those importing cars with the same designs. Increasing the rate of local contribution in production was financially desirable for them. This is especially important because TMFT is a costly policy for SOEs. It usually cost JVs several million USD to import each car model.[19] The permission for car models was usually tied to the agreements of production royalties, equipment procurement, and importing CKD assemblies and components. For example, when FAW (First Automobile Works, China)-Volkswagen was set up in 1991 to import the Jetta-A2 1983 model, a manufacturing line of Volkswagens in Westmoreland, Pennsylvania (US) was

[17] Zhu was the Mayor of Shanghai in 1988–91, the Vice Premier in charge of economy in 1993–8, and the Premier in 1998–2003. He was a major executor of the SOE reform and TMFT framework at the national level.

[18] The Santana model was the first imported model of Shanghai-Volkswagen, imported from Volkswagen.

[19] In most cases, what the JV obtained through paying this fee was only production permission for corresponding models. The Sino-foreign JVs did not have the IPRs of relevant models. Except for special statements, all terms relevant to the importation of product models refer to this interpretation here.

Table 5.3. Initial Expenditures of TMFT Practices

JVs	Initial expenditures for importing models & equipment (RMB/USD, billion, current prices)	Note
Shanghai-Volkswagen	3.58/1.54	In 1984, for importing the Santana model and some production equipment (not including the supply chain)
FAW-Volkswagen	11.13/2.09	In 1991, for importing the Jetta-2 model and some production equipment (not including the supply chain)
DongFeng-Citroën	13/2.36	In 1992, for importing the Citroën-ZX model and some production equipment (not including the supply chain)

Source: Collected by author.

disassembled and moved to China. The machines on this line cost the Chinese side 11.13 billion RMB; even though Volkswagen was already abandoning the Jetta-A2 model[20] in its GPN, as the Westmoreland factory had been retired back in 1988.[21]

Table 5.3 lists the initial expenditures of TMFT practices for model imports and relevant equipment procurement from foreign partners.[22] The Chinese were unable to negotiate bargain prices even with out-of-date models.

Additionally, the JVs still had to raise capital to import the CKD components, complementary equipment, training personnel from the foreign side, technical consultancy, technical certifications, and once again, all the similar fees for importing the subsystem and component designs and realizing the corresponding production localization. In the end, the expenditures for completing the physical construction of production localization were usually several times more than the nominal fee of technical transfer. Take the CKD imports for producing Santana as an example. The Shanghai-Volkswagen planned to purchase 89,000 sets of highly packaged CKD kits of Santana before it realized 60 percent localization, each cost US$10,000 for Shanghai-Volkswagen. When the target of 60 percent localization was achieved in 1990, the real imports of CKD kits had already exceeded 89,000 sets. And even after achieving 60 percent localization, the Shanghai-Volkswagen still had to import assemblies and components from Volkswagen's global network.

[20] Jetta-A2 was officially replaced by Jetta-A3 in 1992 by Volkswagen for its global mainstream market.

[21] This manufacturing line was previously established to produce the Golf. The FAW-Volkswagen cooperation was priced at US$25 m. It was made with additional conditions: the Chinese side had to import 14,500 sets of Audi-100 CKD kits at a set price in 1987.

[22] The equipment procured from foreign partners did not entirely fulfill the basic needs of production. In many cases, in order to save money, SOEs would not import the full set of machines.

Governmental investment accounted for a significant portion of this expenditure. In the 8th FYP (1991–5), Chinese investment in the automobile industry reached 75.61 billion RMB.[23] While the foreign exchange shortage might have prompted the TMFT policy, implementing TMFT in fact increased the demand for foreign exchange, at least at the beginning stage. Among these investments, about 9 billion RMB from state revenues were appropriated for the establishment of TMFT JVs, not including the expenditures of regional governments (Xu and Ouyang 2011). Therefore there were multiple domestic stakeholders, including the central and regional government, banks, local suppliers, and engaged SOEs, looking forward to the return on their investments. In coming up to their expectation, a rapid expansion of production scales of these JVs was treated as a priority.

Foreign firms were attracted by the TMFT due to the favorable policies of Chinese governments, the potential of the Chinese market, and low production costs. However, the MNCs understandably had no interest in cultivating Chinese indigenous technological capability and creating potential competitors for themselves. In fact, MNCs had clear plans to protect their intellectual property rights and minimize such competition.

For example, soon after the establishment of the Shanghai-Volkswagen JV, the Shanghai side found that within the JV there was no institutionalized mechanism to generate technologies related to new product development, in spite of the fact that the German partner assisting the Chinese side to cultivate local capability of product development was written into the JV contract (Liu 1992: 264). Instead, the German side wanted to build this JV into a pure manufacturing base for a single Santana model. Welkener, the assistant managing director and the chief of the German team of Shanghai-Volkswagen at that time, explained Volkswagen is motivated to make Santana the most competitive model as long as the Chinese side continuously improved the design without any major changes, so the Santana model could be produced at USD $5,000 per set around 2000.

The Chinese side insisted on the original contract. The German side cited a series of "practical difficulties." Only after the intervention of the Chinese government did the German side propose three alternatives to fulfill its promise. The first option was to take over a brand new model designed by Volkswagen; the second was to take over a mature model from Volkswagen; and the third was to jointly develop a brand new model suitable for the Chinese market and with potential aims for the global market. The Chinese side chose the third option. However, during the cooperation, the German contingent was reluctant to respond to the requests from Chinese side such as to

[23] Data source: *China Automotive Industry Yearbook* (1999), 9.

send its experienced experts to China, provide consultant services, demonstrate the operation of equipment, and teach the Chinese concrete methods for designing a new car model. Since the Chinese team had not been expected to make technological contributions to the German model, it did not have the ability to produce a jointly designed model and the target for completion was repeatedly postponed. Finally, the Chinese agreed to a "new" project in 1993, which was to develop the Santana-2000 model. The Santana-2000 project was still dominated by the German side, carried out by a Volkswagen subsidiary in Brazil, and was still based on the same Passat-B2 platform as the original version of Santana. Other than face-lifting and regular upgrading of modules from the "weaponry" of Volkswagen, no important change relevant to the product platform had been made.

In other JVs, the situations were similar. In the Beijing-AMC (later Beijing-Chrysler), the establishment of an R&D center was outlined in the original JV contract. However, because of reluctance on the American side, this center was only established in 1995, 10 years later. In 2002, when Beijing-Chrysler moved its factories to another site, this R&D center was "temporarily" dissolved and in fact never re-established.

In addition, MNCs had institutionalized patterns to restrict the routine learning activities in these JVs. In fact, the majority of the knowledge generation process barely went beyond the manufacturing activities of designated products. The information provided to their Chinese partners was strictly controlled and the technical information in the transfer package was often incomplete and fragmented. Generally, the foreign partners were not willing to provide information beyond what was necessary to achieve production localization as outlined in the contract. Therefore, two kinds of data were markedly absent from the blueprint packages that JVs obtained from their foreign partners: (1) data relevant to product design and engineering development; (2) data for manufacturing beyond local activities. MNCs deliberately minimized the information provided to JVs. Therefore, the blueprints that Chinese engineers received were usually incomplete. Chinese engineers found it difficult to comprehend the product and technological systems beyond the manufacturing process. In Richardson's terms (1972: 889), manufacturing knowledge is complementary, but not identical to the developmental activities for generating new products. Without such developmental activities, the Chinese could not build indigenous innovative capabilities.

With JVs, the MNCs also had firm control over activities of incremental technological improvement. The "inspecting and confirming right" was the core of relevant devices. This method was originally developed to enhance the accountability and quality control in developmental processes. But in TMFT practices, it meant the MNCs, as the owners of imported designs and corresponding brands, had the right to control technological changes in their designs. The Sino-foreign JVs only had the permission to produce imported models by paying fees for importing the models. Certainly, they could carry out incremental improvements and claim intellectual property rights. However, their changes would only be acceptable if they worked with the original designs. Since the inspection and confirmation of critical sections were implemented by the foreign sides in their headquarters, the Chinese sides had no control over how long the process of certification should take and what qualitative standards the adjustments ought to reach. It was easy for the foreign side to increase the time requirement or set up unattainable standards if the changes were undesirable. Chinese engineers gradually recognized that they had no real influence over the technological progress and no effective communication channels. In one case, abnormal noises from the hinges of Santana's rear doors were reported by Chinese engineers in 1986, but the German side was slow to respond and Chinese engineers did not have the rights to carry out relevant changes. Not until a Vice-Premier of China spoke to the president of Volkswagen in 1987 was this technical problem solved by Volkswagen (Hahn 2005). This was symptomatic of the unanticipated problems with the power differential between companies in JVs. In his autobiography, Carl Hahn (the president of Volkswagen, 1982–93) admitted that the German side indeed dominated product technologies in all these productive Sino-foreign JVs (Hahn 2005: 119).

To summarize, Chinese policymakers had high hopes for implementing TMFT to generate the "close learning" of Chinese SOEs from their MNC partners. But the results have been disappointing. The MNCs view the Chinese partners as market access and manufacturing facilities, and they were not interested in raising the capability of their Chinese partners for technological innovation. The Chinese side lacked the technological competency and strategic control for information flow and technical adjustments. The technological transfer promised in the negotiations for these JV projects was not achieved.

LEARNING ACTIVITIES AND ORGANIZATIONAL ARRANGEMENTS

It is hardly surprising that MNCs would not actively cultivate the indigenous technological capability of host countries. But the local partners also have little reasons to challenge the technology dominance of MNCs. As described earlier, TMFT policymakers believed that the activities of manufacturing advanced products would gradually and naturally bring about technological capabilities. TMFT therefore focused solely on production localization, not R&D. JVs were founded on the premise that endorsed MNC corporate structure as the model of advanced governance patterns that Chinese companies should respect and imitate. Thus the JV organization of technology is not only dominated by the foreign side, it is also structured bureaucratically, excluding all dissenting opinions from the official plans (Shrivastava 1983).

The goal of learning under TMFT was to produce an exact replication of the imported designs in local production. Thus the decision-makers of firms required their front-line practitioners to stick to imported models rather than develop their own creative skills. They reasoned that the existing designs provided by their MNC partners were top-level technology, so there was no need to make change. This organizational system discouraged any heterogeneous development because potential uncertainties could result in the delay of reaching their manufacturing targets. Product development follows a precisely prescribed path; any deviation was seen as a mistake to be corrected. Organizationally, JVs had no R&D centers for specialized development of products and relevant technologies. Even in cases where local R&D centers had been established according to the JV agreements, these centers were marginalized with only limited budgets and very little power to mobilize the personnel and resources of other departments for meaningful (in scale) tasks of development. Since there were no projects to develop new indigenous product designs or relevant technologies, the R&D centers would have no influence on other departments.

In some cases, the marginal positions of R&D units were even ensured by formal arrangements. For instance, the Pan-Asian Technical Automotive Center, or PATAC, a R&D-oriented company set up by the Shanghai-GM JV, had the authority to access the database of technological information of GM global. However, it was strictly forbidden to apply any proprietary technology of GM to a project unless there was permission from GM. Therefore, the Chinese engineers of the PATAC could work from the database, but they were restrained from any practical activities of product development. In other words, they were isolated into a big laboratory, without any effective interaction with the production division and the real market. The situation for DongFeng's R&D center was similar. It belonged to the DongFeng-Nissan JV and the "information officer" of the center was required to be appointed by the Japanese side. Because the role was filled by manager appointed by Nissan, the "information officer," in charge of property rights, blocked any effective interaction between this center and other divisions in name of protecting the integrity of property technologies. FAW's R&D center had over 2,000 developmental engineers, and it was the largest in the whole domestic automobile industry in the early 2000s. However, according to the relevant JV agreements, this center had no real interaction with the production localization of Volkswagen's model. It mainly focused on the development of trucks and the truck business was not included in the JV. In short, the interactive processes among the manufacturing, market, and development divisions, elementary to the innovative process (von Hippel 1988; Lundvall 1988), did not exist within Chinese JVs.

On the shop floor, each position was appointed to carry out fixed tasks. The knowledge gatekeepers of each unit, namely the heads of working teams, workshops, and factories, were required to have their subordinates follow the existing operation manuals (usually Chinese versions directly translated from the original ones) and achieve the existing targets instructed by the imported blueprints. Therefore, in the SECI (Socialization-Externalization-Combination-Internalization) circles of knowledge conversion, in Nonaka's term (Nonaka and Takeuchi 1995), the knowledge gained through this practice was limited to the manufacturing skills related to the given blueprints. No creative ideas beyond these were allowed. The learning on the Chinese side was mostly focused on studying the dimensions of existing models, operations of equipment, and managerial approaches. Know-how is not the same as know-why (Kogut and Zander 1992: 391). These modes of learning yielded positive outcomes for promoting production capacities, but they did not increase employees' understanding of the product development system.

Overtime, the JVs did engage in incremental improvements, such as the upgrading of processing technologies and the upgrading or adjustment of products. Such projects would be intended for better market acceptance or for adaptation to a local supply chain. However, most of these developmental activities were strictly under the control of MNCs. The Chinese teams were only permitted to carry out projects for appointed modules. In other words, complex projects were exclusively integrated at the MNC headquarters, which would control the system reconfiguration, the change of critical modules, and the systemic testing required to implement changes.

The exceptions to this pattern were few and far between. Some SOEs did not agree to abandon the development of indigenous products. FAW had extended experience developing the "RedFlag" series of car models.[24] When FAW started its cooperation with Audi and transferred the "RedFlag" brand name onto the Audi-100 platform, some senior engineers were very disappointed. As FAW gave up on "RedFlag" development, a mini-car model named "SanKouLe" was spontaneously developed by a dozen FAW engineers without official permission.[25] It progressed to the point of road tests from ChangChun City to Beijing.[26] However, when this project was revealed to the executive managers of FAW, they struck it down as a distraction from the primary tasks of production localization for the Audi-100 model. Participants in the

[24] The "RedFlag" is regarded as the national car of China, created in 1958. As the first mass-produced car model and the protocol car, the "RedFlag" brand is an icon of the Chinese car industry.

[25] These developers used leftover and waste materials to develop this prototype. And two of these leading developers joined Geely when they retired from the FAW and played an important role in the car development in Geely's early years.

[26] The Changchun city is where the headquarters of FAW is located. It is the capital city of Jilin Province, in the northeast of China. The distance from Changchun to Beijing is about 1,000 km.

SanKouLe project received a serious official warning, and no engineer proposed a new car model development thereafter.

A second category of exceptional cases included several developmental projects carried out by marginal sub-companies of the large SOEs involved in TMFT. For example, a regional sub-company of FAW in YunNan tried to develop a "HongTa" car model and a service subsidiary of DongFeng developed a "XiaoWangZi" model. In these cases, the developers were not on the central stage of these large SOEs and they did not receive any significant technical and economic support from those major units involving TMFT. In fact, not much interaction between the sub-branches and major SOEs could be found. Neither model was successful in the long run.

Evaluating Consequences of the TMFT Practices

Measuring against its original goal, the TMFT policy did achieve its designated target for import substitution. The output of the car sector grew from 5,418 sets in 1980 to 703,525 sets in 2001, with a 26 percent average annual growth rate. The market share of the direct import of entire cars was reduced to about 6.6 percent by that year. However, such growth was nothing compared to what happened after 2001. The annual rate of growth from 2002 to 2009 shot up to 40 percent; in 2009, the output of the Chinese car sector reached 10,330,300 sets. The amount of exports has exceeded that of imports since 2005. In short, under TMFT policies, China's production capacity did grow, but not as substantially as it would in the later period when different types of automobile companies emerged.

TMFT practices had disappointing results in terms of capability building for product innovation. No new indigenous car models were developed and successfully mass-produced, and already-existing indigenous product platforms were discarded. For example, the "RedFlag" platform discontinued its upgrades during the mid-1990s. Instead, this prestige brand was transplanted onto the Audi platform. The well-known "Shanghai" and "Beijing" platforms were totally discarded. The only exception was the BJ2020 platform of Beijing Auto. It was regarded as an important asset to the Beijing-AMC JV, and was retained. Because the automobiles based on import models had poor market performance, the entire JV had to be financed by profits made by the BJ2020 until the mid-1990s. Even so, the BJ2020 platform received little further development, and gradually disappeared around 2005.

All in all, two decades of TMFT practices did not cultivate any obvious improvement of technological capability for indigenous product development. Chinese decision-makers tried to adjust the policy by doing basically more of the same: continuing to import foreign designs, extending the duration of JV cooperation, and introducing more MNCs to establish JVs. TMFT

Table 5.4. Product Sequencing of the TMFT JVs (up to 2009)

Firm	Major product platforms manufactured
FAW-VW	Jetta, Audi 100, Audi A6, Audi A4, Golf, Bora, Caddy, etc.
DongFeng-Citroën	Citroën ZX, Citroën Elysée, Citroën Quatre, Citroën Xsara Picasso, Citroën Triomphe, Citroën Visiospace, Citroën C2, Citroën C5, etc.
Shanghai-VW	Santana, Passat, Polo, Gol, Touran, Octavia, etc.
Beijing-Chrysler	BJ2020, Cherokee BJ6420, Cherokee BJ2021, Cherokee Jeep 2500, Cherokee Jeep Star, Chrysler 300C, etc.

measures were crafted to continuously increase production capacity, fulfilling a GDP-oriented strategy of development. Table 5.4 indicates the sequences of product models that were imported from foreign partners.

The JV agreements for FAW-Volkswagen started in 1991, and in 2002 they were extended to 2041. Similarly, the Shanghai-Volkswagen JV agreement went into effect in 1985; in 2002 it was extended to 2030. Meanwhile, the large SOEs also signed more JV agreements to introduce more foreign partners. By the end of 2001, with the exception of Hyundai, BMW, and Mercedes-Benz, all major car companies in the global market had already established their production JVs in China. In 2005, Hyundai, BMW, and Mercedes-Benz also joined in the game. The longest duration of a JVs in the Chinese car sector will be DongFeng-Nissan, which in 2003 was extended until 2053.

The increasing number of JVs has intensified the market competition, but it has not addressed the weak features of TMFT, namely, the lack of strategic control for the Chinese partner and the dominance of MNCs on technology trajectory. Since the Chinese sides had little influence on product upgrading, their negotiation power for importing more and newer models from their foreign partners was highly constrained. Strategic decisions were consistently made by the foreign sides and were based on how to maximize their profits, which in the 1990s led to continued production of outdated models. The "Old Three" (Jetta, Santana, and Fukang) held over 50 percent of market share in the Chinese market until the early 2000s, when the production of two of them had been halted for almost 10 years in the global mainstream market. TMFT helped the Chinese automobile industry grow, but it did not ensure any real catching up.

The Rise of Indigenous Innovation

Two disruptive changes around 2001 led to a turning point for auto innovation in China. First, the new type of automobile companies were gradually approved de facto by the central government; secondly, the Chinese market

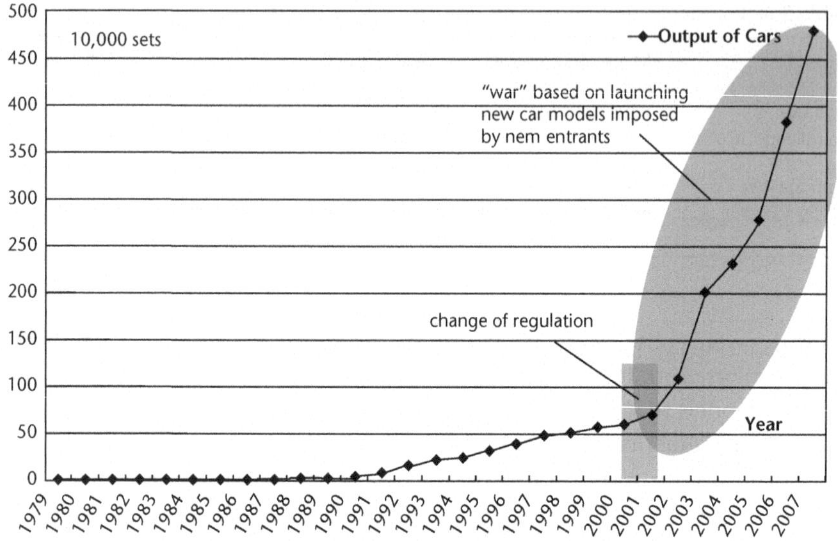

Figure 5.2. Output of the Chinese Car Sector
Source: *China Automotive Industry Yearbook* (2008).

entered a new phase of a more rapid growth in output quantity (see Figure 5.2).

The sudden growth in car-makers also meant that competition was increasingly based on product innovation. By 2001, 12 MNCs had already established their production JVs under the TMFT framework in China. MNCs had preferred to stick to a limited number of product models (and, relevantly, equipment) to maximize their profits in JVs, rather than raising the competition level by emphasizing product innovation. There were only 12 models in 2000 and 13 in 2001 that were launched in the Chinese market, achieving annual production volumes of over 10,000 sets. The emergence of new entrants with indigenous car models created a major challenge to this structure by pushing the competition to become product-oriented. In response to the competition, the MNCs launched more and better models in China. The number of newly launched models increased from 80 to 120 after 2005 and has stayed at a comparatively high level since then (see Table 5.5).

The "Old Three" models began to lose their market share, from 47 percent of the domestic market in 2001, to 34 percent in 2002, and 18 percent in 2004. The sales of Jetta, Santana, and Fukang also decreased, with a loss of 43, 70, and 77 percent respectively in 2004. Prices also experienced steep decline under the competitive pressure. The average price of these three models was about 140,000 RMB in 2002; by 2008, it was forced down to 60,000 RMB. By comparison, the market share of indigenous car-makers sharply increased

Table 5.5. Newly Launched Car Models in the Chinese Market

Years	2001	2002	2003	2004	2005	2006	2007	2008	2009
Newly launched car models	13	28	50	50–	80+	117	90	107	221

Note: These statistics include common cars (sedan, hatchback, and station-wagon), MPVs, and SUVs.
Data source: Collected by author based on annual reports of China Association of Automobile Manufacturers.

from about 5 percent in 2001 to over 20 percent in 2004, reaching a peak of 30.89 percent by 2010.[27] New entrants have a very different make up from the large SOEs, which may explain the differences in their innovative performances in comparison with the TMFT participants. This section of the chapter focuses on the construction of new entrants' organization and core knowledge, which help to explain their success in innovation.

The Origin of New Entrants

China's entry into the WTO finally forced a change in industrial regulation for the car sector. The license regulation system in the Chinese car sector loosened after 2001, with a transition from a strict top-down management system to a review mechanism for projects invested in by existing firms with licenses. For new entrants, the license regulation remained but was also loosened de facto. However, the emergence of new entrants with indigenous products predated the WTO change and in fact contributed to the change in policy.

Before 2001, according to regulations, all cars launched in the Chinese market had to obtain a license from the ministries of central government. The Chinese central planners only granted licenses to those SOEs in their existing plan.[28] There were two main strategies that new entrants mainly used. Some companies used related licenses to produce cars unofficially. HaFei, for example, had a license for producing minibuses, so it transferred in-house technological capability from minibus to car production despite

[27] Data from China Association of Automobile Manufacturers, HongYueXinSi Consultant, and the 2012 Annual Report on Automotive Industry in China.

[28] This structure of central planning for the automobile industry was established in the 1980s. In the "Notice for managing car producers strictly" (关于严格控制轿车生产点的通知) issued by the state council in 1988, the central planners only licensed "three big and three small" firms for car production, namely three leading SOEs (FAW, DongFeng, and SAIC) and three comparatively small-sized SOEs (Beijing Auto, Tianjin Auto, and GuangZhou Auto). However, since the license for car production was made to be an economically beneficial asset, the other governmental departments and regional governments also competed for licenses by mobilizing political forces beyond the industrial regulation. For example, the military and semi-military division of industries won two licenses to produce mini-cars under this 1988 regulation. And more firms carried out automobile manufacturing de facto without official permission from the industrial regulators. But since the automobile industry was a critical one in stimulating local economies, the regional governments usually supported their "underground" business operations.

having no license. ChangAn Auto was part of the division of military (and aerospace) industry so it was allowed to develop vehicles for these purposes, which became their excuse to produce cars.[29]

A second group of firms started entirely outside of China's central planning system. Chery, today's leading indigenous firm in China, for example, received investments from the regional government of Wuhu City. Without license authorization from the central government, they had to build their factories, develop their car models, and sell their products all through underground means. In 2000, Chery gave 20 percent of its stock to Shanghai Auto for free, so that it could share the company's car license.[30] Geely and BYD managed to purchase some bankrupt, small SOEs with licenses to produce cars so that they could make use of these licenses.

Besides licenses, these new entrants also violated another requirement for complete-car production to have an initial production capacity of 300,000 set/year and initial investments over 3 billion RMB. None of the new entrants could meet these requirements. The initial investments of Chery, Geely, and HaFei were about 0.7–0.8, 0.1, and 0.9 billion respectively. To survive, they all publicly announced larger initial investments than were the case. They were often supported by their regional governments since their output could enlarge local economic activity. The central government chose to ignore such violations to some extent. These new entrants managed to survive the harsh policy environment prior to the change in 2001.

The Learning Pattern in New Entrants

STRATEGIC INTENT FOR INDIGENOUS PRODUCT DEVELOPMENT

Without policy supports or external resources, all new entrants were forced to develop an independent system of product development with considerable risk and even bigger commitment. No MNCs wanted to establish JVs with unproven entrepreneurs, as the TMFT framework regulated that each MNC could partner with no more than two JVs in China for complete-car production. The risk was high as they invested in cars before having a license. In fact, all the founders of these firms had to put all of their careers, investment, and even personal reputation on the line. For example, the founders of Chery had stable and promising careers before they joined together and established the firm. Yin Tongyue, the leader of the founding team, was one of the "Top 10

[29] The HaFei belonged to the division of state aviation industry and now it is a part of the AVIC (Aviation Industry of China) Automobile Industry Co., Ltd. The state aviation industry was a semi-military one in China. The ChangAn belonged to the division of state weapon industry and now is a part of the China South Industries Group Corporation, which was characterized as semi-military.

[30] Two years later, when the policy changed, the Chery bought back its share.

Young FAWers" in the early 1990s, which would have given him a secure future with FAW. Chery's other core engineers also came from the chief engineering force of FAW. In order to establish Geely, Li Shufu had to sell his motorcycle business, which had the second largest share in the domestic market for light motorcycles. These entrepreneurs were highly self-motivated and had no other resources than what they had taken the initiative to learn in their previous jobs.

LEARNING ACTIVITIES AND ORGANIZATIONAL ARRANGEMENT

The central task for new entrants during the early stage was to build up a basic team with the necessary internal expertise, and integrate external knowledge for developing the product and technological system of cars. This meant that, from the very beginning, these entrepreneurs positioned themselves as integrators of knowledge and focused on building effective learning systems.

(1) Building the basic core of technological capabilities

The personnel cores of the new entrants were engineers. At the very beginning, the founders had to recruit people from the broad fields of relevant mechanical industries, because the majority of engineers from large automobile SOEs did not want to risk their careers. New entrants recruited engineers from the firms of agricultural vehicles, modified trucks, motorcycles, and other industries. There was, however, a core of engineers who did leave large SOEs, and they became a critical source of initial capability for new entrants. Most of them were retired senior engineers of FAW, who had never developed an indigenous car model under the TMFT policy. After retiring from FAW and securing benefit packages, they joined new entrants. Up until the mid-2000s, FAW did not take such "invisible brain-drain" seriously.

Chery was a major beneficiary of a spillover of veteran expertise from the FAW, and through personal networks it recruited high-quality retired FAW engineers. This team of engineers helped Chery to establish the backbone of its technological system. Some senior engineers even started to develop innovative technologies that they were barred from exploring at FAW. For example, the first-generation engine designs of Chery were accomplished by a team led by one senior engineer who just left FAW. As Chery continued to succeed in the market, it became more attractive for other retired engineers. In the mid-2000s, there were more than 200 veteran engineers playing different roles in Chery.[31] Some might not have contributed to the central task, but

[31] The numbers of retired FAW engineers in Chery and Geely are both based on estimates. Because such relations are quite sensitive and in some cases are informal, there is no formal disclosure of data. The roles they play in these new indigenous firms are also rather diversified. Some of them have acted in critical and standing roles, such as the leading engineers of the former FAW "SanKouLe" project in Geely, while others have acted as consultants.

they were active in giving advice, guiding less experienced engineers, and connecting the firm to relevant suppliers. Another source of experienced engineers for Chery's primary capability building was a group of engineers from DongFeng, the truck maker. The DongFeng-Nissan JV cut down the size of its R&D center, and over 30 developers, who had been trained by the French in an Elysee adjustment project, left DongFeng to join Chery.[32]

Geely was another major beneficiary of retired engineers from FAW. It attracted the senior engineers by appointing them to important posts. This group of engineers included the leading engineers of the former "SanKouLe" project and contributed to the development of Geely's second-generation car models. HaFei was an exception. Because of its ownership as a military-owned company, it was restricted on external recruitment of human resources, so it had to rely on the engineers transferred from its aircraft division.

So indirectly and over time, the TMFT policy has helped build up the automobile industry in China. Experienced engineers trained by TMFT practices were recruited by these emerging firms, some of them played a significant role in the development of the initial core of new entrants' technological capabilities. But these achievements do not address TMFT's dismal record in technological progress. The migration of these engineers to indigenous companies was an unintended consequence of TMFT, not any part of its policy design. In fact, the contribution of these migrated engineers to the development of these new car-makers shows how much organizational difference matters. Because the Chinese partners in JVs consistently lacked strategic control, their capability building was notably stunted.

(2) Integrating external knowledge

For decades under the TMFT framework, there was no real systemic product development, so there was little domestic expertise available when new entrants emerged. For example, Geely did not develop a digital model for its first-generation products, because it did not even have CAD (computer-aided design) capability during its startup stage. As a result, the designs of the two front doors were not even symmetrical in its early models. Among all new entrants, only HaFei had CAD capability. However, since it was transferred from the aircraft division, its engineers still needed to learn how to adapt their knowledge in the new context.

After developing their first generation of car models (or minibus models, in the case of HaFei), the new entrants tried to work as knowledge integrators

[32] As mentioned, the DongFeng has more than one JV with different MNCs. Its JV with Citroen was the first one, established in 1992. In this JV, in order to adjust car models to be more suitable for the Chinese market, the French side trained a group of Chinese engineers. The purpose was to benefit from the low cost of Chinese local engineers. After being trained, these engineers were assigned to carry out specified supporting developmental tasks.

in their cooperation with global partners. They purchased design services from specialized technical firms in Europe. More importantly, they increased the competence of their own engineers through cooperative projects with these design service providers. HaFei was the first mover of this developmental pattern, and its experiences were rapidly diffused to others. Starting with the development of their second-generation car models, both Chery and Geely tried to become integrators and learners of foreign knowledge.

The timing when the indigenous Chinese car-makers entered this industry was helpful. The development of the global automobile industry in the late 20th century was characterized by vertical disintegration, with a group of specialized technological firms working on particular modules, especially in the European automobile production networks (Li 2010). The downturn of the global automobile industry in the West since the 1990s forced some of these specialized technological firms, such as those in bodywork design, engine development, or chassis engineering, to look for potential opportunities from emerging economies. Chinese indigenous firms exploited these global technological assets.[33] For example, Pininfarina, a specialized Italian firm for car body design, began to seek potential cooperators in China from 1993, and found HaFei an excellent candidate because of HaFei's motivation to implement product development and its digital design capability. Pininfarina proposed a project to HaFei in 1995, in which it offered HaFei the service of developing the bodywork for a minibus model. HaFei developed the chassis and other subsystems on its own. The ZhongYi (China-Italy) prototype, the outcome of the first HaFei-Pinifarina cooperative project, was successfully launched in 1998. After that, HaFei employed the services of more external technical firms to develop products and build capability. In developing the Lubo model, Lotus (a UK and Malaysian firm) provided engineering optimization for the chassis. HaFei also hired the Mitsubishi in the SaiMa project, and hired Tjinnova (a Chinese technical firm) in other projects.

Learning from HaFei's experience, other new entrants such as Chery, Geely, and BYD were able to overcome bottlenecks of important technologies. The service of foreign specialists also helped the indigenous car-makers to cultivate their own in-house development teams. Starting from its second-generation car models, Chery purchased at least one technical service project from foreign firms on each major subsystem (e.g. bodywork, chassis, and electronics) for each case of model development. Detailed tactics evolved along with the accumulation of Chery's capability, represented by its purchase of technical services of AVL. Between 2002 and 2008, there were three series and 25 engine models in total developed in this

[33] Referring to the interviews with Takahiro Fujimoto by *DongFang QiYeJia* (Asian Business Leaders), Nov. 2007, 98–101.

Table 5.6. Chery's Cooperation with AVL in Development of Engines (2002–8)

Phase	Models Developed	Location	Role of AVL	Time Period
1	4	Mainly in AVL	Dominant	3 years
2	3	Mainly in AVL	Supervisor	
3	11(18)*	Mainly in Chery	Consultant	4 years

* Another 7 models were added in the third phase mainly by the Chinese side.

Table 5.7. Important External Technical Projects of Chery in 2005

Cooperative Partner	Projects	Location of Cooperation	Start Date
AVL (Austria)	ACTECO Engines (including 3 families, 18 models)	Both sides	2002
Bertone (Italy)	Car configuration	Both sides; mainly in Italy	2002
I.de.A (Italy)	Car configuration	Both sides; mainly in Italy	2004
Pininfarina S.P.A.	Car configuration	Both sides; mainly in Italy	2003
IAT (China)	Car configuration	Both sides	2003
Torino Design (Italy)	Car configuration	Both sides; mainly in Italy	2004
MIRA (UK)	Chassis engineering; testing & adjustment	Both sides	2002
Lotus	Consulting; testing & adjustment	Both sides	2002
Ricardo plc (UK)	Hybrid power tech; Transmission	Both sides	2004
Dürr (Germany)	Paint shops	Both sides; mainly in China	2002
MAG Hüller Hille (Germany)	Production line of engine cylinder	Both sides; mainly in China	2002

Note: Data were collected by the author, and do not represent accurate numbers. In particular, most projects related to the development of components and processing technologies are not included.

partnership.[34] The structure of cooperation changed with Chery's growing in-house capability. Table 5.6 outlines the changes of location and the primary roles of foreign partners in the cooperation.

Table 5.7 presents some important cooperative projects carried out by Chery and its foreign partners in the year 2005. It indicates that such a pattern of learning was generally followed during the rise of indigenous car-makers.

As discussed earlier, external projects assisted by foreign specialized technical company would not necessarily bring about the growth of in-house

[34] Originally, there were only 18 engine models planned to be developed through this cooperation. But in the third phase, both sides decided to engage seven additional projects. The AVL agreed to such an arrangement because, based on Chery's capability, accumulated during the early two phases, its role in the cooperation had become increasingly marginal. Therefore, for the AVL there was not much additional workload in accepting this proposal.

technological capability. Lazonick (2003) in his theory of innovative enterprise argues that three social conditions are essential for innovation: strategic control, organizational integration, and financial commitment. This well describes the new entrants.

The first condition, strategic control, involves the control of intellectual property rights. The experiences of these firms show the critical importance of developing intellectual property rights from their investment. Since they invested in their own projects, these firms owned the IP rights, including the "inspecting and confirming right" in each section of the development process. The financial commitment to cultivating indigenous technological capability ensured these rights and empowered the Chinese side to shape the cooperation with foreign partners.

Of course, even with control over IP, the outcomes of learning still relied on a firm's absorptive capacity and their organization for learning. Absorptive capacity was critical for the recipient to assimilate relevant knowledge (Cohen and Levinthal 1990). In these cases, the absorptive capacity was also critical for latecomer firms to establish a basic understanding of what kind of knowledge they should seek out from their cooperation with advanced partners. It is not surprising that all successful firms had some experience of independent development prior to the cooperation with foreign technical firms. In this sense, it was essential for Chinese firms to gain basic knowledge through reverse-engineering, imitation, and rough integration of domestic expertise in developing their first generation of car models.

Active learning in cooperative projects was only possible if Chinese new entrants were able to mobilize their resources, such as the critical input assets, the rights to approve or disapprove projects. That meant the front-line engineers had to be authorized to integrate knowledge and make key decisions during the cooperation. All the new entrants were engineer-oriented organizations. During early phases of their development, they adopted the project-oriented organizational structure, and gave the development team considerable power for resource mobilization. In fact, the core leaders of new entrants often worked directly in these projects. By comparison, the large SOEs involved in TMFT were all bureaucracy-oriented, and their leaders seldom had hand-on experiences of car design and engineering. So this is consistent with the second condition for innovation outlined by Lazonick, namely organizational integration.

When new entrants did not meet the outlined pre-conditions of learning, failures resulted in employing foreign technologies. The development of the "ZhongHua" model by Brilliant Auto serves as one such example. Brilliant Auto was led by a financier and later his successors, who were famous for their successful financial manipulation but not for any experience in a manufacturing industry. In its project to develop the "ZhongHua" car model during

1997–2005, Brilliant Auto used foreign technological services from top-level international technical firms in each domain. However, its own in-house engineers acted as the auditors and blueprints receivers. They were not supported by higher authorities and had no control over the necessary resources, thus they were unable to work as the integrators of developmental activities. The decision-makers of Brilliant blamed the failure on the quality of domestic engineers and employed more foreign companies that they regarded as high quality. Between 2000 and 2005, after three rounds of significant revisions to the "ZhongHua" model, at a cost 4 billion RMB in development projects, and over 20 foreign partners, this model did not succeed in satisfying customers as a technically appropriate car until important organizational changes happened after 2005.

(3) Effective learning system of organization

There are evident differences in the functions of learning between SOEs under TMFT and the new entrants. In the case of SOEs, the central task of organizations was production localization of imported models and the organizational systems were built to serve this purpose. Firms were organized to stress functions of process technology, workshop management, and accounting, and personal promotions were guided based on such objectives. The new entrants were engineer-led; and many of these leaders had experience with product development in their careers. Geely developed a quasi "duel-leader system"; each subsidiary company or manufacturing base was led by a manager from Li Shufu's intimate circles[35] and an engineering background manager who was recruited externally.

The participation of engineers in strategic decision-making ensured that product development was located at the front and center of resource mobilization. In addition, the authority to mobilize the relevant resources was decentralized and given to front-line development units. It was the development units that decided what to research and what to learn. The structure encouraged the creation of heterogeneous ideas, which was critical for innovative learning (March 1996).

Recent Changes

The rise of indigenous product development in China's car sector was a driving factor behind the policy debates during 2003–5 regarding

[35] Geely was a family-controlled firm before it entered the car sector.

"indigenous innovation (or not)" in Chinese industries. Some policymakers and leaders of large SOEs defended the lack of progress on indigenous capacity, claiming that the transition to in-house product development was time-consuming and difficult. For example, YanFeng Zhu, the president of FAW and the Alternate Member of the Political Bureau of CCP at the time, said that "the preparation to develop indigenously product would take two generations of engineers, and over two decades."[36]

Yet, the rapid market success and technological progress of new indigenous firms—especially Chery, Geely, and HaFei—argued that a shift toward indigenous innovation is feasible. Around 2005–6, "indigenous innovation" became a focus of speeches made by top leaders of China, and this emphasis was echoed explicitly by the "15 year National Medium to Long-Term Plan for Development of Science and Technology (2006–2020)" announced in 2006. In 2007, in the governmental 11th FYP outline for the automobile industry, developing "indigenous technological capability" and "indigenous brands" became one of the central goals for China.

The policy transition brought pressures for large SOEs. They announced their own plans for building up "indigenous brands." However, because of their lack of developmental activities for about 10 or 20 years, it was difficult to shift despite the plentiful resources. For example, FAW tried to restore the market reputation of RedFlag and developed a luxury model for the protocol uses of political leaders and for the top-level market. However, because of the weakness of FAW in in-house technological capability, this project failed to meet its original development scheme. As a consequence, at the 2006 Beijing Auto Show, FAW had to change its prototype and transplant this model onto the 2004 version of Toyota's "Grown-Majesta." This was not well received by the market either.

Large SOEs then turned to their foreign partners for help developing "indigenous brands." They bought existing technological solutions or commissioned new designs from their foreign partners. A series of international mergers by the large SOEs were also motivated by the indigenous innovation mandate. For example, Beijing Auto purchased a package of Saab's technological assets, while Shanghai Auto merged Sangyong and Rover. Beijing Auto even tried to buy product platforms from Chery. In 2012, Guangzhou Auto brought product platforms from Chery through an alliance between these two firms.

[36] Referring to the *Dialogue* program on CCTV (China Central Television), Feb. 15, 2004. This was claimed by YanFeng Zhu, the president of FAW and the Alternate Member of the Political Bureau of CCP at that time. Similar statements had been made by leaders of other large SOEs as well.

As a result, there came a series of so-called "JV indigenous" brands, such as Baojun of Shanghai Auto and LiNian of DongFeng Auto. By forming JVs with homegrown brands, the large SOEs tried to convince policymakers and the Chinese opinion leaders that they would contribute to the development of indigenous technological capability in China. But unless there are fundamental changes in organizational systems and resource allocation within these SOEs, they will continue to rely on outsiders for technology and innovation. The SOEs have so far failed to create a mechanism for acquiring and internalizing new knowledge to develop new products and critical technologies in the car industry.

On the other hand, the new non-SOE firms have also been faced with steep challenges since 2009. The comparative market share of indigenous car brands stalled and even started to decline: it slid from about 30 percent in late 2009 to less than 26 percent by the end of 2012.[37] This can be attributed to a variety of causes. First, MNCs changed their strategies in the Chinese market. Facing competition from new indigenous firms, MNCs accelerated their launch of newer and better car models in China. They reduced prices and even introduced their own low-end car models to the Chinese market. Because foreign brands had a better market reputation among the consumers, such a change of strategy eroded the market share of indigenous firms.

Secondly, indigenous firms thrived in the low-end market but had trouble in the middle or top-level markets where customers cared more about the reputation, quality, and service than price. The newcomers have yet to accumulate sufficient expertise in this area compared with established brands. Since new entrants could not expand their market space upstream, huge R&D investment was becoming burdensome. This reflected the weaknesses in management of new indigenous firms. For the middle and high-end market, coordination between R&D, production, marketing, and service becomes more important in providing a better customer experience. In fact, during the early phases of product development, such interactions involving customer opinions must be introduced. New indigenous firms did well in mobilizing personnel and resources for product and technological development. However, their inexperienced, engineer-led organizations had little expertise in the intricate interactions between different departments for continual subtle improvements, especially when production reached a large scale. According to the J. D. Power report in 2011, in the Chinese market, the amount of faults per 100 new cars (two to six months after purchase) for foreign brands was 131, and only 108 for Japanese brands. But for the Chinese indigenous

[37] Referring to the report of *The Economist*, Apr. 20 2013, "Voting with their Wallets: Chinese Car Buyers Overwhelmingly Prefer Foreign Brands."

brands, there were 232 faults per 100 cars, even higher than the 224 faults in 2010.[38]

Such unfavorable market performance dampened the rather ambitious plans of product development for indigenous car-makers. During 2005–8, Chery had more than 50 car and engine projects in progress. However, it did not have an effective management system to optimize the development input among projects and cultivate close interactions between different departments. When market conditions changed, different projects failed for similar reasons. Increasing competition and the negative feedback of market reputation pushed this firm into a worse situation, although its engineering teams still made steady progress in different domains of technologies.[39]

Today, Chery is still struggling. Confronted with management problems, it has lost confidence and hired professional managers to guide strategy-making, usually executive managers from Sino-foreign JVs and engineer leaders from MNCs. But these managers or engineer leaders had not experienced anything like Chery's rapid progress and intensive learning. Instead, an extreme reform was led by these managers, and it caused a huge brain-drain at Chery. Many experienced engineers were forced to leave Chery, and the reform has not helped Chery revive.

Geely has been going through a similar process since 2009. Since it merged with Volvo in 2010, it has been financially vulnerable.[40] However, in terms of quality control and management efficiency, the management team of Geely achieved better performance with incremental improvement. It has kept investing in workshops and tried to improve the interactions between product development, manufacturing, and marketing. All these improve the quality and market reputation of its cars. In 2012, it initiated a joint car development project with Volvo. The goal of this plan was not simply to introduce Volvo technologies to China, but also to apply Geely experiences in the low-end compact car market to a new joint platform with Volvo technologies. The products created from this platform will be aimed at the global market.

Given the lack of experience of these new entrants and the intensified competition in the Chinese market, their bumpy ride is not surprising. Regardless of their difficulties with certain aspects of management, they were indispensable in introducing a more dynamic and innovation oriented auto-industry in China.

[38] Source of data: *China Business News*, Nov. 7, 2011 (referring to <http://jingji.cntv.cn/20111107/104896.shtml>).

[39] For example, in recent years, Chery built up the best facilities for safety research in the Chinese automobile industry, and had good performance in developing hybrid power cars. However, these individual technological advances would not be appreciated by the market.

[40] For this, the Chinese government gave Geely financial support.

Summary

Over the past three decades, the growth of China's car market has attracted different types of players, with differences in fundamental organizational systems and learning mechanisms. TMFT policy successfully expanded China's production capacity, but it created passive SOEs as partners of foreign firms. These SOEs have little control of the technological trajectory and little incentive to learn and innovate. The indigenous newcomers represent a distinct group which emerged in the 2000s and were pushed forward by forces outside the government central planning. These new entrants catalyzed rapid growth of the Chinese market and intensified product-based competition. Instead of being integrated into global production networks, these firms were integrators that mobilized knowledge within the Chinese and global industrial community, and invested intensively in in-house capability building. The Chinese governments at the central or regional level did not lead in initiating such a transition. Rather, the central government played a reactive role by launching the "indigenous innovation" policy in the mid-2000s after examining this development in industry.

This dual development of the automobile industry in China makes it difficult to assess the overall innovative capability of the Chinese car sector. It is similarly difficult to forecast where and when the Chinese car sector can be globally competitive in terms of product and technology development. The JVs, to some extent, are changing their strategies and introducing more in-house development of products and critical technologies. They have brand reputation, a well-trained labor force, and competent manufacturing skills to ground their innovative learning. However, they will have to make fundamental reforms to their organization in order to become more self-directed in technological changes.

The new indigenous firms are also confronted with challenges. They have already developed certain competences and gained certain innovative capacities, as shown by the example of BYD's entry into the electric car market. But visible gaps remain between the current capabilities of Chinese car-makers and the international frontier in design and manufacturing quality. For these firms, the more urgent challenge is in management facing a new market reality.

Overall, it is indisputable that the Chinese car industry has already changed greatly because of the rise of these new indigenous firms and their efforts in developing technologies. In 2003, when the author of this chapter started his investigation into the Chinese automobile industry, some industrial regulators and researchers in Beijing insisted that there was little hope for indigenous innovation because "we can't even develop an engine or even a car door assembly." Within only a decade, the situation has changed dramatically.

Leading indigenous firms like Chery and Geely are a source for learning and training that inspires other indigenous innovators. In fact, a new generation of indigenous firms, such as ZhongTai and HuaTai have emerged, specializing in clean technology, often staffed by the talents of former Chery workers.

In short, the landscape of the Chinese automobile industry has been thoroughly changed and continues to evolve. In the future, some of today's leading indigenous firms might fail, but their experiences have already upgraded the Chinese intellectual and human resource base for this industry.

References

Amsden, A. (1989), *Asia's Next Giant: South Korea and Late Industrialization*, New York: Oxford University Press.

Chu, W. (2011), "How the Chinese Government Promoted a Global Automobile Industry," *Industrial and Corporate Change*, 20(5): 1235–76.

Clark, K. B., and T. Fujimoto (1991), *Product Development Performance: Strategy, Organization, and Management in the World Automobile Industry*. Boston, MA: Harvard Business School Press.

Cohen, W. M., and D. A. Levinthal (1990), "Absorptive Capacity: A New Perspective on Learning and Innovation," *Administrative Science Quarterly*, 35(1): 128–52.

Drucker, P. F. (1946), *Concept of the Corporation*, New York: John Day.

Editorial Office of China Automotive Industry Yearbook, eds. (1999), *ZhongGuo QiChe GongYe NianJian (1999) (China Automotive Industry Yearbook (1999)*, Changchun: Editorial Office of China Automotive Industry Yearbook.

Editorial Office of China Automotive Industry Yearbook, eds. (2008), *ZhongGuo QiChe GongYe NianJian (2008) (China Automotive Industry Yearbook (2008)*, Changchun: Editorial Office of China Automotive Industry Yearbook.

Fujimoto, T. (2007), *NengLi GouZhu JingZheng* (in Chinese, tr. from *Noryoku Kochiku Kyoso*, Tokyo: Chuokoron-Shinsha, Inc., 2003), Beijing: China Citic Press.

Hahn, C. H. (2007), *Wo Zai DaZhong QiChe SiShi Nian* (My 40 Years in Volkswagen) (tr. by Liuhua Zhu from *Meine Jahre mit Volkswagen*, Munich: Signum Verlag, 2005), Shanghai: Shanghai Far East Publishers.

Kim, L. (1997), *Imitation to Innovation: the Dynamics of Korea's Technological Learning*, Boston: Harvard Business School Press.

Kim, L. (1998), "Crisis Construction and Organizational Learning: Capability Building in Catching-up at Hyundai Motor," *Organization Science*, 9(4): 506–21.

Kogut, B. and U. Zander (1992), "Knowledge of the Firm, Combinative Capabilities, and the Replication of Technology," *Organization Science*, 3(3): 383–97.

Lazonick, W. (2003), "The Theory of the Market Economy and the Social Foundations of Innovative Enterprise," *Economic and Industrial Democracy*, 24(1): 9–44.

Li, D. (2010), "ZhongGuo QiChe ChanYe de ZiZhu KaiFa (Indigenous innovation in Chinese automobile industry)," *ZhongGuo Qiche ChanYe (Chinese Automobile Industry)*, S. Yamazaki. Beijing: China Commercial Publishing House, 9–59.

Liu, W. (1992), *Technology Transfer, Technological Capability and Late Entry into the International Automobile Industry: A Case Study of Shanghai-Volkswagen Automotive Corporation in China*. SPRU, University of Sussex, Brighton. D.Phil. dissertation.

Liu, W., and P. Dicken (2006), "Transnational Corporations and 'Obligated Embeddedness': Foreign Direct Investment in China's Automobile Industry," *Environment and Planning A*, 38: 1229–47.

Lu, F., and K. Feng (2005), *The Policy to Develop the Indigenous Automobile Industry* (FaZhan WoGuo ZiZhu ZhiShi ChanQuan QiChe GongYe De ZhengCe XuanZe). Beijing: Peking University Press.

Lundvall, B.-Å. (1988), "Innovation as an Interactive Process: From User–Producer Interaction to the National System of Innovation," in G. Dosi, C. Freeman, R. Nelson, and L. Soete (eds), *Technical Change and Economic Theory*, 349–69. London: Pinter.

March, J. (1996), "Exploration and Exploitation in Organizational Learning," in M. Cohen and L. Sproull (eds), *Organizational Learning*, 101–23. London: Sage.

Ming, H. (2006), "DengXiaoPing yu ZhongGuo QiChe GongYe de FaZhan" (Xiaoping Deng and the Development of Chinese Automobile Industry), *DangShi ZongHeng* (Over the Party History), 12: 13–17.

Naughton, Barry (2007), "China's State Sector, Industrial Policies and the 11th Five Year Plan," testimony before the US-China Economic and Security Review Commission Hearing on the "Extent of Government's Control of China's Economy and Implications for the US," May 24.

Nonaka, I., and H. Takeuchi (1995), *The Knowledge-Creating Company: How Japanese Companies Create the Dynamics of Innovation*. New York and Oxford: Oxford University Press.

Richardson, G. B. (1972), The Organization of Industry. *The Economic Journal*, 82(327): 883–96.

Shrivastava, P. (1983), "A Typology of Organizational Learning Systems," *Journal of Management Studies*, 20(1): 7–28.

Thun, E. (2006), *Changing Lanes in China: Foreign Direct Investment, Local Governments, and Auto Sector Development*. Cambridge: Cambridge University Press.

von Hippel, E. (1988), *The Sources of Innovation*. New York: Oxford University Press.

Womack, J., D. Jones, et al. (1990), *The Machine that Changed the World*. New York: Maxwell Macmillan International.

Xia, D., D. Shi, et al. (2002), *QiChe GongYe: JiShu JinBu Yu ChanYe ZuZhi* (Automobile Industry: Technological Progress and Industrial Organisation). Shanghai: Shanghai University of Finance and Economics Press.

Xu, B., and M. Ouyang (2011), *ZhongGuo JiaoChe FengYun: 1953–2010* (Chinese Car Sector: 1953–2010). Beijing: Enterprise Management Publishing House.

Zhang, R., and H. Gao (2001), *ShiJie QiChe GongYe: DaoLu, QuShi, MaoDun, DuiCe* (World Automobile Industry: Path, Trend, Challenge and Strategy). Beijing: China Economic Publishing House.

6

High Speed Rail Development in China

A Case Study of State-Guided Technology Transfer

Rongfang (Rachel) Liu, Liu (Willow) Lv, and Zhaodong (Tony) Huang

Introduction

The High Speed Rail (HSR) network in China has grown from zero to more than 13,000 km in the past decade (Sheng 2013), representing a feat few countries in the world could accomplish. The explosive growth of the HSR system has reconfigured Chinese geography in profound ways. It has not only propelled China into a leadership position in HSR development but also attracted global scrutiny in terms of the roles of the state, technological transfer and innovations, business models, and investment returns in large infrastructural development.

Many view China's HSR as a triumph of technology, but controversies abound over whether China's HSR is primarily a result of technological transfer, as claimed by foreign companies, or home-grown improvement or innovation beyond purchased technology, as claimed by the Chinese government. The most exceptional aspect of the technological development in HSR is the fact that the entire system was orchestrated and implemented by one governmental entity, the Ministry of Railroad (MoR). The MoR is the centralized governmental body for railroad construction, operation, and regulations (Chan and Aldhaban 2009). This single-handed feat raises the crucial question about the role of state in successful technological acquisition and indigenous innovation.

While technology advances are visible and important, it is crucial not to credit them as the only enabling factor for an HSR system. Contemporary China has a number of unique conditions that fostered the development of HSR: capital investment, travel demand and supply, economic capabilities, political will, and technology capacity. All of these combined made HSR a reality within a short period of time over a vast geographic territory.

The first section of this chapter provides an overview of the high-speed rail development process in China. The second section documents the key factors that made the Chinese HSR network a success. The third section highlights the unique aspects of China's HSR development and technology transfer processes. The last section provides an analysis of HSR technology transfer practices in China, and compares findings with other industries, and with HSR development in other countries.

Historical Development of HSR in China

The idea of HSR development in China was conceived in the early 1990s, almost 30 years after the opening of the Shinkansen (SKS) HSR in Japan (Perl 2002). Running through one of the most developed economic and transportation corridors in China, the Beijing–Shanghai Passenger Dedicated (BJ-SH) HSR Line exemplifies the overall development of the HSR in China and can be used as a case study to demonstrate many facets of HSR development (Liu and Deng 2004).

There is no single standard definition of high speed rail: many transportation organizations use various speed thresholds ranging from 200 kmph (125 mph) to 380 kmph (240 mph) (Leber 2013; Liu and Deng 2004). In general, HSR is a combination of all the following elements that constitute the "system": fixed guideway infrastructure, rolling stock, communication and control, and operating processes. As shown in Figure 6.1, rail infrastructure development as measured by total kilometers of track has been growing steadily over the past two decades while China's per capita Gross Domestic Product (GDP) growth accelerated.

The overall development for HSR in China can be divided into four distinctive phases.

Exploration: 1991–97
Evaluation: 1998–2004
Adaptation: 2005–2010
Assimilation: 2011–present

Exploration: 1991–1997

The first phase of HSR development, exploration, was initiated in the 1990s. The driving force behind this early development was largely the Ministry of Railways (MoR). A recent reorganization in March 2013 placed the

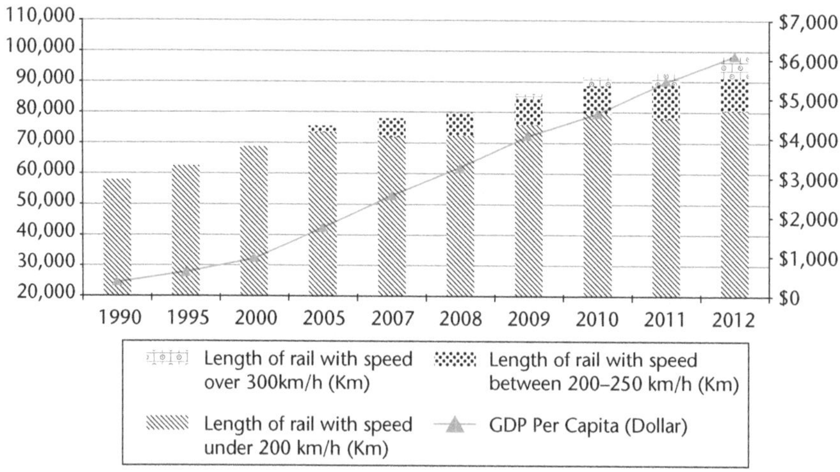

Figure 6.1. Historical Development of High Speed Rail in China
Sources: MoR 2012; XinHua Net 2002.

Table 6.1. Main Events during the HSR Exploration Phase

Year	Events
1991	Feasibility study started by the Ministry of Railways (MoR).
1992	Feasibility report finished and submitted to the government.
1994	Beijing–Shanghai project office is set up by MoR.
1996	Preliminary Study on "HSR experimental engineering" and "200 km/h electric passenger train set and force scattered AC driven EMU."
1997	MoR started speed increase of national rail networks.

Source: Li 2006.

construction and operation of railways under the China Railway Corporation and regulation of railways under the Department of Transportation (DOT) (China's State Council 2013).

As shown in Table 6.1, the HSR exploration effort was officially launched in 1994 by the establishment of the Beijing–Shanghai HSR Project Office under the MoR. The Chinese government's commitment to HSR was signaled by not only planning and policy development, but also by the initiation of speed increases across all existing rail networks in the country. The first rail-speed increase started in 1997 and was a precursor to five more speed increases and, eventually, high-speed rail implementation.

As part of its effort to increase operating speeds of existing railway trains, the MoR started to purchase and license foreign rail equipment in the 1990s, which served as an introduction to advanced knowledge of HSR technologies.

The initial HSR exploration was conducted in a low-key fashion when the six speed increases were applied to already existing rail services. This spared the MoR and railway engineers decisions between domestic or foreign technology in the early stages of development. At this stage, domestic and foreign approaches were explored simultaneously.

Evaluation: 1998–2004

The second phase of HSR development, evaluation, was dominated by the debate between Maglev and "steel wheel on steel rail" HSR technologies. Since the most extensive and oldest HSR in the world was located in neighboring Japan, Shinkansen was naturally selected for evaluation as a candidate for implementation along the BJ-SH Corridor. Unexpectedly, however, very strong anti-Japanese sentiment rooted in historical grievances and contemporary territorial disputes spilled into the technological transfer discussion, and hundreds of thousands of Chinese signed a petition against using Japanese technology (Feng 2004). As a result, the MoR was forced to examine different HSR technologies such as Maglev.

As documented in an earlier study (Liu and Deng 2004), there are basically two distinctive branches within the family of high-speed ground transportation technology: conventional High Speed Rail (HSR) and Magnetic Levitation (Maglev). Both conventional HSR and Maglev provide higher operating speed and passenger riding comfort, but only conventional HSR has existing commercial applications, such as Shinkansen (SKS) in Japan, Inter City Express (ICE) in Germany, and Train à Grande Vitesse (TGV) in France. It is difficult to compare conventional HSR and Maglev in technical merits, since both are complex technologies. Many studies have assessed the potential trade-offs of implementing either Maglev or conventional HSR systems in the BJ-SH Corridor. The proven commercial applications, excellent safety record, and cost advantages favored conventional HSR, while operating speed, energy consumption, riding comfort, and environmental impact favored Maglev technology.

Several variations of Maglev technology have been developed in various parts of the world. A low-speed Maglev pilot project was installed in the United Kingdom in the 1980s (Maglev UK 2013). A 2,000 foot, elevated testing track for Maglev vehicles was built in the 1990s in America (AMT Inc. 2013), while Germany, Japan, and China have all developed and tested high-speed Maglev technologies. Transrapid International, Inc., a joint venture between Siemens and Thyssenkrupp of Germany, led the pack in implementing and promoting Maglev technology.

Maglev technology gained momentum at the end of the last century, as Germany, America, and China all explored the possibilities of implementing

Table 6.2. Timeline for the Maglev and HSR Comparison

Year	Events
1998	Government shows interest on Maglev for Beijing-Shanghai line.
1998	MoR began its experiments with the Guangzhou-Shenzhen Railway 200 km/h Technology.
2000	Shanghai Maglev Project was proposed.
2002	Shanghai Maglev Service inaugurated.
2003	Wen, Jiabao replaced Zhu, Rongji as Premier and Liu, Zhijun appointed as the Secretary of the MoR.
2004	Maglev option officially abandoned by Beijing-Shanghai project. Conventional HSR Technology was chosen. "Medium/Long-Term Railway Network Plan" Released.

large-scale high-speed Maglev projects. Driven by the green movement and earlier technology transfer efforts, Deutsche Bahn (DB), the German national railway company, adopted an ambitious plan to implement Maglev in several German transport corridors (BBC News 1999). Meanwhile, the US Department of Transportation (USDOT) initiated the "Maglev Program" with an $8 million research fund to study application feasibility along seven corridors in the US (Parsons Brinckerhoff 2011). China jumped onto the international Maglev bandwagon at the end of 2000 by initiating a feasibility study of Maglev implementation in Shanghai, as shown in Table 6.2.

In Germany, the 1998 elections ushered in a new governing party, and the Berlin-Hamburg Maglev Corridor Project was canceled as a result. High-level German politicians were interested in locating international markets for its technology developers (Vuchic and Casello 2002). While Chinese sentiment against Japan had pushed the pendulum away from SKS HSR, China's application to the World Trade Organization (WTO) may have subsequently pulled it toward German Maglev technology instead. As one of the founding members of the WTO, Germany had great influence over the status of China's application: a contract was awarded to a German consortium led by Transrapid International, Inc. barely two months after the Chinese Maglev feasibility study started.

The 30-km Maglev line between Shanghai Pu Dong International Airport and downtown Shanghai was inaugurated into commercial operation on January 22, 2003 (Shanghai Maglev Transportation Development Co. Ltd, 2013). The Chinese Premier, Zhu Rongji, and German Chancellor, Gerhard Schroder, both attended the inauguration ceremony, which may have signaled strong support for further use of Maglev technology in China's HSR development. Another Maglev pilot project from Shanghai to Hangzhou was seriously considered but did not move forward.

The international fanfare for Maglev died down right after the new millennium. Changes in the Chinese government, especially the retirement of Premier Zhu, Rongji in 2003, the most powerful proponent of Maglev technology, drove the last nail into the coffin. The plan to use Maglev technology for the BJ-SH Corridor was officially abandoned in 2004, due to lack of commercial application, limited technical staff for large-scale implementation, and higher construction costs than conventional HSR.

The negotiations on Maglev also taught participants a lesson on technology transfer, one that may explain alternative strategies used by the MoR in later negotiations regarding HSR application. Transrapid International's exclusive and tight grip on the core propulsion technology in Maglev trains dictated that all Maglev vehicles for the Shanghai system be made in Germany and shipped to China. This practice did not help sell more Maglev applications in China, such as for the proposed Shanghai to Hangzhou HSR line. In any case, it would have been difficult to implement Maglev on the large scale desired by the Chinese government, given the limited capabilities of Transrapid as the sole vehicle provider.

The debate on Maglev technology delayed the implementation of any HSR system in China for at least seven years, if not more. Once the decision on HSR technology was made, the increased interests from abroad and intense lobbying efforts to promote conventional HSR brought many more potential HSR technology suppliers and partners to Beijing.

During the same period, a second effort to increase the speed of the national railway service quietly took place. As documented in Figure 6.2, the average speed for conventional rail services increased from 55 km/h in 1997 to 66 km/h in 2004 and the highest operating speed changed from 140 km/h to 250 km/h during the same period. This progress is commendable given the

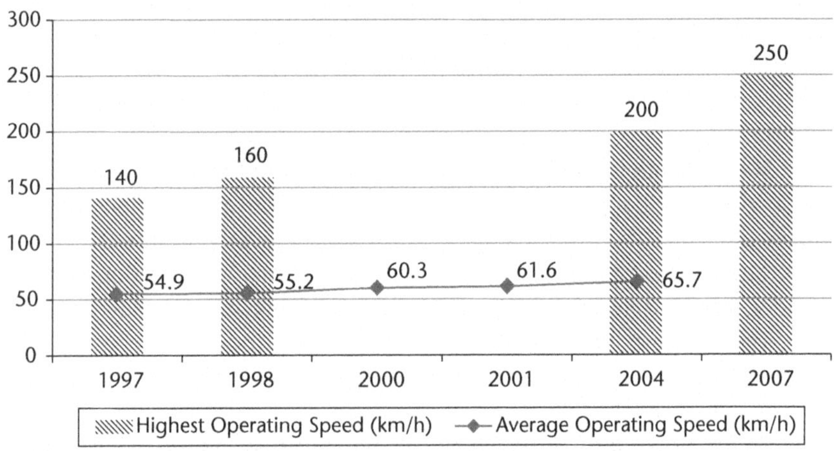

Figure 6.2. Speed Increases for National Rail Networks
Source: Zhou 2007.

massive scale of the conventional rail network and the very low speed of conventional rail services.

In the meantime, domestic development of HSR technology in China expanded. By 2003, several HSR train models had been developed—the process of technological transfer will be discussed later in this chapter. The domestic models can be grouped into two large families: locomotive-hauled HSR trains and Electric Multiple Units (EMU). The "Blue Arrow" and "China Star" models represent the centrally powered, locomotive-hauled HSR trains, while the "Pioneer" and "Changbai Mountain" models fit into the EMU category, where propulsion power is decentralized (China Institute of Railway Research and Technology 2009). These indigenous developments indicated an accumulation of some understanding and technological expertise regarding HSR.

Adaptation: 2004–2010

The transportation sector is often credited as a public utility and associated with pioneering such technological innovations as the steam engine, airplane, wheel, and streetcar. However, introducing new technology is risky for the public sector. Many public bodies favor "commercial off-the-shelf" (COTS) products when massive infrastructure is needed to satisfy large demand. COTS products are commercially available, can be leased, licensed, or sold to the general public and require no special modification or maintenance over their life cycle. For example, immediately after the 9/11 terrorist attacks, the newly established US Transportation Security Administration (TSA) launched a global search for "commercial off-the-shelf" explosive/trace detection technologies, since the critical importance of national security did not allow any time for research and development, or for the high risks associated with trial and error testing of technology (Transportation Security Administration 2003).

China's plan to develop a large-scale HSR network over a short period of time, especially in a sector in which safety concerns are paramount, meant that it faced a similar scenario favoring commercial off-the-shelf technology. MoR eventually rejected Maglev technology due to its lack of commercial, large-scale application. Similar concerns also arose regarding domestic HSR development. For example, the operating speeds for the most advanced domestic models ranged from 100 to 125 mph, which was far slower than international standards for high speed rail operations (often above 200 mph). Chinese policymakers quickly realized that domestic HSR technology alone would not be able to provide the safe and efficient HSR system demanded by China's speed of economic, political, and military advancement. By the mid-2000s, China had entered a developmental period during which rising income and changing economic activities led to strong demand for rapid, reliable, and mass means of transportation. Large-scale migration from the interior provinces to the coastal areas created huge congestions; the acute shortage of rail capacity during the Chinese New Year time created a predictable public outcry year after year. A rapid deployment of HSR was seen as the solution.

The MoR's decision to pursue off-the-shelf technology instead of indigenous R&D also reflected political calculation of individual leaders. The political appointment system in China consists of five-year terms combined with age limits for high-level posts (such as the Secretary of the MoR). This gives leaders a fairly short time to create and solidify their legacies—certainly not enough time to nurture the developing domestic HSR technology sector to maturity. In 2003, the ambitious Liu Zhijun (nicknamed "Crazy Liu") took the helm of the MoR. Liu was determined to push the HSR projects through to completion and was supported by the central government, especially with stimulus spending after 2008. Even though foreign HSR technology was more expensive up front, it promised safety standards, durability, and technical certainty that were lacking from indigenous models. Pursuing the transfer of HSR technology actually promised to be cheaper in the long run, given uncertainties about domestic technology and the time associated with developing an alternative approach.

Since Liu was determined to build the world's largest HSR system within his tenure, the MoR evaluated and acquired all four existing HSR technologies from Canada, France, Germany, and Japan respectively. As shown in Table 6.3, each foreign transferee was matched up with a domestic partner, initially under the auspices of satisfying the 75 percent localization rate, but with the eventual goal of training domestic custodians on the technologies once the transfer was completed. These models are named CRH1, CRH2, CRH3, and CRH5 respectively.

In a striking contrast to the technological transfer agreements in the auto-industry (see Chapter 5), in which individual companies negotiated their own deals on joint ventures under the guidance and support of the central government, the MoR was the sole negotiating party on the Chinese side, representing the largest HSR market in the world, with multiple foreign suppliers. This arrangement heightened the MoR's bargaining power to negotiate favorable deals with whoever eyed the vast HSR market in China. As a result, the MoR entertained offers from the world's four HSR suppliers. Instead of "the turn-key operation [that] characterized the auto-industry," foreign firms supplied various components, while China's own railroad enterprises and research institutes were tasked with integrating various parts and local implementation. This way, MoR positioned itself as the sole party with strategic control of the project.

The principal tasks during the adaptation stage were to integrate and adapt foreign technology for different applications. Some incremental improvements, such as body weight reduction, reduced energy consumption, and

Table 6.3. Imported HSR Technology and Their Chinese Partners

Series	CRH1	CRH2	CRH3	CRH5
Original model	Regina C2008; ZEFIRO 250	E2-1000	ICE-3	Pendolino + Sm3
Manufacturers	Bombardier	Kawasaki	Siemens	Alstom
Country of Origins	Canada	Japan	Germany	French
Domestic Partners	CSR Qingdao Sifang/BST	CSR Qingdao Sifang	CNR Tangshan	CNR Changchun
Production Period	2006–13	2005–11	2007–10	2006–12
Inauguration	2007.2.1	2007.1.28	2008.8.1	2007.4.18
Format	5M3T, 10M6T	4M4T/6M2T/8M8T	4M4T	5M3T
Capacity	1299/673/642/ 612/597	588/610/630/1230	557	662/587
Power	5300/11000 kw	4800/7200/8760/ 9600 kw	8800 kw	5500 kw
Speed (Running/Highest speed)	200/250 km/h	200/250 km/h, 350/ 380 km/h	350 km/h	250/250 km/h
Train Sets	120	170	80	110
Train Value (Million Dollar)	2,842	3,585	2,417	1,939
Adaptations to Chinese Market	Aluminum body	DSA 250 Pantograph		
Initiation of Second Generation Models	2012	2010	2011	2011

Source: BST-Transportation 2013.

alternative pantograph connections, had to be made during development. Other minor interior configuration changes—such as converting regular seats to sleepers—were made to fit the unique demands of the Chinese HSR market. The highlights of this developing phase were the inaugurations of various short-distance HSR services, such as the Beijing–Tianjin and Shijiazhuang–Taiyuan corridors.

Assimilation: 2011—Present

The latest (and current) phase of HSR development in China can be labeled or anticipated as assimilation. Indigenous HSR development was bumped to the back burner when foreign technologies were purchased during the last phase, but it was never abandoned. On the contrary, the domestic partners who had the HSR developmental experience were selected from the very beginning in the adaptation processes.

The technical expertise and production capabilities acquired during the exploration stage came in handy when sufficient imported technical know-how was accumulated. The assimilation process was marked by the second generation of HSR trains produced in China: CR380A, B, C, D, and CRH 6. As noted in Table 6.4, each of the CR380 models originates from the earlier CRH1, 2, 3, and 5 designs, but has higher operating speeds and improved key features such as aluminum bodies, enhanced traction systems, and optimized air tightness. While the improvements are likely to be achieved in collaboration with the foreign partners, it signals the tremendous growth of local HSR expertise in a few years.

Key Factors for Success of HSR in China

While technological barriers to HSR are high, it is economic and political calculations that prevent most countries from pursuing large-scale HSR development. It is in these areas that China presented a uniquely favorable case. China has the world's largest share of existing and expected rail travel demand. China also has the ability to act as a powerful financial backer due to decades of successful development and domestic accumulation of capital. Its political structure enabled fast and relatively inexpensive land acquisition, allowing right of way for extensive and straight tracks, thereby removing a major stumbling block to HSR development. As a result, Chinese HSR has always had a reasonably optimistic return on investment, which further encouraged its construction.

Large Share of Rail Travel Demand

Benefitting from a growing economy and huge population base, China's demand for travel has grown rapidly in the last two decades. Intercity travel in China increased dramatically, and railway was the dominant mode of travel, especially since the economic reform of the 1980s. As depicted in Figure 6.3, railway travel accounted for around 60 percent of overall travel in 1980. Due to increased highway construction and airline services, the railway's share dropped to 37 percent in 2000 but remains a major force in passenger transport, especially along intercity corridors. The total amount of long-distance travel has increased tremendously during this time interval.

While expanded highway construction and airline services might have taken some passenger traffic away from railways, new demand for HSR services was generated by three major shifts. First, previously nonexistent routes and trips were made possible by HSR development, drawing in a new customer base. According to a World Bank study (Olivier et al. 2014), more than 50 percent of HSR trips in China result from such newly harnessed demand. The remaining passenger traffic is made up of customers who have switched over from conventional rail and other modes of transport.

Table 6.4. Second Generation HSR Models and their Key Features

Series	CRH380A		CRH380B		CRH380C	CRH380D
Advanced Models	CRH380A	CRH380AL	CRH380B	CRH380BL	CRH380 CL	
Manufacturer	CSR Qingdao Sifang		CNR Changchun		CNR Changchun	BST
Original Model	CRH2C Phase2		CRH3		CRH3C & CRH380BL	ZEFIRO 380
Production Time	2010–11	2010–13	2012	2010–13	2011–13	2012–14
Inauguration	Sept. 10	June 11	Oct. 12	Jan. 11	2011	2013
Consist	6M2T	14M2T	4M4T	8M8T	8M8T	4M4T
Capacity	480	1028/1061	450	1043	1053	
Power	9600 kW	20440 kW	9600 kW	18400 kW	19200 kW	
Speed	350/380 km/h		350/380 km/h		350/380 km/h	350/380 km/h
Train Sets	41	95	41	102	25	70

Source: BST-Transportation 2013, etc.

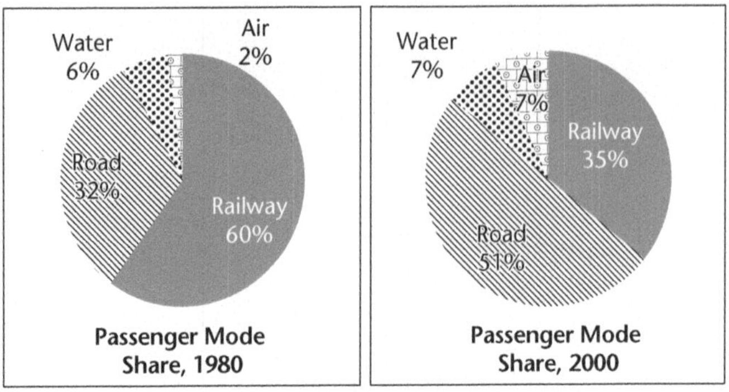

Figure 6.3. Mode Shares in China
Source: National Bureau of Statistics of China 2011.

Table 6.5. Highway Comparison

Nation	Highway Length (km)	Population (million)	Density (km/ hundred sq. km)	Length Per Capita (meters)
China	25,600	1,354.04	0.26	0.02
USA	88,700	315.25	0.96	0.28
Germany	12,000	81.75	3.25	0.15
Japan	6,114	127.95	1.62	0.05

Source: China Association Highway and Waterway Engineering Consultants 2005.

If the long history of railway development in conjunction with the habit of railway use amongst travelers has provided a solid base for the development of HSR in China, then the infancy of air travel and late development of the highway system has presented the perfect void for HSR to fill. As of 2004, there were only 25,000 km of highways in China, less than one-third of the distance covered by interstate highways in the US. Given the similar sizes of territory and large differences in total population, it is clear that highway system development in China lags far behind its Western counterparts. As shown in Table 6.5, total highway length, density, and per capita distribution in China are only a fraction of what they are in the United States, Germany, and Japan. In addition, private vehicle ownership is still low and highway toll charges are high, which also discourages highway travel.

Intercity highways in China primarily connect capital cities of large provinces and large cities with more than half a million people. Only 60 percent of medium-sized cities, with populations between 200,000 and 500,000 people, are connected to highway networks. According to a recent estimate (China Association of Highway and Waterway Engineers 2005), an additional 80,000 km of highway would be needed to connect all Chinese cities with more than 200,000 persons.

Similarly, the popularity of air travel only took off long after the economic reform. The passenger share of air travel in China was under 10 percent in 2010, a far cry from the 14 percent of the intercity travel share held by commercial air carriers in the US (Lv 2013). The total number of airports and airline routes is still very low, even by the standards of developing countries. The total Passenger Miles Travelled (PMT) in China via commercial airlines totaled around 1.88 billion in 2004, which translates to less than two miles per capita. In contrast, the per capita PMT by air in the US is around 2,500 miles.

In short, as Chinese income grows, demands for travel both for work and pleasure will continue to rise. Fast and reliable HSR would probably attract a large share of these passengers.

Financial Capabilities

One of the major barriers to HSR development in most countries is the prohibitive cost. The Chinese state, however, made HSR a priority and lent considerable financial support. Buoyed by the explosive economic development and rapid increases in GDP, HSR development in China has accounted for a large portion, almost 60 percent, of the government's growing infrastructural investment since the mid-2000s (Liu and Lv 2012). As shown in Figure 6.4, the total capital investment in rail infrastructure jumped from \$8.4 billion in 2004 to \$82 billion in 2012.

Funding for China's HSR development project comes from three main sources. The first is the Railway Construction Fund, which is extracted from the income of operating railways and amounts to approximately \$16

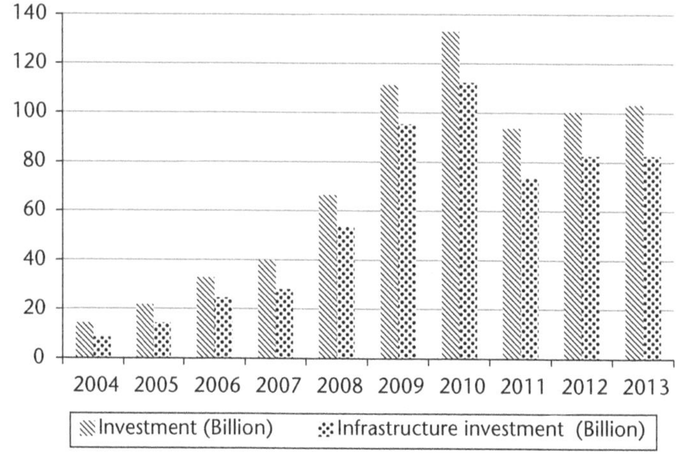

Figure 6.4. Capital Investment in Rail Infrastructure
Source: MoR 2012, etc.

billion per year. Railway Enterprise Bonds that are issued by specific railway construction projects but backed by the central government provide the second source. Third, direct loans and budget allocations from the central government that are earmarked for specific construction projects also finance rail infrastructure development. For example, about half of the capital investment in the Beijing–Tianjin HSR line—$1.46 billion—came from bank loans with interest rates of 6.65 percent. While the life spans of the loans are not specified, and most can be negotiated depending on the health of the HSR operation, the HSR operator is only required to pay interest during the three years immediately after the loan was initiated. This is the case largely because all the banks in China are owned and operated by the central government, and because the Chinese accounting system is significantly different from that of the US, especially in regard to treatment of debt (Liu and Lv 2012).

The Chinese government, like its US counterpart, initiated stimulus funding when the world economy went into recession in 2008. The total value of the Chinese stimulus package was $645 billion, which is comparable to the $787 billion stimulus package authorized by the American Recovery and Reinvestment Act of 2009. In contrast to the US practice of handing money to failing banks, however, the Chinese government channeled 40 percent of the stimulus fund toward investment into transportation infrastructure such as railways, roads, airports, and waterways (Liu 2010). The HSR program benefitted greatly from this stimulus spending (Zhou 2012).

Political Will

Besides the benefits derived from a long tradition of rail technology development, the large share of the travel market held by China's rail system, and the strong backing of China's financial capabilities, national policy and political will proved to be other key factors that facilitated the successful implementation of HSR.

As the regulatory agency responsible for railway operations, the MoR pioneered early HSR development by establishing the "BJ-SH HSR Development Program" in 1990. In April 2001, the MoR issued an "HSR Train Design Task Book," which called for an annual output of 15 trains with operating speeds higher than 200 kmph. As a result of this "Task Book," the "China Star" HSR vehicles were implemented along the Qinhuangdao–Shenyang Passenger Line in 2003.

Another key milestone in China's HSR development journey, "The Medium to Long Term Railway Development Plan," was released in 2004. The first version of the plan anticipated the construction of 12,000 km of passenger-accessible HSR lines by 2020. A 2008 revision of the plan raised this projection to a nationwide total of 16,000 km. Judging by the pace of construction and operation thus far, the total kilometers of operating HSR track in China will easily surpass this projection before 2020. HSR development has also been incorporated into several "Five Year Plans," national development plans, and budgeting blueprints. This priority treatment allows China's political machine to allocate substantial attention and resources to HSR projects.

Due to a high operating speed of HSR vehicles, HSR tracks must be planned to have a minimum number of curves. For example, the minimum turning radius along the Wuhan–Guangzhou HSR line is seven kilometers, a straight line to the naked eye, which facilitates the maintenance of high speeds. Ensuring right of way (ROW) is a thorny issue because local residents often resist railroad construction and expenses can be prohibitive.

Political will from various levels of government was essential to acquiring ROW for HSR development in China. Since the state owns all land in China, the central government only delegates its usage rights to citizens, government agencies, and independent enterprises. There is no lengthy "due process" or "eminent domain" when the central government wants to acquire or take back land. Provincial governments are very enthusiastic about HSR development in their provinces, since it is viewed as the most powerful and enduring economic stimulus for the local economy. It is also largely paid for by the central government. This is especially true for provinces such as Fujian, which benefitted little from prior railway development. Interviews with railway staff members revealed that some provincial representatives literally lived in the Beijing MoR compound in order to lobby for railroad development in their respective provinces in the decision-making period.

Local and municipal governments are also major supporters of HSR development, particularly when stations are planned in their territories. Many HSR stations are located relatively far from existing central cities in order to reduce land and relocation costs, minimize resistance from existing residents, and open new sites for urban land development, something local municipal governments typically crave. The state's sole ownership of land avoids years of litigation and market appraisal of land intended for HSR ROW, which usually contributes to long delays in the US or Europe. The ability to acquire land fit for HSR construction is arguably the single most important factor in the quick deployment of the technology.

Positive Returns on HSR Investment

One of the key concerns about HSR development is the return on investment, mainly due to the general unprofitability of public utilities such as public transportation. Some researchers (Liu 2011; Wen 2010) are concerned about the large amount of debt and high debt/asset ratio accumulated by the

Table 6.6. Land Use Planning along Beijing–Shanghai HSR

City	Developing Areas (sq. km)	Planned Investment (in millions of dollars)
Tianjin	10	3,200
Cangzhou	28	560
Dezhou	56	608
Jinan	26	15,200
Taian	4	480
Qufu	6	N/A
Tengzhou	49	N/A
Zaozhuang	26	N/A
Xuzhou	26	512
Suzhou	50	780
Bengbu	9	424
Tuzhou	45	1,280
Nanjing	46	6,400
Changzhou	2	3,200
Wuxi	125	2,936
Suzhou	29	416
Total	536	35,580

Source: Lv 2013.

MoR. The situation is, again, rather unique in China. A systematic analysis of HSR lifecycle costs and travel demand/supply balances provided assurances about the potential long-term economic and social health of HSR development in China. In fact, two of the authors of this chapter predicted elsewhere that the BJ-TJ HSR line had the potential to break even—or even produce positive returns—some 36–48 months before its realization in 2014 (Liu and Lv 2012).

Furthermore, HSR development achieved not only investment returns for some early passenger lines but also brought immediate economic benefits for the areas in close proximity to the railway stations. As shown in Table 6.6, HSR stations along the BJ-SH corridor have attracted more than $35 billion in investment since the initiation of the HSR line.

Conversations with MoR staff revealed that previously skeptical local governments and residents now largely support HSR development or even demand it in their area, after witnessing the positive economic impact that the technology can bring. Similar to the value added to real estate developments around subway or light rail transit (LRT) stations, properties located in proximity to HSR stations also see increased values due to the increased flow of goods and people through the area.

The creation of more than 13,000 km of HSR dedicated to passenger transport has freed up a large portion of conventional rail for freight movement, a utility that had previously shared tracks with passenger trains. The separation of freight and passenger movement in turn has increased the speed and efficiency of freight movement, which is vital to China's economic development and stimulation of the market economy.

The Characteristics of Technology Transfer

To the general public, the most recognizable components of a HSR system may be the slick bullet trains and straight, endless rails. To a trained transportation engineer, the key to a successful HSR service is not a single piece of technology, but rather the seamless integration of civil and track works, HSR trains propelled by high tractive power, and modern train control and communication systems that are inspected and maintained continuously. Unlike most other technological transfer projects, in which whole, "turnkey" operations are the norm, HSR development in China was sourced from different suppliers and leaves significant integrative work to domestic enterprises.

For imported components, there are also aggressive localization requirements and schedules for the purpose of mastering the technology thoroughly and quickly. The fact that China seems to have largely accomplished its goals and become a leader and international provider of HSR technology suggests a highly successful transfer process. The rapid gain of internal capacity contrasted sharply with the experience of the auto sector (discussed in Chapter 3), which also used joint venture as a primary technological transfer vehicle. The distinctions of the HSR story are discussed in the following sections.

Long Tradition of Rail Development

First, China did not start from scratch with railroad development. The first railroad in China, the Tang–Xu Railway, was established in 1881 (Crush 2005). Another long stretch of early railway development, the Jinghan Railway, was surveyed in 1937 by Mr. Parsons. Parsons was the founder of Parsons Brinkerhoff Inc., the largest transportation engineering company in North America until it was bought by a British company a few years ago.

As shown in Figure 6.5, overall railway development in China has experienced several distinct stages since the PRC was established in 1949. Total track miles expanded slowly during the early years of the People's Republic, while operating speed stagnated and ridership crept upwards. Compared to the overall development of rail services around the world, China steadily built and expanded passenger and freight rail services during this period, in stark contrast to the shrinking of US railroad networks since the 1950s.

While often slow, development of railway systems and technology in China, especially in terms of civil and track work, has been continuous. For example, ballastless tracks, railway surveys aided by remote sensing and artificial intelligence, and switching and dispatching technologies have kept pace with international developments. Supported by a unique Railway Corp—a military branch that specialized in railway construction for more than three

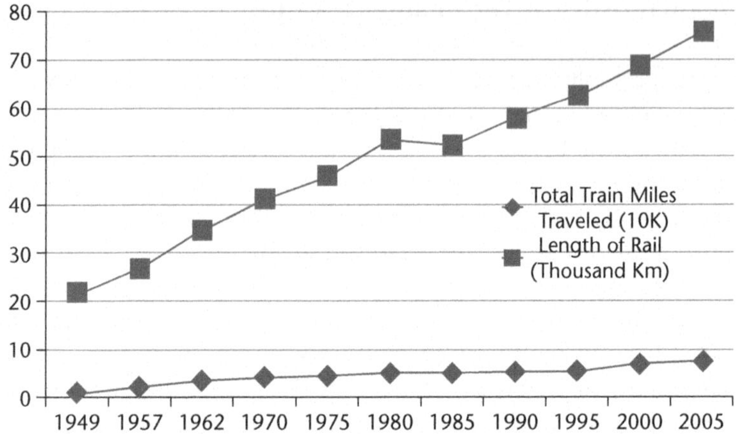

Figure 6.5. Historical Railway Track Kilometers in China
Source: United Nations 2006.

decades until it was absorbed into the MoR in 1984—the Chinese railway workforce developed a military structure of project execution, which was the foundation for the rapid implementation of the large number of HSR lines. Compared to the rapid growth trajectories of other high-tech sectors such as IT or automobile, international development of HSR was relatively gradual during the last 30 years, which also allowed China to catch up.

As discussed in Chapter 2 of this book, China has engaged in technological transfer for a long time in many sectors. Successful assimilation distinguishes China's HSR development program from most sectors with recent experience in imported technology. China has achieved not only a high localization rate but also improved existing technology and become an international HSR provider. All of this happened in approximately a decade, a comparatively short time. Some scholars argue that the success of the HSR technology transfer program proves the superiority of state-led R&D over market-oriented models. But before we accept such a sweeping conclusion, it is important to understand the unique features of HSR systems and the case-specific technological transfer arrangement. Some of the conditions witnessed in the transfer of HSR technology may not be applicable to other sectors. The following factors deserve consideration.

Singular Massive HSR Market with Central Authority

Sheer scale and speed of deployment are the most remarkable features of the Chinese HSR network thus far. But there is more to come. The plan to develop an HSR network consisting of four horizontal (west to east) and four vertical (north to south) lines will yield the eighth Wonder of the World. The HSR network will eventually connect almost all major metropolitan areas across Chinese provinces except Tibet. In 2020, the high-speed rail network is projected to connect all 30 provincial capitals and all cities with more than five million people to one another. As a result, more than 90 percent of the Chinese population will have access to the HSR network (He 2010).

The scale of developments has different implications for different players. For foreign suppliers, it promises a long-term and bottomless contract for construction and maintenance at a scale unimaginable elsewhere. There are only a handful of manufacturers from four countries that possess HSR technology, but some of them do not even have any applications in their own country, such as Bombardier in Canada. This is very unlike the automobile industry, which many countries have developed and for which virtually all countries have a market. Few countries in the world, including the United States, have or plan to have an HSR system. The Chinese market is practically singular and irresistible to HSR technology providers. However, for Chinese policymakers, the large scale means that China has an imperative to master and develop as much HSR technology as possible at home. Otherwise, the financial and security risks might be unacceptable. The National Development and Reform Commission (NDRC) and MoR determined that the technology transfer system must also foster indigenous technological innovation (Development and Reform Commission and Ministry of Railways 2004). They reasoned that China has a better chance of producing viable and competitive HSR technology than automobile or telecommunication technology, given the country's long history of rail development, its extensive existing railway infrastructure, and the privileged position of the rail sector in the Chinese economy.

Regarding the question whether centralized power is beneficial to technological development: the conclusion is likely to be case specific. In the case of HSR, China's centralized authority seems to have played a pivotal and beneficial role, although concentration of power also led to corruption. The arrests of Liu Zhijun and a few of his associates in 2011 for corruption make this amply clear (CCTV 2013).

The MoR is a unique ministry, even by China's standards. It is a semi-military organization that unilaterally developed national policy on HSR and selected contractors to design, build, and operate all the HSR lines in China. Once the decision was made to acquire HSR technology from international developers, the MoR acted as the single broker and entertained all offers from potential technology providers.

China's market potential enticed all the main HSR technology developers—Alstom, Bombardier, Kawasaki, and Siemens—to work with the MoR. This allowed the Ministry to negotiate contracts favorable to the Chinese, especially in terms of technological transfers to Chinese partner companies, Given that there is no market that comes anywhere close to China's scale of HSR implementation, it is an opportunity that no major suppliers could afford to miss. Siemens learned a bitter lesson from the negotiations: it was dropped during the first round of train set orders because it tried to resist the technological transfer requirements. Obviously, the MoR held enormous bargaining power in negotiating HSR contracts.

On June 17, 2004, the MoR announced the first round of high-speed EMU tenders, or bids, which clearly specified four general principles:

- Tender entities must come from a domestic enterprise that will be supported by international partners.
- Key technologies must be fully transferable.
- The price must be the lowest in the world.
- The model must be a Chinese brand.

As a result, all four HSR developers (Siemens of Germany, Alstom of France, Japan's Kawasaki Heavy Industries, and Bombardier of Canada) teamed up with a Chinese enterprise partner. Each won a bid to build 60 or 140 trains using their own HSR technology. Through this bidding, China obtained not only various HSR trains at the lowest possible price but also direct exposure to all HSR technologies around the world.

Continuous Research and Development Efforts

Joint ventures are a typical institutional arrangement for technological transfer agreements in China. They have precedents in other industries, most notably in the automobile sector. But the Chinese partners in other industries eventually lose out because they lack the motivation and competence to strategically control the changing technology or corporate operation, and fast-moving frontiers quickly leave them behind. In the case of HSR technology transfer, MoR was put into a strategically controlling position as a system integrator. China's domestic parties were also better positioned to absorb foreign expertise because of prior indigenous technological accumulation and considerable state investment into the research necessary to master the new technology.

A common misconception amongst foreign observers is that Chinese HSR trains accelerated from 60 to 300 km/h overnight. Rather, the six speed increases applied to all Chinese rail operations represented an incremental approach to Chinese rail industry transformation and user preparation. These early transition efforts did not grab headlines, since they were carried out without much fanfare, but they helped the system integration effort of the railway sector.

Chinese HSR train technology development was in progress for many years before it was prioritized as a national goal during the last few years of the twentieth century. Many ongoing experiments and evaluations were supported and carried out by the giant MoR and Ministry of Science and Technology (MOST), another central government agency that studies technology.

During the late 1980s and early 1990s, the MoR initiated preparatory HSR key technology studies and scientific research, which explored high-power AC traction converters, high-power AC induction motors, high-speed bogies, straight-through braking systems, and microcomputer control systems (China Institute of Railway Research and Technology 2009). Around the new millennium, two projects—one on the experimental engineering of a 200 km/h electric passenger train set, and another on force-scattered AC drive EMU—were incorporated into the 9th Five Year Plan, the official plan that guides the direction of the Chinese government. All of those efforts paved the way for the rapid deployment of HSR in China.

MoR and MOST established a joint program during the same period to pursue HSR research that provided funding to top national research institutes, experimentation, and demonstrations. The joint program focused on 10 individual topics, involved 25 leading universities, employed more than 10,000 researchers, and invested more than $1.58 billion during its tenure, the largest sum in MOST's history.

Several domestic HSR train models have been developed as a result of such research and experimentation. As mentioned above, the "Blue Arrow" was introduced in 2000 and tested at operating speeds up to 270 km/h. Eight "Blue Arrow" HSR trains were tested along the Guangzhou–Shenzhen corridor, and four more sets were ordered for the Shenyang–Dalian Line before the decision to switch to foreign HSR technology.

In 2001, a consortium made of four top research institutes, two national universities, and four manufacturers received $130 million to support the development of the "China Star" HSR train model. By 2005, the "China Star" HSR model was put in operation along the Qinhuangdao–Shenyang Passenger Dedicated Line. It accumulated more than a quarter million passenger kilometers before foreign models replaced it.

Once the MoR made the decision to utilize imported HSR technologies for the newly created HSR passenger lines in 2004, all the domestic HSR train projects were put aside. In the meantime, the MoR designated three Chinese train manufacturers as business partners for the four foreign technology suppliers. As recipients of transferred HSR technology, these manufacturers were instructed to host, adapt, and improve the foreign technology once they received a batch of HSR trains from their respective partners. They were also instructed to produce more train sets to supply rolling stock for corresponding line operations. For example, CSR received two trains from Siemens in 2008 for the Beijing–Tianjin Intercity Passenger Line; 78 more train sets were then produced on site in China in the following years.

Liu (2011) conducted a detailed study about the roles of basic research in HSR technological absorption and improvement in China. Even though the technology was imported, considerable adjustments were necessary to make it work in China. First, China has different geography and geology from the source countries. China's diverse landscape and changing climate zones presented problems for HSR tracks. For example, northeast China has permafrost and cold temperatures. Ensuring track stability in such a climate is not something Japan or Germany need to deal with. China also resorted to elevated tracks for most of the route in order to cope with the region's complicated geology, ranging from high elevation plateaus, to river deltas, to Karst landscape. Secondly, China operates the world longest HSR route, from Guangzhou to Beijing. The length requires great understanding of fatigue damage to tracks and cars that operate at such high speeds for long periods of time. Thirdly, in European countries and Japan, it is not necessary to have trains capable of speeds higher than 200–300km/h due to the size of the territory. China's HSR network covers a far larger territory and therefore needs to have higher operating speeds.

Second-generation Chinese HSR train models were designed to run at 350 and 380km/h, but were later reduced to 300 km/h operating speeds even though test runs reached 500km/h. Increasing speed presents problems with coupling between wheel and rail, their combined vibration, train body design, ground effects, and many other repercussions. Even if such high-speed trains never go into service, research and testing will help Chinese engineers understand various technical issues embedded in the imported technology.

For these adjustments and system integration to work, Chinese engineers and researchers needed to have not only knowhow, but also know-why—specifically, the detailed rationale underneath the design of imported technology. Liu et al.'s research tabulated various research projects funded by the National Science Foundation that targeted HSR adjustments. Most projects had to do with technology adjustment, training, and structures integration (bridges, dynamics, tracks, roadbed, and vibration). It is clear that considerable work and innovation went into China's effort to implement an HSR system.

By 2010, a new generation HSR train, the CRH 380, achieved an operating speed of 380 km/h. The claimed 87 percent localization rate and application for wholly owned intellectual property rights was evaluated and certified according to US Patent and Trademark Office (USPTO) guidelines (Xin 2011).

In short, with a favorable technological transfer agreement and indigenous technological accumulation, Chinese partners were able to absorb the imported technology and hasten the pace of localization. The conditions for successful absorption are unique and likely to be inapplicable in sectors where a virtual monopoly of the market does not exist. For example, in the automobile and renewable energy sectors, governmental and corporate stakeholders are much more diffused, and international technological changes happen much more swiftly. The bargaining positions of Chinese governments and corporations are far less formidable, and thus technological transfer agreements are unlikely to be as generous. But the example of HSR does demonstrate to other sectors the critical role of internal capacity building as an integral part of successful technological transfer.

China's rapid deployment of HSR was not without its glitches, the most serious of which was the fatal crash on July 23, 2011 (Osnos 2012). Forty people died and nearly 200 were injured when one HSR train hit another stopped HSR train on the same track just outside Wenzhou City in Zhejiang province. It is known that one of the key causes was a faulty signal system. The exhausted and inexperienced dispatchers were also to blame. It is unclear whether the accident had anything to do with the technological transfer agreement; there have been speculations that Chinese companies did not fully understand imported signal designs, but this is not confirmed (Areddy and Shirouzu 2011). The government report blamed Liu Zhijun and chief engineer Zhang Shuguang, who, also facing corruption charges, for the accident, as they had undercut some safety measures in favor of higher speeds. After Sheng Guangzu was appointed the new minister of the MoR, the overall speed of China's HSR was reduced for safety and energy conservation reasons (Anderlini 2011). The construction of new HSR lines also slowed down after the crash.

Despite all the setbacks, it is indisputable that China has managed to implement the world's largest, longest, and fastest HSR system over the course of a decade. As the Chinese HSR network continues to operate and expand, Chinese engineers have the opportunity to make further adjustments and incremental improvements. It is likely that China's HSR application capabilities will eventually lead the world, even if it continues to rely on the purchase of certain key components for some time.

Summary

By combining "commercial off-the-shelf" HSR technologies from around the world with the operating and marketing expertise of various international consortia, China acquired, adopted, and inaugurated 13,000 km of HSR track within a five-year span. The Beijing–Tianjin HSR line was inaugurated in 2008 to celebrate the Beijing Olympics, and the commissioning of the Nanjing–Hangzhou Passenger Dedicated Line occurred in 2013. The greatest achievement for the transportation community is its provision of HSR services to more than a billion people scattered throughout China's vast territory.

The Chinese process of technology transfer is not exactly indigenous innovation, but improvement and innovation based on purchased technology. It was probably the fastest and cheapest way for China to build an HSR infrastructure and provide services that could keep pace with its economic, social, and political evolution.

Nevertheless, this process was not without its risks and costs. Due to the differing goals and strategies of involved parties, international technology transfer is complicated and influenced by many domestic and contingent factors (Chan and Idhaban 2009). The technology transfer process documented here was a high-risk endeavor for all stakeholders. For technology developers, the sale of technology had the potential to render the developer irrelevant before sufficient returns on initial investment had been made. On the other hand, there was no guarantee for the technology receiver that the technology could be localized quickly and completely. The tug-of-war between technology developers and receivers can be long-lasting and tricky (Dorf and Worthington 1990; Eldred and McGrath 1997).

The transfer of HSR technology to China is generally considered successful. The sophistication of HSR technology dictated that indigenous innovation would not be able to catch up overnight; therefore, it was less expensive for the MoR to acquire HSR technology than to nurture innovation and wait for technology to mature. Similarly, intricate technology adjustment and process improvement ensures long-lasting partnerships between foreign technology developers and their Chinese counterparts.

If technology exporters were initially motivated by the high profits from selling their HSR technologies to China, they may now be concerned about the loss of the competitive advantage in the long run once China assimilates and absorbs the technology and is able to improve upon it (Simon 1997). There is no doubt that some technology exporters have tried to address the concern by retaining core technologies and assets. On the other hand, some have decided to embrace the market and capture the potential to be derived from production by becoming the manufacturers of train sets and other elements of transportation infrastructure.

The key objective of the technology importers is to assimilate the imported technologies and make improvements for real world applications. The MoR, the central controlling authority over China's railways, was able to negotiate favorable deals on technical transfers from a number of developed countries largely due to the unique scale of projected HSR development. No other country could have supported such a large-scale infrastructure project, and limited suppliers of technology would not have similar or even close market potential in countries other than China. While some of the lessons on technological transfer from the development of Chinese HSR may not be applicable to other sectors, it is certain that continuous research and the technology competency of Chinese manufacturers will be the necessary condition for any successful adaptation and assimilation of technology.

References

Anderlini, J. (2011), "China Acts on High-Speed Rail Safety Fears," *Financial Times*, Apr. 14, <http://www.ft.com/intl/cms/s/0/4cd5723e-6685-11e0-ac4d-00144feab49a.html#axzz2UY2mWMhc>.

AMT Inc. (2013), "Zero Emission Transportation," <http://american-maglev.com/maglev> [accessed Apr. 2013].

Areddy, J., and N. Shirouzu (2011), "China Bullet Trains Trip on Technology," *Wall Street Journal*, Oct. 3.

BBC News (1999), "Sci/Tech: the Magnetic Attraction of Trains," Nov. 9, <http://news.bbc.co.uk/2/hi/science/nature/488394.stm>.

BST-Transportation (2013), "Product," <http://www.bst-transportation.com/products_list/&pmcId=153a5296-f66d-4a47-a55a-2da02f931f0b.html>.

CCTV (2013), "Liu Zhijun in Court," June 3.

Chan, L., and F. Aldhaban (2009), "Technology Transfer to China: With Case Studies in the High-Speed Rail Industry," Portland International Conference Management of Engineering & Technology (PICMET).

China Association Highway and Waterway Engineering Consultants (2005), "Expressway in China and the Developed Countries are the Gaps," <http://www.cahwec.com/article.php/599>, Mar. 2.

China Institute of Railway Research and Technology (2009), "Overview of International HSR Development," Presentation, Beijing.

China's State Council (2013), *Notice of the State Council on the Institutional Settings*. Beijing: National Development and Reform Commission, no. 4.

Crush (2005), "The First Self-Built Railway in China," <http://lishi.huisongshu.com/HisMain/5611> accessed Mar. 2013.

Development and Reform Commission and Ministry of Railways (2004), "*Implementation Plan about the High-Power AC Drive Electric Locomotive Technology Introduction and Localization*, July 29. Beijing.

Dorf, R.C., and K. K. F. Worthington (1990), "Technology Transfer from Universities and Research Laboratories," *Technological Forecasting and Social Change*, 37, 251–66.

Eldred, E., and M. McGrath (1997), "Commercializing New Technology-II," *Research Technology Management* (Mar.–Apr.), 40(2): 29–33.

Feng, S. (2004), "Talking about the Current Sino-Japanese Relations," *People's Daily*, Dec. 17.

He, H. (2010), Press Conference of the Ministry of Railways, July 28.

Leber, J. (2013), "Waiting for the Hyperloop? Here is Where you can Actually Travel the Fastest Right Now," Co.Exist, Fast Company, Aug. 13.

Li, S. D. (2006), "Study Course and Main Construction Survey of Beijing–Shanghai High Speed Railway," *China Railway Standard Design*, Zl (Aug.), 1–5.

Liu, R. (2010), session organizer and presenter. "Transportation Revitalization Plan: Chinese Stimulus Package in Responding to the Economic Crisis (p10-0140)," Transportation Research Board (TRB), 89th Annual Meeting, Jan. 11, Washington, DC.

Liu, R., and L. Lv (2012), "Investment Returns for High Speed Rail in China: A Life Cycle Cost Analysis," Proceedings of Transportation Research Board (TRB) 91st Annual Meeting, Jan. 2012, Washington, DC.

Liu, R., and Y. Deng (2004), "Comparing the Operating Characteristics of High-Speed Rail and Maglev Systems: A Case Study of Beijing–Shanghai Corridor," *Transportation Research Record*, 1863 (Journal of Transportation Research Board), 19–27.

Liu, Z. (2011), "High-Speed Rail, Little Hope to Recover the Investment," <http://blog.ifeng.com/article/12513046.html>, July.

Lv, L. (2013), "Social Economic Impact of HSR for Beijing Shanghai Corridor," TRB 2013 Annual Meeting, Washington DC, Jan.

Maglev UK (2013), "Birmingham Maglev," <https://sites.google.com/site/maglevunited/birmingham-maglev> accessed Apr. 2013.

Ministry of Railways (2012), *2011 Annual Report of the Ministry of Railways*. Beijing: Ministry of Railways.

National Bureau of Statistics (2011), *The People's Republic of China 2000 National Economic and Social Development Statistics Bulletin*. Feb. 28.

Olivier, G., et al. (2014), "High-Speed Railways in China: A Look at Traffic," *China Transport Topic*, 11 (Dec.).

Osnos, E. (2012), "Boss Rail," *The New Yorker*, Oct. 22.

Parsons Brinckerhoff, Inc. (2011), "High Speed Rail," Company Internal Report No. 73, Sept.

Perl, A. (2002), *New Departures: Rethinking Rail Passenger Policy in the Twenty-First Century*. Lexington, KY: University Press of Kentucky.

Shanghai Maglev Transportation Development. (2013), "Important Events," <http://baike.baidu.com/link?url=xw0FqsvxSdE0UMUeJG4Hf4JS_Gb624Pu1C93JAhH4RF31nXP1_SCuc-vl836g0NxRTBY108pKvX1wdusCx6WJK#2> accessed Mar. 2013.

Sheng, G. (2013), "Pioneering a New Situation of Railway Scientific Development, and Making New Contributions to a Moderately Prosperous Society," National Railway Work Conference, Jan. 17, Beijing.

Simon, D. (1997), *Techno-Security in an Age of Globalization: Perspectives from the Pacific Rim*. Armonk, NY: M. E. Sharpe.

Transportation Security Administration (2003), *Conducting Global Market Survey of Commercially Available Explosive/Trace Detection Systems*. Washington, DC: US Department of Transportation.

United Nations (2006), *Monthly Bulletin of Statistics*.

Vuchic, V. R., and J. M. Casello (2002), "An Evaluation of Maglev Technology and its Comparison with High Speed Rail," *Transportation Quarterly*, 56(2): 33–49.

Wen, S. (2010), "Great Leap Forward Economic Calculations, the Beijing–Tianjin High-Speed Rail Operating Loss of 700 Million a Year," *Economic Observer*, Apr. 5.

Xin, D. (2011), "High-Speed Technology Eyes US Patents," *China Daily*, June 23.

Xinhua Net (2002), "Shanghai Maglev Demonstration Line Construction has Created a Miracle," Dec. 31.

Zhou, D. (2007), "China's Railway Experienced Five Large Area Speed up," Xinhuanet, Apr. 13.

Zhou, X. (2012), "China Echoes 2009 Stimulus with Railway Spending Boost," *Bloomberg News*, July 17.

7

State, Market, and Business Enterprise

Development of the Chinese Integrated Circuit Foundries

Yin Li

Introduction

In 2001, the influential semiconductor technology executive Morris Chang, a long-time skeptic of China, declared, "The future of the world's chip industry lies not in Taiwan or the United States, but in China" (Landler 2001). This comment served as a remarkable indicator that China is becoming a significant force in the global semiconductor industry in the twenty-first century. Many analysts have believed that China's ascension in this industry would be inevitable due to the enormous Chinese market and strong governmental support for local producers (see e.g. PricewaterhouseCoopers 2004).

Indeed, on the market side, China became the world's largest semiconductor market in 2005. The Chinese market is estimated to have made up half of world consumption in 2012[1] (PricewaterhouseCoopers 2013). As the market expanded, the gap between semiconductor production and consumption in China widened; in 1999, it was already at $5.7 billion (PricewaterhouseCoopers 2004). Industrial analysts predicted that this huge production–consumption gap could provide ample opportunities for local producers, and fuel a successful semiconductor production system.

[1] PricewaterhouseCoopers (2013) estimates that about 20% to 40% of semiconductors consumed in China are "dislocated purchasing," i.e. those semiconductors consumed in China are purchased outside of China. Yet after excluding "dislocated purchasing," China is still the largest single market, larger than the second largest market North America with less than 20% of world market.

On the policy side, the objective of building an advanced semiconductor industry has been a focus of the government since the 1970s. The history of the Chinese semiconductor industry reflects the changing ways in which the Chinese government has promoted industrial development. In a Soviet-style system between the 1950s and 1970s, China emphasized self-reliant development. After Economic Reform in the late 1970s, they began to favor technology imports. In the 1990s, hoping to gain leverage from the massive domestic market, they promoted international joint ventures between multinational corporations and China's leading state-owned enterprises. Since 2000, they have handed down policy initiatives to local governments in the hope of incentivizing at-home development. The shifting policy regimes trace policymakers' search for ways to promote industrial development, with the consistent goal of empowering Chinese semiconductor companies to become leading global innovators.

A decade after Chang's grand statement, the Chinese chip industry has experienced both remarkable progress and severe setbacks. Today, China is among a handful of nations with the ability to manufacture advanced integrated circuits (ICs) with millions of tiny transistors as small as nanometer scales. Chinese-made semiconductors, including processors, memory chips, and different kinds of sensors, are powering personal computers, tablets, mobile phones and a wide array of high-tech electronic devices that are exported around the globe. Such capability barely existed in the 1990s. But China has strikingly not become the center of world chip production, yet. In 2005, China reached one-tenth of world semiconductor production, but not much further progress has been made since then. Instead, the consumption–production gap has increased to more than $100 billion since 2012 (PricewaterhouseCoopers 2013). Among the top 30 semiconductor producers supplying the Chinese market, none of them are local Chinese companies. In major segments of the semiconductor industry, including IC design and IC chip fabrication, China's leading semiconductor corporations are still catching up with—rather than challenging—the dominant multinational corporations (MNCs).

The semiconductor industry is one of the most dynamic, competitive, and globalized industrial segments of the economy. The experience of the Chinese semiconductor industry raises important questions about the power and the limitations of the state and market in enabling industrial growth. To explain the progress and setbacks of the Chinese semiconductor industry, I contend that an analysis that includes only the state and market forces is insufficient. It is necessary to understand how these industrial corporations innovate. More specifically, we must examine the innovation process that enables firms to generate higher quality and lower unit cost semiconductors (Lazonick 2004b, 2009, 2010). In this chapter, I analyze the process by which major Chinese semiconductor corporations develop and utilize their innovative capabilities, with a focus on the fabrication of IC chips. Since 2000, this industrial activity has evolved into a separate IC foundry sector in China. Yifei Sun's chapter in this volume addresses the spectacular growth of the IC design sector, where technology, market, and competition are quite distinct from the IC foundries. My analysis uses a firm-centered perspective with a historical lens in the hope of revealing the dynamics of industry growth and technology development. Such a lens deals not only with the forces of market and state, but also with how specific companies developed capabilities over a long period of time.

In carrying out the analysis, I utilized the "social conditions of innovative enterprise" framework developed by Lazonick and colleagues (Lazonick 2004a, 2004b, 2009, 2010; Lazonick and O'Sullivan 2000; O'Sullivan 2000; Lu 2000). This framework examines the social and historical process in which innovative enterprises use strategy, organization, and finance to overcome the challenges of innovation. Specifically, it asks three key questions about the social process in a given business enterprise: *Who* has the power to allocate resources and make investments in the company? *What* types of investments are made? *How* are these investments financed and sustained? Answers to those questions prescribe whether a business enterprise can or cannot respond to challenges in developing and utilizing innovative capabilities, which determines its ability to compete and grow in the long run. In this chapter, I ask the same "Who, What, and How" questions about the process of developing major Chinese semiconductor companies and analyze their achievements in terms of innovation and competition.

The following sections analyze the history of the Chinese semiconductor industry and models to develop Chinese IC foundries. The first section follows the development of the state-owned semiconductor industry, including the various national projects to build advanced IC companies before 2000. The following section illustrates the transition to a globalized industry led by non-state companies. Then there is an analysis of the relationship between the industry and the state, particularly local government, under the new conditions of non-state ownership and globalization. The final section summarizes lessons from the history of the Chinese semiconductor industry and engages in a discussion about the role of the state, market, and business enterprise in industrial innovation and development.

The State Industry

China's development of semiconductor technology can be traced back to as early as 1956, when the country's first semiconductor was made in a state lab (Dewey-Ballantine 2013). In 1964 the Chinese Academy of Sciences (CAS) developed the country's first integrated circuit (IC) (Simon and Rehn 1987: 261). The semiconductor technology developed in the state labs started at small-scale integration (SSI) in 1965, to medium-scale integration (MSI), and further to large-scale integration (LSI) in 1972. From SSI, MSI, to LSI, the number of transistors integrated on one IC increased from less than 100 to more than 1,000. CAS had developed 4K random access memory (RAM) in 1979 and 16K RAM in 1980, enabling these indigenously made ICs to power mainframe computers developed in Chinese universities and state labs.

Being isolated from both the West and the Soviet Bloc, China followed a strategy of self-reliance in developing semiconductor technology for military use during this period (1956–80). Under a Soviet-style regime of central planning, the semiconductor industry was organized into two parts. The research and development (R&D) activities were conducted in state labs, mainly the Chinese Academy of Sciences Institute of Semiconductors and regional semiconductor labs. The manufacture of chips was undertaken by two major state-owned factories, one in Beijing (No. 878 Factory) and one in Shanghai (No. 19 Shanghai Wireless Electronics Factory). R&D and manufacturing were originally separated for the purpose of efficient planning. This separation, however, became a severe barrier for technology development. Transferring technology developed in state labs to industrial production was difficult, if not impossible, in this planned system. When an American scholar Denis Fred Simon toured the semiconductor factory in Shanghai in 1985, the factory was producing semiconductors based on technology that had been developed by state labs 10 to 15 years earlier (Simon and Rehn 1987: 269).

While the separation of R&D and manufacturing made it difficult to integrate technology from different parts of the industrial system, the state industry also suffered from inefficiency as the result of the managers' inability to motivate efforts and integrate skills on the shop floor.[2] In the 1980s, the aforementioned Shanghai semiconductor factory had yields as poor as 20 to 40 percent—this meant that 60 to 80 percent of the semiconductors produced were defective (Simon and Rehn 1987: 268). During the same period, the Japanese producers had achieved yields of 70 to 80 percent, demonstrating both higher quality and reliability.

By the time China embraced economic reform and opening up in the late 1970s, its semiconductor technology was severely trailing that of the international mainstream. For example, in 1985, CAS finally developed 64K RAM

[2] Feng Lu (1999) explained that in China's Soviet industry system, the party-state has a social contract with the working class, in which the state provides permanent employment and social provision through "work unit" in exchange for political legitimacy. In this arrangement, managers at state-owned factories were deprived of the ability to motivate workers through incentives.

in its lab, but the more advanced 256K RAM was already accessible in the international mass market. Being backward and incompatible with the mainstream standard, the indigenously developed technology was gradually abandoned.

In 1982, as part of the 6th Five-Year Plan (1981–5), the State Council created the "Computer and Large Scale IC Lead Group" chaired by Vice-Premier Wan Li[3] to promote the development of a modern semiconductor industry. The involvement of the state at such a high level signaled that the Chinese government considered ICs to be an industry of great strategic importance. The Lead Group adopted an import-substitution strategy, emphasizing production based on imported technology. Provincial governments were allocated quotas to import semiconductor production lines. By 1985, 33 state-owned factories imported 24 IC manufacturing lines. Due to the high cost of new lines, all these imported lines were second-hand lines with out-of-date technology. Even so, most factories involved in localization failed to realize the full capacity of the imported lines. Only one factory, Wuxi Jiangnan Wireless Device Factory (No. 742 Factory) managed to meet its production target by successfully launching the 3-inch wafer line[4] imported from Toshiba in 1984.

Around that time, a semiconductor market in China began to emerge. IC consumption in China was between 350 million and 400 million units annually in 1989, while domestic production totaled 114 million units (Simon 1992). Foreign electronics manufacturers showed interest in this emerging market, and the Chinese government allowed them to establish joint ventures (JV) with Chinese state-owned enterprises (SOEs). Joint ventures with Belgium's ITT (JV's name: Shanghai Belling), with Netherlands' Philips (JV: Shanghai Philips, later ASMC when Nortel was involved), Canada's Nortel (JV: ASMC), and with Japan's NEC (JV: Shougang-NEC) became the largest semiconductor producers in China. NEC even introduced China's first 6-inch wafer line to its JV in Beijing in 1991.

Besides using JVs to access foreign capital and technology, the Chinese state also mobilized resources to build its own indigenous companies. The 8th Five-Year Plan (1991–5) highlighted the microelectronics/semiconductor industry as a "Pillar Industry" to be promoted, meaning an industry that would support a number of related industries and promote long-term growth of the national economy. In 1990, the Ministry of Electronics Industry (MEI) selected No. 742

[3] In this chapter, Chinese names are written with family name first, followed by given name.

[4] Semiconductor fabrication plants (also known as fabs) are defined by the diameter of wafers, or layers of semiconductor material, that they are tooled to produce. A 3-inch wafer line or fab means a fabrication line that processes 3-inch wafers. The diameter of wafers has gradually increased over the years along with more advanced fabrication technology and heavier capital expenditure at each upgrade. A fab with a larger diameter of wafers almost definitely means more advanced technology.

Factory, the only successful factory in using imported technology in the 1980s, as China's semiconductor national champion. No. 742 Factory became the state-owned Huajing Group, an integrated device manufacturer (IDM) with the ability to design, fabricate, package, and test ICs. MEI allocated a budget of 2 billion RMB to upgrade Huajing with 6-inch fab and submicron (0.8–1.2 micron) process,[5] allowing the company to compete with international mainstream technology. Lucent Technologies was recruited in 1994 to transfer process technology, train engineers, and provide an intellectual property (IP) library for IC design to Huajing. The upgrade of Huajing was officially coded as State Project 908.

Huajing, however, failed to meet the state's target this time. The 6-inch fab did not enter production until 1997, by which time its technology had already fallen behind the global standard. The causes of the delays were multifold, but the most critical ones were a lack of autonomy to make investment decisions and a lack of sufficient financial resources as a state-owned company. According to Hu Qili's reflections (Hu 2006), when deciding to make any investment in production (e.g. to import a lithography machine), Huajing had to go through MEI's bureaucratic procedures, which typically took months, if not years. MEI then negotiated the investment budget with the Ministry of Finance for another lengthy period. When the investment requirements evolved beyond the initial plan, the negotiation had to begin over again.

Huajing's managers also failed to develop organizational capabilities that could utilize imported technologies and access markets. Huajing's engineers were well known for being good at reverse-engineering ICs that were established in the market (Fuller 2005: 253–4), and the company relied heavily on such self-taught skills. But when Huajing acquired process technology and the IP library from Lucent, Huajing engineers' routines of reverse-engineering were inadequate for incorporating these new technologies. Lucent's engineers reported that the transferred IP library was not used at all, and Huajing continued to manufacture ICs based on reverse-engineering (Fuller 2005: 254). As a result, Huajing produced merely 800 wafers per month in 1997 on its new 6-inch fab with the capacity of 12,000 wafers per month. Huajing recorded a loss of 240 million RMB in 1997 and was taken over by a Hong Kong-based startup Central Semiconductor Manufacturing Corporation (CSMC) in 1998.

By 1995, China's IC consumption-production gap had increased to 4.5 billion units, or $1.7 billion annually (Simon 1996: 9). In this period, South

[5] Besides wafer size, the other measurement of semiconductor fabrication technology is the line width of lithographic node. Smaller line width can enable more transistors to be made on one chip. For example, a 0.8um process means that the lithographic machine can draw semiconductor feature as small as 0.8 micrometer. The measurement of line width is usually in micrometer or nanometer.

Korea and Taiwan became significant players in the world electronics industry. Chinese leaders decided to emulate the Korean business groups, or *chaebol*, in their role as backbone enterprises of the national economy. In December 1995, MEI launched a national project coded 909 as the successor of Project 908. Project 909 targeted the creation of a multi-divisional semiconductor business group that integrated production and distribution. The group would be capable of producing ICs on state-of-the-art 8-inch wafer lines with 0.35- to 0.5-um node process, far beyond the prevailing 6-inch fab technologies in China. Rather than upgrading an existing SOE, Project 909 chose to build a brand-new organization from scratch: the Shanghai Huahong Group. Founded in April 1996, Huahong registered a capital of $604 million with a 60:40 split in shares between MEI and the Shanghai municipal government. Hu Qili, the Minister of Electronics Industry as well as a high-rank cadre in the Communist Party, became Huahong's chairman of the board to exercise direct supervision over the project.

At the time, Project 909 was China's largest state project in the semiconductor industry, involving capital investment above 10 billion RMB, more than the sum of all prior state investment in semiconductors (Hu 2006: 6). Funded directly from the Premier's Fund, Huahong gained unusual autonomy in allocating the project budget. As Hu admitted later, such autonomy allowed Huahong to pursue its own investment strategy. This was exemplified by the company's sustained investment in production facilities during the world market downturn (Hu 2006).

The centerpiece of Project 909 was, as intended, the construction of an 8-inch fab. It was undertaken by Huahong-NEC, a joint venture established in 1997 between Huahong and Japan's NEC.[6] Huahong contributed $500 million for a 71.4 percent stake in the JV while NEC put up $200 million for the remaining 28.6 percent. Both companies had two seats on the JV's board of directors. Even though Huahong had a controlling share, NEC was asked to manage the fab, purchase outputs, and train the workforce for the first five years. The fab's construction was based on the blueprint from NEC's Hiroshima plant, which provided Japanese engineers to Huahong-NEC to implement NEC's proprietary production process. With NEC's help, Huahong-NEC quickly finished building its fab within two years and entered pilot production in early 1999. Within three months of production ramp-up, the fab improved its yields from 50 percent to more than 90 percent, an efficiency level

[6] Huahong approached other multinational semiconductor companies including IBM, Intel, AMD, TI, Rockwell, Fijistu, Toshiba, Siemens, and Philips. But it was rejected partly due to the stringent requirements of entering into a JV (i.e. technology transfer and guaranteed purchase of at least 35% of output), and partly due to the tight US control over the export of advanced semiconductor devices to China (*Business China* 1997). NEC at that time already had a large stake in China after investing in a JV with Shougang in Beijing.

competitive on the world market. NEC distributed the output, mainly 64 Megabit DRAM (dynamic random access memory), under the NEC brand. As a result, Huahong-NEC made a profit of 350 million RMB in 2000 after its first full year of production, a record in the history of China's state industry projects.

However, reliance on NEC's expertise had its drawbacks. As NEC engineers and managers assumed the role of process management and skill development, Huahong's managers were deprived of the opportunity to develop management capabilities. According to Fuller (2005), an internal report by the Ministry of Science and Technology accused NEC of excluding the Chinese from core operations, though it is unclear whether NEC had done so intentionally. Others found that Huahong engineers lacked the knowledge of the whole production process beyond their specific tasks. In one case, Huahong managers could not even confirm to customers whether the fab had the capability to produce to certain specifications without consulting NEC engineers in Japan (Fuller 2005: 261–2).

With NEC taking strategic control over the fab, Huahong's ambition to build more integrated business operations fell apart. Huahong had sent engineers to the IMEC (Interuniversity Microelectronics Centre), the European semiconductor research center in Leuven, Belgium, to learn skills for the 0.18-micron process. But without a solid understanding of its existing production process, the new technology could not be transferred successfully. Without access to the fab's advanced manufacturing process, Huahong's two IC design houses in Beijing and Shanghai could not develop sophisticated application processor chips as planned. And Huahong's distribution arm, Shanghai Huahong International Electronics, played no role in distributing Huahong-NEC's products. Instead, it transformed into a commercial trading firm serving its own profit goals. In short, the multi-divisional structure of the Huahong Group ended up in organizational disintegration.

In 2002, Huahong-NEC was hit by a severe market downturn in DRAMs and recorded a loss of 700 million RMB in a single year (Hu 2006: 199). Under political pressure, Huahong-NEC terminated its management contract with NEC, making NEC a passive stockholder. But Huahong still lacked the capabilities to access markets on its own. Its design houses could barely ramp up the fabs to fulfill orders of smart card ICs for government procurement. By the end of 2002, Huahong-NEC decided to restructure itself completely by bringing new returnee management, taking on the US-based Jazz Semiconductor as a new foreign partner, and adopting the pure-play foundry business model (explained in the next section).

From the 1950s to the 2000s, the powerful Chinese state was the main driver behind the evolution of the nation's semiconductor industry. The state provided capital investment, designed organizational forms for the state-owned companies, and acquired technology. The technology was acquired through various channels, including indigenous R&D, purchases from abroad, and technology transfers from foreign companies. Yet the state still had to rely on individual firms, primarily SOEs, to transform these inputs into productive capabilities. The SOEs failed to do so. Among many causes, a critical one was the lack of autonomy and sufficient financial resources within SOEs to make developmental investments in organizational capabilities to utilize technologies provided by the state and transform them into successful products. The control over company strategies to invest and develop was segregated in the state bureaucracy in the planned economy, and later on, it was ceded to multinational partners in international joint ventures. Strategic control was rarely in the hands of capable people whose interests were aligned with the growth of companies. In other words, these enterprises lacked strategic control—a critical social condition for innovative enterprise (Lazonick 2009).

The lack of strategic control made investment in organizational capabilities difficult, if not impossible, and hence posed a barrier to a second social condition of innovative enterprise, organizational integration. Learning to transform technology and access markets is a complicated organizational process involving the integration of efforts and skills of people in different hierarchical and functional roles to achieve shared goals. The lack of organizational integration has resulted in the inability of Chinese state industry to mobilize the workforce for technology learning (e.g. Huajing) or to coordinate operations across business units (e.g. Huahong). Under Japanese management, Huahong-NEC has achieved organizational integration in its production process that resulted in high-yield chip fabrication. Unfortunately, such achievement on the shop floor could not be integrated to serve the strategic goal of the Huahong group, at least from the Chinese point of view. The combination of a lack of strategic control and organizational integration thus might explain how, even with powerful state involvement and strong market incentives, the state industry never evolved into independent and competitive business operations.

The Non-State Companies

China's accession to the World Trade Organization (WTO) in 2001 necessitated the liberalization of the semiconductor industry. This meant it would need to have zero or low trade barriers and free entry for foreign firms. In anticipation of this change, China began to adjust industry policy in 2000. On June 24, 2000, the State Council released a new industry policy under the name "Several policies for encouraging the development of software industry and integrated circuit industry," better known as Circular 18 (in Chinese, 鼓励软件产业和集成电路产业发展的若干政策). Shifting away from the previous strategy of building IC industrial companies by the state, Circular 18 provides preferential tax schemes and other incentive programs to encourage semiconductor-manufacturing activities in China.[7] In particular, Circular 18 extended benefits to all firms regardless of ownership, only requiring that operations be undertaken in China.

Circular 18 was followed by major investments in semiconductors from the non-state industry. In 2001, the largest two semiconductor companies in China, Semiconductor Manufacturing International Corporation (SMIC) and Grace Semiconductor, were established in Shanghai's Zhangjiang High-Tech Park. SMIC was founded by a team of Chinese repatriates[8] previously working in leading semiconductor companies in the United States, Taiwan, and Singapore. Richard Chang (or Chang Ru Gin), previously a senior Taiwanese manager at Texas Instruments with extensive experience in fab construction and operations, led the team (Iritani 2002). SMIC raised an initial capital of $1.48 billion from global venture capital firms, including Walden International (Silicon Valley), Vertex Venture Holdings (Silicon Valley), H&Q Asia Pacific (Taiwan), and Goldman Sachs (USA), as well as the Shanghai municipal government (SMIC Annual Report 2004). Grace was co-founded by Jiang Mianheng, the son of the president of China Jiang Zemin, and Winston Wang, the son of the Taiwanese industrial tycoon Wang Yongqin. Grace drew from Taiwan's deep industrial expertise in semiconductors, and received strong backing from the Chinese state banks, raising an initial capital of $1.6 billion.

With substantial capital in hand, SMIC and Grace invested in advanced 8-inch wafer fabs with the 0.25- to 0.18-micron node technology. SMIC's growth strategy was particularly aggressive. After establishing operations in Shanghai, SMIC quickly expanded to Tianjin by acquiring Motorola's fabs. By 2004, SMIC had already expanded to Beijing and constructed China's first 12-inch fab. In less than three years, SMIC controlled more than half of the foundry capacity in China.

The newborn non-state industry adopted a new business model, separate from the integrated device manufacturing (IDM) model that the state industry adopted in the 1990s. In order to participate in the global production networks,

[7] Circular 18 imposed a reduced 3% of value-added-tax (VAT) in effect for domestically produced semiconductors while maintaining 17% for overseas manufacturers, in addition to tax-free imports of equipment and raw materials as well as preferred status for investment. The VAT rebate was withdrawn later in 2005 after the US challenged China at the WTO for its discrimination against foreign manufacturers.

[8] This chapter distinguishes between repatriates and returnees. Returnees refer to mainland-born Chinese returning to China. Repatriates refer to a broader group of ethnic Chinese who might originate from all over the world and migrate to mainland China.

SMIC and Grace followed the "pure-play foundry" model, which meant they specialized in providing IC fabrication services to system companies and "fabless" design companies (i.e. semiconductor design companies without fabrication facilities). The foundry strategy allowed SMIC and Grace to access markets in the existing global production chain and acquire large numbers of orders from established multinational corporations in a relatively short period of time. By 2005, SMIC surpassed Singapore's Chartered to become the third largest pure-play foundry in the world, after Taiwan's TSMC and UMC, while Grace had made its way to the seventh.

With access to foundry providers at home and abroad, an industry ecosystem comprising a large number of semiconductor design houses began to emerge in the early 2000s in China. The number of semiconductor design companies in China increased from 76 in 1999 to 463 in 2003. In the single year of 2002, there were 189 new entrants. The majority of entrants were fabless design companies with less than 250 employees relying on foundries for IC fabrication, distinct from those design houses linked with system companies or IDMs dominating China in the 1990s (Chesbrough 2005).

The rapid growth of non-state IC foundries followed China's new industrial policy, but its deeper cause should be understood within the context of technological change, globalization, and China's economic institutions under reform. Since the 1980s, the rising capital requirements for IC wafer fabrication drove greater vertical specialization in the global semiconductor industry. Pure-play foundries, a business model pioneered by Taiwan Semiconductor Manufacturing Corporation (TSMC), gradually gained market shares over more integrated IDMs. Multinational semiconductor companies responded by outsourcing and offshoring to lower cost locations, especially Asia (Brown and Linden 2009).

China emerged as the next destination for semiconductor production sourcing when Circular 18 liberalized foreign entry and favored local production. In the early 2000s, major semiconductor multinational corporations, including Motorola, Intel, Texas Instruments, TSMC and UMC, established IC fabrication operations in China. The early growth of the Chinese IC foundries significantly benefitted from being part of the supply chain in access to technology and markets. SMIC, for example, acquired technology licensing from IBM, Fujistu, Infineon, Elpida, and Toshiba, and sourcing orders from Broadcom, Elite, and Marvell, which were three long-term customers of TSMC. The US semiconductor equipment suppliers even helped SMIC obtain a special import/export license from the US government in 2003, allowing it to import the most advanced fab tools.[9]

[9] Under the Wassenaar Arrangement, the US government requires semiconductor technology exported to China to be at least two generations older than the state-of-the-art in the United States.

The other benefit of being part of the global supply chain was the ability to raise funds from the international capital market. The IC foundry industry is a high-tech, high-risk business that requires a huge amount of capital over a long period of time. Nevertheless, as noted above, SMIC was able to receive early investment from foreign banks and VC firms, and the company launched an IPO within three years. Importantly, public listing on the American stock market was much more than simply raising funds; it was closely linked to SMIC's organizational strategy.

SMIC's human resource and organization strategy tapped into the flows of high skilled engineers and managers moving between Silicon Valley and emerging technology centers in Asia, which is often described as "transnational talent circulation" (Saxenian 2005). Among the 1,043 engineers that started SMIC, 393 were Chinese repatriates (SMIC Annual Report 2005). Many of them were seasoned engineers and managers hitting the glass ceiling of career advancement in established Western companies (Iritani 2002; Saxenian 2007). While SMIC was only paying wages at the prevailing Chinese rate (which for senior managers were 25 to 30 percent of US salaries in 2001), SMIC also offered the classic Silicon Valley bargain—"a chance to be on the ground floor of a pioneering venture, with stock options" (Iritani 2002). While it might be an established practice in Silicon Valley, SMIC pioneered employee stock option plans among large industrial corporations in China. Essentially, stock options offer employees the ability to purchase company shares at low prices under a contract (often requiring the options to be vested over an extended period of time). It allows employees to reap a profit in selling these shares when the company stock is publicly traded at higher prices. Thus stock options attract, motivate, and retain important employees, without compromising the scarce capital of the startup company. The effectiveness of stock option plans is tied to public trading of company stocks, which appears to be one reason why SMIC pushed for an early IPO.

In addition to the employee stock option plan, one other notable characteristic of SMIC's recruitment strategy was investing millions of dollars in its Shanghai campus (provided by the municipal government at low cost) to offer low-cost, high-quality housing to employees and education to their children. Backed by the company, the SMIC School became one of the best primary and secondary schools in East Asia using an American curriculum. As the

Officially, the US Department of Commerce and US Department of Defense stated that export licenses were granted or denied on a case-to-case basis, and thus there is no official "two generation gap" policy. But "the generation gap" view was held widely by the industry, policymakers, and academics (see Hu 2006).

result of these organizational strategies, SMIC successfully lured hundreds of managers and engineers from the rival foundries of TSMC, UMC, and Chartered. From TSMC alone, SMIC hired more than 140 production experts (Clendenin 2004).

By 2004, hiring repatriate executives became the new norm for management of semiconductor companies in China. Not only did repatriates operate the rising non-state semiconductor companies, but they were also hired as top executives in the state industry. Huahong-NEC, the largest foundry in the state industry, hired several returnees when restructuring from a NEC-captive fab to a pure-play foundry competing for sourcing orders. In another example, the smaller state-owned foundry Shanghai ASMC hired a former SMIC executive and returnee as the CEO in its restructuring. By the end of the decade, nearly all major semiconductor companies in the state industry employed repatriate executives.

Repatriate executives used their intimate knowledge of production and management skills gained from overseas experience to improve industry performance by implementing competitive strategy and developing effective organization. In one notable case, CSMC, formerly Huajing (i.e. Project 908), employed the same workers, engineers, and production facilities after they were taken over by the Hong Kong-based firm with Taiwanese managers. Even so, because of the capabilities and knowledge of the new Taiwanese and repatriate managers, CSMC was now able to fabricate semiconductors at competitive prices and access new markets. Subsequently, the company was able to make investments in newer fabs using accumulated surplus (Li 2011: 90–1). At the leading SMIC and Grace, the repatriate expertise was translated into high technology capabilities. The initial process qualification of the 0.18 um node technology took Grace and SMIC only 21 and 12 months, respectively. Rival foundry TSMC described this astonishing speed of SMIC as "an implausibly quick ramp-up of its production facilities and fabrication processes" (TSMC court statement, quoted in Clendenin 2004). Repatriate engineers played central roles in speeding up this process development (Saxenian 2007).

While non-state semiconductor companies grew out of globalization, they were still deeply embedded in Chinese economic institutions. After 2001, the WTO agreements largely stymied China's ability to protect national champions with trade barriers in their industrial policy (such as Projects 908 and 909). It was the Chinese local governments, which were institutionalized as the local developmental states throughout the economic reform, who became the new sponsors of the semiconductor industry. Since the beginning of 2001, the Beijing, Shanghai, and Shenzhen municipal governments competed fiercely to attract investments in the semiconductor industry. Local implementations of Circular 18 in the three cities were known as the "S+1" and "B+1" policies

(*Economic Observer* 2003).[10] These policies established explicit competition between cities—"S+1" signified that the Beijing government would offer subsidies for one more loan interest point than whatever Shanghai would offer. Shenzhen targeted Beijing in the same way with their "B+1" policy. The typical policy package for attracting the semiconductor firms might include cheap or free land, government financing for up to 15 percent of the costs of fab construction,[11] loan interest subsidies, and corporate income tax breaks. For giant companies like SMIC, local governments might even make larger investments in the company as well as in infrastructure. For example, in China's largest semiconductor industry cluster in Shanghai, $8.5 billion was invested in fabrication lines by public and private actors between 2000 and 2003 (*Economic Observer* 2003).

Using the ample capital provided by global investors, Chinese banks, and local governments, non-state IC foundries strove to compete with global players by investing in state-of-the-art, high-fixed-cost technology with the hope of achieving the necessary economies of scale and technology sophistication. SMIC and Grace started with 8-inch fabs with a monthly capacity of over 100,000 wafers and used the mainstream 0.18 um node process. By 2004, SMIC began to construct China's first 12-inch fab in Beijing, pushing into 0.09- to 0.13-um node process and lagging the global technology leader by only one generation (GAO 2008). Fearing these growing challenges from SMIC, TSMC, the Taiwanese foundry leading the global semiconductor industry, charged SMIC with industrial espionage that gave them an illegal edge in market competition (Clendenin 2004).

Financing the Non-State Industry

The Chinese semiconductor industry, however, soon encountered setbacks in its rapid expansion. By 2006, the world semiconductor industry fell into a downturn as the traditional PC market matured and relevant demand weakened. Historically, latecomer semiconductor producers might have taken advantage of such crisis time to leapfrog incumbents by outspending on capital investment and accessing new markets (Kim 1997). Unfortunately,

[10] Policies from the three cities are: Shanghai Circular 54 of Dec. 1, 2000, "Some Policy Guidelines of This Municipality for Encouraging the Development of the Software Industry and the Integrated Circuit Industry," *Shanghai Gazette*, Jan. 2001; Beijing Circular 2001–4, "Measures for Implementing 'Policies for Encouraging the Development of Software Industry and Integrated Circuit Industry'," *Jing Zhen Fa*, 2001–4 (Feb. 6, 2001); and Shenzhen Circular 2001–11, "Several Policies of Shenzhen City for Encouraging the Development of the Software Industry and the Integrated Circuit Industry," *Shen Fu*, 2001–11 (Jan. 19, 2001).

[11] Such investment took the form of shares held by the government, but the firm paid no dividends and government officials usually did not sit on the board of directors.

this was not the case for Chinese companies in the late 2000s, even though SMIC as the leading company might have had that intention. It appears that China's financial institutions were not ready to take on the task of sustaining the finance of an advanced semiconductor industry.

As already mentioned, IC fabrication is an extremely capital-intensive business. A state-of-the-art 12-inch fab already cost $5 billion to build in 2007, and it is typically unprofitable in the first five to seven years of operation (Brown and Linden 2009: table 2.1). Nevertheless, multinational semiconductor companies spend billions of dollars annually on capital expenditures in order to keep their fabs up to date. As startup foundries, Chinese fabs carried significant losses during the downturn, and they often responded by cutting capital expenditure. However, SMIC decided to maintain a high investment level so that it could compete with multinationals (Shih 2009).

Financing this aggressive investment strategy was no easy task. While struggling for profitability (partly due to capital depreciation expenses as high as $700–800 million annually and fines lost to lawsuits brought by TSMC), SMIC found it difficult to raise large amounts of funds from the stock market and Chinese banks. It was, again, the local governments in China that sponsored the industry. At its startup stage, SMIC had leveraged the competition between Beijing and Shanghai to raise capital for fab construction in both cities. By 2005, SMIC engaged with Chengdu and Wuhan, two metropolitan cities located in the interior of China that promised to finance SMIC's operations in their jurisdictions. SMIC's joint venture with local governments led to the establishment of Chengdu Cension Semiconductor Manufacturing Corporation and Wuhan Xinxin Semiconductor Manufacturing Corporation in 2005 and 2006, respectively. In the two joint ventures, local governments financed the construction of plants, while SMIC came in as contracted management. In practice, SMIC ran the two factories in the same way as its own subsidiaries, even though local governments claimed ownership (Shih 2009). The Chengdu factory built an 8-inch fab with used equipment purchased from SMIC's Shanghai and Tianjin fabs, while the Wuhan factory made an investment of $1.5–3.0 billion to construct brand new 12-inch fabs (SMIC 2009). The main advantage of these joint ventures was to permit SMIC to add capacity while keeping its own capital expenditure low.

Thus the relationship with local governments became a critical part of SMIC's strategy. With the additional capacities from joint ventures as buffers, SMIC sought to upgrade its own operations to access higher value-added markets. Initially, SMIC kept its custom IC foundry business in the Shanghai fab and devoted the Beijing fab to commodity DRAM[12] manufacturing. DRAM helped

[12] Dynamic Random Access Memory.

SMIC ramp up production, train workers, and generate revenue quickly, but it also exposed the company to risks in a low-margin, highly volatile commodity market. Between 2007 and 2009, SMIC transferred this memory chip manufacturing to Chengdu and Wuhan and retooled the Beijing fab to fabricate logic ICs (i.e. application processors) for foundry operations (Shih 2009).

Local governments might have decided to cooperate with SMIC, but they did not expect the joint venture to be so difficult. Starting in 2005, SMIC carried operational losses for eight consecutive years. Shortly after the resignation of SMIC founder Richard Chang in 2009, the deals with Chengdu and Wuhan went sour. Chengdu sold its Cension Semiconductor to Texas Instruments in 2010, while Wuhan hired new management to remake its Xinxin Semiconductor as an independent company in 2013. A new CEO was installed at SMIC in 2011 and shifted the company toward a conservative investment strategy, cutting back capital expenditure to preserve profitability.

SMIC's relationship with Wuhan and Chengdu was a snapshot of the deeply intertwined ties between China's local governments and the finance of semiconductor companies. Beginning in 2003, the semiconductor industry diffused to second-tier cities without any experience in advanced semiconductor industry. Chengdu, Dalian, Wuhan, and Xi'an had all made ambitious plans to become chip manufacturing centers. While larger cities could offer enough incentives to lure multinationals or large domestic firms (i.e. SMIC), smaller cities could only bet on new entrants. The first local initiative of the later sort was Ningbao Zhongwei founded in 2002. The mid-coastal city of Ningbao hoped its joint venture with a group of former middle managers from TSMC could forge the "flagship" semiconductor company of Zhejiang province. While claiming an investment of $150 million, the Taiwanese management brought second-hand 6-inch fab equipment from TSMC for a much reduced cost. The factory could never overcome the limitations of the out-of-date equipment, and it was never able to process more than 10,000 wafers per month, insufficient to sustain itself. In 2008, the factory was taken over by a local car-maker.

The other notable case is Green Mountain IC, a poster child of the "returnee plus local government" business model (Wang 2011). Founded in 2004, the firm is the only local initiative supported by a county-level government.[13] Hai'an County is a small county with an agriculture-based economy on the periphery of the industrialized Yangtze Delta region. With limited resources, the local officials were eager to build high-tech industries at low cost. The returnee founder of Green Mountain IC convinced the county to make initial investments, which were relatively cheap since they were buying used

[13] In China, county is the jurisdiction level between a city and a town.

equipment. The company promised to bring in foreign investment through an overseas IPO and talent through the founder's personal networks. Hai'an County, thus convinced, invested in one-fifth of the new firm, constructed factories and facilities, provided virtually free utilities (water, electricity, etc.), and organized loans from local banks. But the firm failed to bring in money or a market. After a brief pilot period of production in 2006, the factory was completely shut down in 2009. A business journalist has shown that the founder might have contributed only $2 million of the overall $110 million investment, with the remainder coming from local government investment and state-owned bank lending (Wang 2011).

Ningbao Zhongwei and Green Mountain IC are just two examples among the numerous failed local initiatives in the 2000s. Analysts estimated that at one point there was local investment in 130 fabrication sites across more than 15 provinces, none of which succeeded in reaching the market (Orr and Thomas 2014). These flops of local-government-led fabrication projects reflected a profound institutional failure of the Chinese political economy in managing the financing of the capital-hungry semiconductor industry. The political economic system failed in two important aspects here. First, it is difficult to provide capital to the semiconductor industry efficiently through a system of competing local governments; regional competition critically fragmented investment capabilities. This stunted the growth of larger foundries, which couldn't reach sufficient economies of scales to innovate, and offered insufficient investment to help smaller foundries overcome entry barriers. Second, local governments generally lack the resources and capabilities to manage the capital requirements of chip manufacturing, even though they were well incentivized to sponsor economic development. There was no guarantee that the inexperienced local governments could allocate capital to semiconductor companies with the greatest potential.

Since the financial crisis, the Chinese state has responded to the difficulties of the IC foundries in various ways. The state-owned telecommunication technology group, Datang Telecom, and the sovereign wealth fund, China Investment Corporation (CIC) have both provided capital and acquired substantial stakes in SMIC. The CIC investment in particular has shown the state's commitment to maintaining the independence of SMIC, as opposed to allowing it to be captured by the interests of one integrated company Datang. Sustained by the patient capital from the state, SMIC was able to make it through its toughest time. By the end of 2012, SMIC regained profitability, thanks to a booming smartphone market led by indigenous system companies and IC design houses. Huahong received a renewal of state Project 909 (also named Project 909 Upgrade) in 2010, acquired Grace Semiconductor's 8-inch fabs, and invested in 12-inch fabs under the new company, Huali Microelectronics Corporation.

At the time of writing this chapter, the political and economic institutions for financing the Chinese semiconductor industry are going through another round of reforms. In June 2014, a new industrial policy was institutionalized in the *National Framework for Development of the Integrated Circuit Industry* ("the Framework"). The Framework reestablished the "National IC Industry Lead Group" headed by the President of China, Mr Xi Jingping. A similar Lead Group was last seen in the early 1980s, but the latest one has even higher authority. The Framework includes plans for China's Government Funds for National IC Industry Support with 120 billion RMB (~$19 billion) to be invested in 2014–17 and managed by China Development Bank. About 40 percent of the national funds will be allocated to IC wafer manufacturing, especially targeting leading indigenous companies such as SMIC.

The Framework also calls for local governments and private equity funds to invest another 600 billion RMB (~$98 billion) in semiconductor companies. Under this mandate, the city of Beijing announced the first "regional" fund, the Beijing IC Industry Equity Investment Funds.[14] The first project funded by Beijing's 30 billion RMB is the construction of SMIC's new 12-inch fabs with advanced 40–28nm process, which is a joint venture between SMIC and Beijing. Other cities and provinces, such as Shanghai, Wuhan, and Shenzhen soon followed Beijing's lead. Meanwhile, Chinese private equity funds are on the lookout to acquire successful IC companies in the globe, as seen in the recent proposal of a group of Chinese investors to buy the leading US digital imaging chipmaker OmniVision (Yoshida 2014). All these indicate that China's new industry strategy is to re-emphasize the role of the state while fully embracing non-state industry, globalization, and competing regions.

State, Market, and Business Enterprise in Industrial Development

This analysis of the Chinese semiconductor industry's evolution reveals why the Chinese IC foundries remain in a catching-up stage, behind the growth of Chinese markets. That being said, the Chinese semiconductor industry has been through remarkable transformations, evolving from a small state-owned, domestic-oriented sector to a truly globalized and diverse industrial ecosystem over the last 30 years. Today, as the Chinese semiconductor industry is going through another critical phase of transformation, the historical lessons could be valuable.

[14] Beijing Circular 2014-6, "Beijing Policies to Further Promote Software and IC Industry Development" (*Jing Zheng Fa*, 2014-6).

Based on the structures of dominant business and economic institutions, this chapter analyzes the Chinese semiconductor industry through two stages of development: a first stage of state industry from the 1950s to the end of the 1990s and a second stage of non-state industry since 2000. During the period of state industry, the Chinese state led development by carrying out at least three large-scale national projects to build national champion semiconductor companies. The role of the state was all-encompassing: it mobilized capital to invest, it leveraged access to the large Chinese market for technology transfer from MNCs, and it initiated the restructuring of state-owned companies to organizational forms that it deemed efficient. Nevertheless, the powerful Chinese state still depended on individual business enterprise to realize its technological and economic ambitions. As this chapter has shown, the lack of strategic control in the state industry, both in state-owned companies and in joint-ventures with MNCs, critically hindered these companies from developing organizational capabilities to carry out the state's tasks. But the state industry did gain and learn from its trials. State investment in industry improved the infrastructure and skill base that formed the foundation for the next stage of development. It also crucially informed the Chinese state that technological progress must be embedded in innovative enterprises, whether state-owned or not, that can succeed in market competition.

The second stage of the Chinese semiconductor industry began in 2000 when non-state companies took over as a major force. SMIC, the non-state foundry founded by expatriate professionals with seed capital from Silicon Valley, led the local industry in global competition and narrowed the gap between China's progress and global technological frontiers. The success of SMIC and other innovative non-state foundries over the state industry can be explained by their strategic control by experienced expatriate professionals. These companies were able to strategize entry into the global production chain by positioning themselves as pure-play foundries, build up organizational capabilities by attracting returnee talent, and finance aggressive investment strategies by tapping into domestic and foreign sources of funding.

In the second stage, the top-down, state-heavy system of industrial policy was substantially decentralized. With the rise of the non-state sector, the role of industrial promotion and finance has been transferred to lower-level governments, the local developmental states that have fueled the growth of the Chinese economy since the reform. However, this structure of capital provision does not match the needs of the semiconductor industry very well, as the industry functions in large economies of scale, broad scope, and clustering. Largely due to the fragmentation of the industrial financing system, the Chinese foundries were not able to sustain high growth in the late 2000s.

At the beginning of this chapter, it was noted that the market and state are often regarded by analysts as two great sources of potential for the development of an advanced Chinese semiconductor industry. The analysis of the industry's history shows, however, that the roles of market and state are not straightforward. China's production-consumption gap of semiconductors might have indicated a large market, but the consideration of this market must be complemented by the fact that this industry is subject to global competition and the fragmented production chain. Since production is fragmented, IC foundries must harness innovative capabilities at the upper end of the supply chain in China, including system integrators, IC design capabilities, and others, if they are to realize their potential. Over the 2000s, these subsectors were in a catch-up phase as well. Moreover, even when the supply chain was better developed in China, local foundries faced more fierce competition from the MNCs.

The ever-increasing production–consumption gap and the risks of overreliance on foreign semiconductor capabilities certainly provide strong incentives for the Chinese state to intervene for import substitution. But the Chinese state's capacity for intervention is limited by the nature of globalized production in the industry on one hand, and the state's reliance on competing regions for resource channeling that provide incentives for local development on the other hand. The recent policy changes might have strengthened the role of the state and its position in negotiating with foreign players, but it still follows a path dependence of globalization, regional competition, and non-state industry.

The historical evolution of the Chinese semiconductor industry demonstrates that industrial growth and technological progress ultimately depend on the development of innovative business enterprises that invest in the accumulative process of innovation. Innovative enterprises require strategic control, organizational integration, and financial commitment (Lazonick 2004a, 2004b, 2010). The successful transformation of the Chinese IC foundry industry occurred when the state allowed the strategic control of its key enterprises by experienced professional managers and engineers who could direct resources and invest smartly in organizational capabilities. But external financial institutions were misconfigured to sustain the accumulative process of technological catch-up and innovation in the companies. Whether the Chinese IC foundries will emerge as global technological and market leaders in the next decade or so will depend on China's success in reforming its financial institutions for innovation and development.

Acknowledgments

Research for this chapter was supported by the Ford Foundation project, Financial Institutions for Innovation and Development, directed by William Lazonick, and the

National Science Foundation under grant SES-0964907, directed by Dan Breznitz. The author would like to give special thanks to Dan Breznitz and Michael Murphree for supporting field research, and to Bill Lazonick and Yu Zhou for their comments in developing the chapter.

References

Brown, C., and G. Linden (2009), *Chips and Change: How Crisis Reshapes the Semiconductor Industry*. Cambridge, MA: MIT Press.

Business China (1997), "908, 909 . . . NEC and the National Semiconductor Project." 1–2.

Chesbrough, H. W. (2005), *The Globalization of R&D in the Chinese Semiconductor Industry*. Berkeley, CA: Sloan Foundation Research Report.

Clendenin, M. (2004), "Analysis: Background on TSMC, SMIC Lawsuits," *EE Times*, May 1, <http://www.eetimes.com/electronics-news/4120176/-b-Analysis–b-background-on-TSMC-SMIC-lawsuit>.

Dewey-Ballantine. (2003), *China's Emerging Semiconductor Industry: The Impact of China's Preferential Value-Added Tax on Current Investment Trends*. Washington, DC: Dewey-Ballantine LLP.

Economic Observer, (2003), "中国造芯运动预警" (Early warnings to the Chinese chip making movement). 经济观察报 *(Economic Observer)*, June 22.

Fuller, D. B. (2005), "Creating Ladders out of Chains: China's Technological Development in a World of Global Production." PhD Dissertation, Department of Political Science, MIT.

Hu, Q. (2006), "Memoirs on National 909 Project on Very Large Scale Integrated Circuits" (芯路历程：909 超大规模集成电路工程纪实). Beijing: Publishing House of Electronics Industry.

Iritani, E. (2002), "China's Next Challenge: Mastering the Microchip," *Los Angeles Times*, Oct. 22. <http://articles.latimes.com/2002/oct/22/business/fi-china3> accessed Aug. 2015.

Kim, L. (1997), *Imitation to Innovation: The Dynamics of Korea's Technological Learning*. Cambridge, MA: Harvard Business Press.

Landler, M. (2001), "From Taiwan, a Fear of China Technology," *New York Times*. Oct. 3, <http://www.nytimes.com/2001/10/03/business/from-taiwan-a-fear-of-china-technology.html> [accessed Sept. 2014].

Lazonick, W. (2004a), "Indigenous Innovation and Economic Development: Lessons from China's Leap into the Information Age," *Industry and Innovation*, 11(4): 273–97.

Lazonick, W. (2004b), "The Innovative Firm," in J. Fagerberg, D. C. Mowery, and R. R. Nelson (eds), *Oxford Handbook of Innovation*, 29–55. Oxford: Oxford University Press.

Lazonick, W. (2009), *Sustainable Prosperity in the New Economy? Business Organization and High-Tech Employment in the United States*. Kalamazoo, MI: Upjohn Institute for Employment Research.

Lazonick, W. (2010), "The Chandlerian Corporation and the Theory of Innovative Enterprise," *Industrial and Corporate Change*, 19(2): 317–50.

Lazonick, W., and M. O'Sullivan (2000), "Maximizing Shareholder Value: A New Ideology of Corporate Governance," *Economy and Society*, 29(1): 13–35.

Li, Y. (2011), "From Classic Failures to Global Competitors: Business Organization and Economic Development in the Chinese Semiconductor Industry." Master's Thesis, Department of Regional Economic and Social Development, University of Massachusetts Lowell.

Lu, F. (1999), "State, Market, and Enterprise: The Transformation of Chinese State Industry." PhD Dissertation, Department of Political Science, Columbia University.

Lu, Q. (2000), *China's Leap into the Information Age: Innovation and Organization in the Computer Industry*. New York: Oxford University Press.

Orr, G., and C. Thomas (2014), *Semiconductors in China: Brave New World or Same Old Story?* McKinsey & Co., Aug., <http://www.mckinsey.com/insights/high_tech_telecoms_internet/semiconductors_in_china_brave_new_world_or_same_old_story> accessed July 2015.

O'Sullivan, M. (2000), "The Innovative Enterprise and Corporate Governance," *Cambridge Journal of Economics*, 24(4): 393–461.

PricewaterhouseCoopers (2004), *China's Impact on the Semiconductor Industry*. New York: PricewaterhouseCoopers.

PricewaterhouseCoopers (2013), *Continuing to Grow: China's Impact on the Semiconductor Industry 2013 Update*. New York: PricewaterhouseCoopers.

Saxenian, A. (2005), "From Brain Drain to Brain Circulation: Transnational Communities and Regional Upgrading in India and China," *Studies in Comparative International Development*, 40(2): 35–61.

Saxenian, A. (2007), "Brain Circulation and Regional Innovation: The Silicon Valley-Hsinchu-Shanghai Triangle," in K. R. Polenske (ed.), *The Economic Geography of Innovation*, 190–211. Cambridge: Cambridge University Press.

Shih, W. (2009), *Semiconductor Manufacturing International Corporation: "Reverse BOT."* Cambridge, MA: Harvard Business School.

Simon, D. (1992), "Sparking the Electronics Industry," *China Business Review*, 19(1): 22–8.

Simon, D. (1996), "From Cold to Hot: China Struggles to Protect and Develop a World Class Electronics Industry," *China Business Review*, 23(6): 8–16.

Simon, D., and D. Rehn (1987), "Innovation in China's Semiconductor Components Industry: The Case of Shanghai," *Research Policy*, 16(5): 259–77.

SMIC (2004), "Announcements and Notices: Issuance and Repurchase of Shares under Share Option Plans of the Company," Dec. 10, http://www.smics.com.hk/attachment/2011012018125617_en.pdf.

SMIC (2005), "Announcement of 2005 Annual Results," Mar. 30. http://www.smics.com.hk/attachment/2011012012331117_en.pdf.

SMIC (2009), "Annual Report 2008," Apr. 27. <http://www.smics.com/attachment/2011011717413117_en.pdf>.

United States Government Accountability Office (GAO) (2008), *Export Controls: Challenges with Commerce's Validated End-User Program May Limit its Ability to Ensure that Semiconductor Equipment Exported to China is Used as Intended*. Washington, DC: GAO.

Wang, R. (2011), "南通绿山 1.04 亿整体挂牌让 (Nantong Green Mountain IC on Sale for RMB 104M)," *First Finance and Business Daily (第一财经日报)*, Jan. 25.

Yoshida, J. (2014), "China to Blow $10B a Year on Chips," *EE Times*, May 25, <http://www.eetimes.com/document.asp?doc_id=1321616&_mc=RSS_EET_EDT&cid=NL_EET_Daily_20140325&elq=e228683f4577461c814e25ff8a7c904a&elqCampaignId=16143> [accessed May 2014].

8

Catching Up and Developing Innovation Capabilities in China's Telecommunication Equipment Industry

Peilei Fan and Xudong Gao

Introduction

On October 8, 2012, the US House Permanent Select Committee on Intelligence issued an investigative report on "The US National Issues Posed by Chinese Telecommunications Companies Huawei and ZTE." The report alleged that Huawei and ZTE, China's two largest communication equipment makers, provided opportunities for Chinese intelligence services to tamper with US telecommunications networks to engage in spying. The report provided five recommendations to the US government and companies, including that the US government should block acquisitions or mergers by Huawei and ZTE, that government agencies and contractors should not use equipment from the companies, and that US intelligence agencies should "remain vigilant and focused on this threat." Both companies denied the allegations, and Huawei released a statement saying that the committee had predetermined the outcome of its investigation and that the report was based on "rumors and speculations to prove the non-existent accusations."[1] The recent Snowden case indicated that the US and some European governments have engaged in various global surveillance programs with the cooperation of Western-based telecommunication corporations, and Huawei was among the targets. While the release of the report and wide media coverage have largely barred high-end telecommunication equipment of Huawei from the American market, the

[1] <http://www.bloomberg.com/news/2012-10-07/huawei-zte-provide-opening-for-china-spying-report-says.html>, accessed Nov. 2012.

immediate effect of the publicity actually served to increase the name recognition of Huawei and ZTE in the United States and worldwide. Established in the 1980s in Shenzhen, Huawei and ZTE are two of the leading Chinese telecom-equipment companies that have transformed themselves from local producers into global giants over the past three decades. The hearing also showed that the US Congress not only considers Huawei and ZTE potential national security concerns due to the fact that telecom equipment is critical national infrastructure in today's knowledge economy, but that it may also view Huawei and ZTE as serious competitors to US companies in its own market.

Exactly one month later, on November 8, 2012, in a report delivered by the Chinese President Hu Jintao at the 18th National Congress of the Communist Party, China announced that it had determined to move towards an "innovation-driven economy." This decision was intended to address the slowing of its past growth model, which relied heavily on investments, exports, and its huge low-cost labor force. Although China as a nation still lags behind most OECD countries in innovation capability, some of its high-tech industries, such as telecommunications, and some of its domestic companies such as Huawei, ZTE, and Datang, can offer insights on how a latecomer can grow on the basis of indigenous innovation and high-tech multinational corporations. This chapter analyzes the development of China's telecommunication equipment industry since the 1980s, with a focus on three innovative companies, Huawei, ZTE, and Datang, the developer of China's 3G standard. There is disagreement on whether these companies can be considered legitimate models of innovation, as they are still followers in many aspects of leading technology. However, Huawei and ZTE have evolved to be leading global players whose technological capability allows them to compete on equal footing with other telecom multinationals (Seekingalpha.com 2012). Datang has pioneered the successful technological development of China's 3G, which has been a landmark achievement in China's technological standard-setting. Before we relate the stories of these companies, we will first briefly introduce the related literature on innovation and economic development, innovation strategies, the theory of innovative enterprise, and institutional factors affecting innovation for latecomers.

While innovation capacity has been widely acknowledged as a critical force in national economic growth for developed countries (Nelson 1993; Porter 1990), scholars have divergent views on the contribution of innovation capacity to economic development for latecomers such as China. Some have viewed the growth of China as the result of its advantages in the abundance of low-cost production factors such as labor and land, in addition to foreign investments and technology transfer (Huang 2003; Naughton 2007). Others have attributed China's economic development to technological progress and indigenous innovation (Chu 2011; Fan 2006a; Fan and Watanabe 2006; Feng 2010; Lazonick 2004; Liu and White 2001; Lu 2000; Lu and Lazonick 2001; Motohashi and Xiao 2007).

We define innovation broadly as not only the creation of new processes and products, but also the development of higher quality and/or lower cost processes and products than previously available. How can companies in latecomer countries develop innovative capability? The scholarship on catching-up and technological upgrading strategies of such countries (e.g. Hobday 1995; Leonard-Barton 1995; Kim 1997) provides different models. For instance, the "technology capability ladder" perspective argues that firms in developing nations can follow the sequence of (1) assembly or turn-key operations, (2) adaptation and localization of components, (3) product redesign, and (4) independent design of products to catch up (Leonard-Barton 1995; Austin 1990). According to Kim (1997), while technology development in advanced countries goes through stages of emergence, consolidation, and maturity, firms in catching-up countries follow a process from imitation to innovation. They begin by acquiring, assimilating, and improving mature technologies from advanced countries, and they repeat the process with higher level technology in the consolidation stage in advanced countries. Then, if successful, they may eventually accumulate indigenous capability to innovate—that is, generate higher quality, lower cost products than their global competitors.

The transformation from imitation to innovation at the company level can be best explained by the theory of innovative enterprise (Lazonick 2003, 2006, 2010). The theory focuses on three social conditions that can transform strategy, organization, and finance into innovation: strategic control, organizational integration, and financial commitment. *Strategic controls* refers to "a set of relations that gives decision-makers the power to allocate the firm's resources to confront the technological, market, and competitive uncertainties that are inherent in the innovation process" (Lazonick 2010: 331). *Organizational integration* is "a set of relations that creates incentives for people to apply their skills and efforts to strategic objectives" (Lazonick 2010: 331). Financial commitment includes "a set of relations that ensures the allocation of funds to sustain the cumulative innovation process until it generates financial returns" (Lazonick 2010: 331). Lazonick and his colleagues have compared the social conditions of innovative enterprises in different national settings (e.g. Lazonick 2007), while Lu's work (Lu 2000; Lu and Lazonick 2001) has directly applied the theory to examine the rise of four innovative computer companies in China in the early years of the reform era. While this chapter does not use the innovation enterprise theory to examine the development of innovation capability of each company, we would like to stress that transformations of the three social conditions are important dimensions that should be linked with the latecomers' catching up in technological capability.

The institutional factor, especially the role of the government, in industrialization and innovation has been heavily debated. This is illustrated by the scholarly work on the role of the government in economic growth of East Asia economies, particularly Korea and Japan (Amsden 1989; Johnson 1982; OECD 1972; Rosovsky 1972; Sumiya 2000; Sakakibara and Cho 2002). In fact, the government has also played an essential role in industrial development and innovation in Western countries. The development of several high-tech industrial areas, such as semiconductor, software, and data communication, in the US are obvious cases (Wilson et al. 1980; Lazonick 2011). China has a transitional economy with its particular institutional characteristics and large and dynamic markets. It raises questions of the roles and effectiveness of the government in influencing the innovation capability of domestic companies. More broadly, China's experience also has implications for other late-industrializing countries.

In this chapter, we will focus on how Chinese companies have developed innovation capability over time and how the government has influenced that process. We organize the rest of the chapter as follows. The next section provides an introduction to China's telecom industry and highlights the development of major telecommunications companies. The following section discusses the role of the government in influencing firms' innovative capability. Then findings are presented on innovative capabilities relating to Chinese telecom companies, focusing on options for developing innovative capability, the advantages of domestic versus multinational corporations (or MNCs), and the dynamic process of catching up. The chapter concludes by highlighting that the government affects the technological learning of innovative enterprises through modulated support corresponding to the development levels of domestic firms.

China's Telecommunication Industry

China's telecom-equipment market is among the most dynamic in the world because of its rapid growth and intense market competition (Figure 8.1). Surpassing the United States, China now leads the world in mobile phone and internet usage, with over 1 billion mobile phone subscribers and over 500 million internet users in 2012 (Reuters 2012; *China Daily* 2012; World Bank 2014). Its telecom infrastructure has improved dramatically, and the consumption of telecom products has skyrocketed, especially since the mid-1990s. Fixed-line connections were the first to grow; while only 0.2 percent of China's population had access to fixed-line connections in 1980, this figure

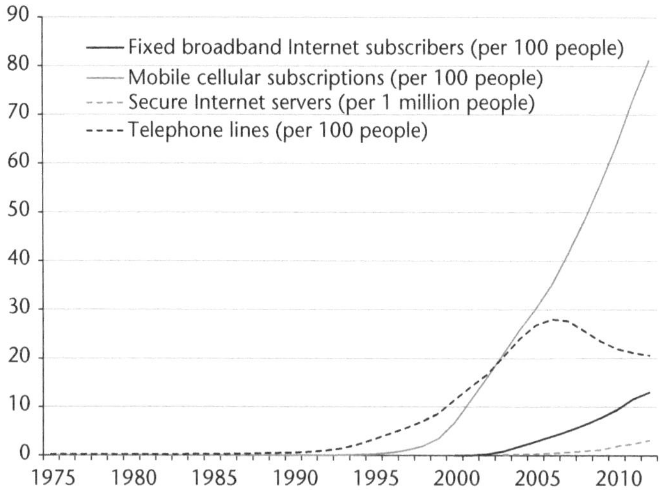

Figure 8.1. China's Telecom Infrastructure.
Source: World Bank. 2014. World Development Indicator. Washington D.C.: World Bank.

increased dramatically to 28 percent in 2006 before declining to 21 percent in 2012 as subscribers opted for mobile phone connections. Mobile phone and internet connections have also multiplied quickly since the late 1990s. Mobile cellular subscriptions reached 81 percent of the population in 2012, compared to 0.2 percent in 1990 and 6.7 percent in 2000. Similarly, fixed broadband internet subscriptions increased impressively from 0.2 percent of the total population in 2000 to 13 percent in 2012, whereas secure internet servers per million population increased from 0.1 in 2001 to 3.1 in 2012 (Figure 8.1). Meanwhile, China's telecom industry has grown rapidly to meet the exploding demand. This is exemplified by the production of mobile phones, which increased 10 times within a short span of a decade, from 0.12 billion in 2003 to 1.2 billion units (Figure 8.2).

Meanwhile, the global telecom-equipment industry has experienced a major shakeout since the middle of 2000s. In 2004, Ericsson, Alcatel, Lucent, Nokia, Siemens, Motorola, and Nortel were the seven global leaders. But this ranking has been dramatically reshuffled in less than a decade: Ericsson has retained its leading position as an independent company; four companies have merged into two, creating Alcatel-Lucent and Nokia-Siemens; and two have exited the sector completely with Nortel filing for bankruptcy in 2009 and Motorola selling its network business in 2011 (Seekingalpha.com 2012). During this time, Chinese telecom multinationals such as Huawei and ZTE have emerged to become industrial leaders of the newly shuffled telecommunication sector. Measured by 2011 revenues, the new global leaders are Ericsson, Huawei, Alcatel-Lucent, Nokia-Siemens Networks, and ZTE. Some argue that external factors, including the global financial crisis, centralized purchasing among operators, and technology

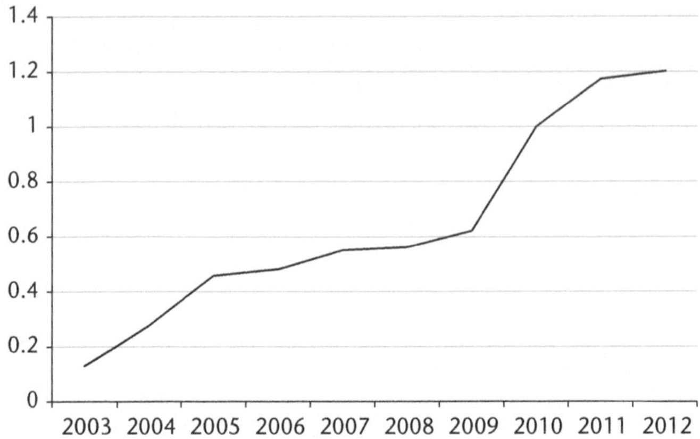

Figure 8.2. China's Mobile Phone Production (in billion units)
Source: Various dataset from the Ministry of Industry and Information Technology (MIIT) of China. 2014. www.miit.gov.cn.

shifts, have contributed to the restructuring of the industry (Seekingalpha.com 2012), while others have pointed out that managerial problems, such as the use of financialization strategies, have plagued companies such as Cisco, Nortel, Lucent, and Motorola. These companies manipulated their corporate stock, either as a currency to acquire technology companies or attract talented people to accumulate innovative capability (such as the acquisitions conducted by Lucent and Nortel in 1999 and 2000). Lazonick (2009) criticizes the financialization strategies of major corporations who allocate resources simply to boost their stock prices. The practices actually have weakened the financial condition of some of these companies, such as Motorola, since diverting the financial resources from investment in more sophisticated technologies made them more vulnerable to collapse of the stock market and the slowdown of the industry (Carpenter et al. 2003; Lazonick and March 2012; Bell et al. 2013).

Before the 1980s, foreign MNCs had little involvement in the China telecom-equipment industry due to stringent government control. After the economic reform, they were able to establish joint ventures, and after China joined the WTO, wholly owned foreign subsidiaries were permitted to produce telecom equipment in China. The Chinese government stipulated that offering technological help was the precondition for foreign producers to enter the Chinese market for telecom equipment. This happened both through joint ventures, which created technological spillovers to the Chinese partners, and through foreign subsidiaries, which gave technological assistance to telecom service providers. Several global MNCs, some with the assistance of their respective national governments, secured deals with the Chinese national government in the 1980s and 1990s, such as Nortel (1984), Nokia (1985), Motorola (1987), Lucent (1988), Cisco (1994), and Qualcomm and

Siemens (1999), to set up their wholly owned manufacturing bases in China. China thus has become an important revenue base for foreign MNCs. For instance, in 2006 Motorola generated nearly a quarter of its global sales from China, while Nokia, Qualcomm, and Lucent obtained over 10 percent of their sales from China (Fan 2010b). While they facilitated the growth of China's telecom infrastructure, foreign MNCs also posed significant challenges for the development of domestic companies.

Despite the early dominance of foreign firms in the telecommunication sector, domestic companies soon caught up in switch equipment and access equipment markets in the middle and late 1990s from a far behind position in every sub-sector in the 1980s. For example, China was totally dependent on foreign-made telephone switch equipment in 1980. It produced jointly with foreign manufacturers from the late 1980s to early 1990s, and finally supplied the market with almost 100 percent self-developed switches in the later 1990s (Fan 2006b). In the 2000s, Chinese companies became diversified and competitive in the markets of optical transport, data communications, and mobile communications (Fan 2011).

Roles of the Government

While it is generally agreed that the Chinese government has played an essential role in domestic firms' catching up and innovation in the telecom-equipment industry (see Chapter 2 of this book; Fan 2006b; Feng 2010), the detailed approaches and strategies are not well understood. In the following paragraphs, we first illustrate the nature of government support in the telecommunication sector, showing that it evolved over time in response to the development of domestic firms. We then examine two particular types of indirect support that have affected the innovation capability of domestic firms: state intervention in demand conditions and the influence of the state in domestic firms' participation in international standards. These experiences can provide policy implications for other emerging economies that also hope to rely on their large domestic markets to facilitate the growth and technological learning of their domestic firms.

Dynamic Support for Domestic Innovation

At the beginning of China's economic reform in the late 1970s, the telecommunications industry in China was backward in technology, small in scale, inefficient in management, and without sufficient investment capital (Wan 2001). In addition, the industry was totally controlled by the government through its monopoly agency, the Ministry of Industry and Information

Technology (MIIT), formerly Ministry of Posts and Telecommunications (MPT). The government had three distinct roles simultaneously: regulator, service provider, and owner of equipment manufacturing enterprises. The structure has been reformed over time, but the influence of the MIIT remains.

As mentioned, governmental support has gone through different stages. The initial policy measures (from the early 1980s to the early 1990s) relied on imports of foreign switch products and the setting up of joint ventures with multinational corporations at the initial stage of the development. The measures did not bring the desired transfer of switch technology to China (similar cases can also be found in the semiconductor and automobile sectors, discussed elsewhere in this book). However, after several domestic firms developed their own large-scale switches in the early 1990s, the government began to offer direct support in finance, marketing, and exports. In particular, the government helped domestic customers and producers by providing them with financial assistance and by encouraging domestic customers to purchase from capable national brands, such as the Customer–Producer Coord-nation conferences organized by the MPT, to be described in detail shortly. In these ways, the state effectively supported the early stage of development of several domestic firms, including Huawei, ZTE, Datang, and Great Dragon (Fan 2006b). When domestic firms had grown strong enough and able to compete with multinational firms in several fields in the middle and late 1990s, the government reduced their direct support but remained active in building infrastructure in the field, such as pushing for the basic technology development for mobile and data communication.

The technological support from the government has evolved as well. Since the mid- to late 1990s, the government has offered technical support indirectly by selecting capable domestic firms for national science and technology programs, such as the "863 Program" and "973 Program." This was the point when domestic firms were dominating China's switch equipment market and starting to expand their product range to other sub-sectors of telecommunications, such as optical transmission and access systems. Since 2000, the government has completely withdrawn direct financial aid for domestic firms and instead focused on supporting the development of cutting-edge technologies and standards in the fields of mobile and data communications. The most prominent example of this has been the heavy state investment in TD-SCDMA (Time Division-Synchronous Code Division Multiple Access), the Chinese 3G standard developed by Datang.

The criticism of governmental support is often that it props up weak, non-innovative firms. But in fact, the Chinese government's support for telecom-equipment firms seems to be dynamic and selective, with the goal being to nurture technologically advanced domestic firms. Firms with the greatest innovation capability have received the most support from government intervention, which in turn enabled them to become even more innovative. For instance, at an early stage, Huawei, ZTE, Datang, and Great Dragon have all benefitted from government support in financial subsidy and demand assistance due to their self-developed technologies in large-scale switch equipment. Even though government policy to support domestic switches seemed to benefit all domestic producers, in reality, only those who have built up R&D capability in switches can take advantage of it. The increased revenue brought by the switch equipment offered these domestic firms more financial resources for R&D.

Companies that take the long-term view and continue to invest in their R&D, such as Huawei, ZTE, and Datang, have flourished, while others, such as Great Dragon, have faded away. Due to their increased technological capability, Huawei and ZTE received further research support from the government by participating in key national S&T programs to develop cutting-edge telecom-equipment technologies, such as optical transmission systems and 3G, high-speed data communications, integrated access systems, and optical transmissions, while Datang has been instrumental in developing TD-SCDMA. Great Dragon is an example that failed despite the government support. It was the first domestic producer of large-scale switch equipment, but could not sustain its success when it missed the chance to diversify into other subsectors. Additionally, it experienced various problems with corporate management and finance (Fan 2006b). A key lesson is that, although government intervention can provide a benign environment to nurture innovation capability, ultimately it is the firm itself that has to have the vision and commit resources for innovation capability through strategic control, organizational integration, and financial commitment.

Intervening in Demand Condition

One type of indirect support that the Chinese government has used is its intervention in domestic demand, both through creating direct linkages between the market and domestic suppliers, and through institutional reform to foster a competitive domestic market. Fast growth and competitive domestic market has fueled demand-stimulated innovation for domestic companies.

First, the Chinese government has played matchmaker by linking the domestic market to domestic producers, and thereby significantly helping the development of these firms in their early years. In the early and middle 1990s, the domestic market had been dominated by MNCs despite the fact that Huawei, ZTE, GDT, and Datang had all developed their own public digital switch systems. It was very hard for local firms to find buyers in China's domestic market. The government encouraged domestic telecom service providers, particularly the state-owned entities at the provincial and city-level telecommunications bureaus, to give priority in purchasing to domestic products when the function/price ratio was similar. This helped domestic companies' business, even though most national-level purchases continued to be dominated by foreign brand products due to the higher requirements on quality. The MPT created platforms and channels to link service providers and domestic producers, such as two conferences it organized in 1996 and 1997 on switch and mobile equipment. At these conferences, producers met representatives from provincial and city telecommunication bureaus and were given the chance to market their products. A significant amount of equipment units were sold through this government-created channel. For instance, the second Domestic Switch Customer–Producer Coordination Conference organized by the MPT helped GDT to obtain orders for about 3.9 million equipment units (Fan 2006b). Domestic companies who were able to develop their own large-scale switch equipment through reverse-engineering were rewarded by orders from these domestic customers. As a result, several of the companies decided that their future development path would be based on self-developed technologies (Fan et al. 2007).

Second, the Chinese government has engineered major reform in organization and competition among the state-owned telecommunication service providers, which liberalized the demand market and further promoted demand-stimulated innovation. The Chinese communication industry has gone through significant reforms to better serve the market. To date, the telecommunications infrastructure of China is well developed compared with other countries that are large in scale, competitive in service provision, and at a similar level of economic development. From 1980 to 1993, the Chinese government primarily focused on improving the telecommunications infrastructure. The state gave companies in the telecommunication industry several advantages during this period: they could charge high installation fees and surcharges and invest the profit, speed up capital investment depreciation, and pay reduced and exempted taxes. In the middle of the 1990s, the government intervened to encourage competition by opening non-backbone telecommunications services to new entrants and adopting new investment and financing vehicles such as debt and equity financing (Wan 2001). Since then, extensive competition in liberalized services resulted in the sharp reduction of prices and improvement in the quality of services. The most dramatic reform occurred in 2008, when the government mandated the consolidation of then six separate companies into three: China Telecom, China Unicom, and China Netcom. Each had a fixed-line business, a wireless business, and a different 3G standard license, and the consolidation strengthened and focused competition between them (Bernstein Research 2010).

The fierce competition among telecom service providers as a result of the governmental restructuring made China's telecommunication market the most competitive in the world (Bernstein Research 2010). This not only benefitted consumers, but it also pushed service providers to seek more innovative, cost-saving solutions. They then passed on this competitive pressure to their suppliers, the telecom-equipment producers, who themselves were driven to innovate in order to satisfy increasingly challenging requests. For instance, in 1999, China Mobile proposed a plan of prepaid fees for cell phones to attract price-sensitive customers. While most MNCs did not act quickly at the opportunity, having distant relationships with their service provider, Huawei reacted immediately and quickly developed an innovative solution for China Mobile. This solution, called the mobile intelligent network, can be conveniently overlaid with existing networks and is based on Huawei's accumulated R&D experience in the Global System for Mobiles (GSM) (Fan 2006a). This product is a typical demand-stimulated solution; while it does not have the characteristics of radical innovation, it improves the current solution by incremental innovation. Importantly, this required firms to have a strong R&D capability in GSM equipment. The mobile intelligence network provided a practical solution in China's market and quickly gained over 30 million users, achieving coverage of 80–90 percent of China's domestic market even in 2002 and later becoming popular in the broader Southeast Asia market. Intensified competition between the service providers, resulting from the government's decision to restructure the telecommunication sector, thus provided opportunities for demand-stimulated innovation of domestic companies. These opportunities for development were especially available to those who have already accumulated R&D capability and were ready to jump at every viable opportunity.

Influencing Local Firms' Participation in International Standards

The development of TD-SCDMA, one of the international standards of 3G mobile communications, illustrates how government can influence technological standards and therefore affect the participation of local firms in the entire production chain. Although there are much to debate on whether TD-SCDMA should be treated as a successful case, we would like to underscore that developing China's own standard offered the opportunity for domestic innovation not only in TD-SCDMA but also in many other sectors where 3G and 4G standards are relevant (see also Chapters 2, 9, and 10). Not only did Chinese manufacturers not have to pay patent licensing fees for TD-SCDMA, but they were also able to negotiate lower fees than foreign manufacturers for other standards (Fan 2006a). Strong government support has pushed TD-SCDMA in the Chinese market and also played a critical role in helping TD-SCDMA to become one of the international standards (Gao and Liu 2012; Kshetri et al., 2011; Zhan and Tan 2010).

However, it is also important to acknowledge that the government support of TD-SCDMA exposed many problems with relying upon the state to coordinate a complex technological and industrial network. The decision to support TD-SCDMA was spurred by a letter from three noted scientists in September 2005. A very high-level government official ordered that the letter be carefully treated because of the potential contributions of TD-SCDMA to the indigenous innovation policy. Acting accordingly, various government agencies knew that they needed to support TD-SCDMA, but their actions were not well-coordinated, proactive, or consistent (Gao and Liu 2012). By 2005, even though all technical tests had proved that TD-SCDMA could be used to build up large-scale telecom networks, service providers still were reluctant to use the new standard, since the former MIIT had not effectively involved the service providers in promoting TD-SCDMA.

The government endorsed the strong lobby of Datang in choosing this standard, but it did not determine which service provider would be the key adopter of the standard for many years (Gao and Liu 2012). This was because of concern that the appointed providers might suffer competitive disadvantage due to the immaturities of the standard. The uncertainty here had a hugely negative impact, delaying the deployment of TD-SCDMA. Firms along the TD-SCDMA value chain were reluctant to make a full commitment of resources to TD-SCDMA for several years, until China Mobile was finally selected to use the TD-SCDMA standard in January 2009 (Gao and Liu 2012). This was the result of the government's inexperience, lack of preparation, and control of the entire production system surrounding 3G. The debate is still raging in China whether the TD-SCDMA decision is a right one. At the end of the 2014, a leading Chinese economic website Caixin released a special report on Chinese innovation in TD style (TD 式创新), characterizing the TD decision as a failure with close to 200 billion RMB investment only producingd the shortest lived 3G network (Tan 2014). Those who defend TD responded that TD has provided a strategic opportunity to upgrade China's telecommunication industry which allowed Chinese domestic players become the leading players in the world (*New Beijing Times*, Dec. 25, 2014). It is indeed arguable that if China had decided to adopt the existing international 3G standards as it had done with 2G, China's 3G development would have been earlier, cheaper, and smoother. However, China would also have lost the opportunity to encourage a host of domestic firms, including Huawei and ZTE, to develop indigenous solutions and products to solve the problems that TD-SCDMA encountered. This process vastly increased the innovative capacity of these firms and of the industry in general. In the end, the bumpy ride of TD-SCDMA is by no means surprising as China's first domestically generated technology standards in a landscape dominated by foreign giants. The government and associated businesses have learned a great deal. In all likelihood, this experience should help China's upgrade to 4G run more smoothly.

Developing Innovation Capability

In this section, we discuss the trajectory of three of the most successful companies, Huawei, ZTE, and Datang, as cases that illustrate how Chinese domestic firms developed innovation capabilities. After a brief introduction describing the growth of the three companies, we will highlight three common characteristics of domestic firms in developing innovation capability: internal development, local advantages vs. the MNCs, and the dynamic process of technological learning.

Brief Introduction of Huawei, ZTE, and Datang

HUAWEI

Started in 1988 in Shenzhen, with a registered capital of a mere RMB 20,000, Huawei spent its first two years selling telephony switches and fire alarms, providing consultancy to various machinery producers, and acting as a distributor of the HAX switch produced by a Hong Kong company. Earnings from these markets funded the entry of Huawei into telecom manufacturing as it started to develop a small-scale switch system—mostly used in rural areas in the late 1980s—by reverse-engineering. This "rural-encircling central cities" strategy was also used by ZTE and 200 other domestic firms in the same field. Rural telecom markets were entry points into the greater domestic market because they needed affordable, small-scale switch equipment. Most foreign firms concentrated on the highly profitable business of supplying large-scale switch equipment to urban markets, ignoring the rural telecom bureaus due to distribution barriers and the lower profit margins on small-scale switches. To survive and to distinguish itself from others, Huawei put significant effort into research and successfully developed its own large-scale switch, C&C08, in 1993, achieving the first significant success in its technological learning. Huawei used in-house R&D to repeat the same imitation process in catching up with higher-level technologies in other subsectors, such as access equipment, optical transmission, and data networking. Based on the experience and innovative capabilities gained from imitation, Huawei progressed to higher-level technology in the consolidation stage of advanced countries.

Huawei's consistent drive to more R&D and innovation today can be illustrated by the growth of its impressive number of patent applications. Not only has it been the domestic leader for patent applications in China since 2002, it also became the fourth-largest global patent applicant in 2007, following only

Matsushita, Philips Electronics NV, and Siemens, bypassing telecom giants such as Nokia, Qualcomm, Ericsson, Samsung, and Cisco. In the emerging area of cloud computing technologies, Huawei possessed 685 patents in China, 226 in Europe, and 107 in the US. Huawai has also played a major role in standard development by being one of the 14 board members of the Distributed Management Task Force (DMTF).[2]

Huawei moved to the international market in the mid-1990s. Similar to its strategy of "rural encircling cities" used in domestic market for small-scale switch equipment, Huawei's international exploration started with developing markets such as Russia, Ethiopia, South Africa, Yugoslavia, and Brazil in the mid-1990s, though initially without much success (Fan 2010a). Through persistence, high-value but low-cost products, and strategic alliances with distributors in the host countries, Huawei was able to make some progress in these developing country markets by the late 1990s. Its advancement into established markets such as Western Europe and North America has proved to be more challenging. It eventually became one of the main suppliers to the global service providers British Telecom and Vodafone in the mid-2000s, but its inroads into the US has been hampered by the security concerns described at the beginning of the chapter. Nevertheless, Huawei's overseas strategies have borne fruit, as its overseas sales increased from 27 percent in 2003 to 66 percent in 2012 of its total revenue. Along with this market expansion, it has established global R&D centers in the United States, India, Russia, and Sweden.

ZTE

Also located in Shenzhen, ZTE was established in 1985 as a *Guoyou-Minying* (state-owned, but privately managed) company and it initially produced watches, electronic keyboards, and telephones (Huawei is a *Minying*, or private, company, by contrast). ZTE entered into telecom manufacturing by investing most of its accumulated financial resources in R&D for switch technology. Like Huawei, in the late 1980s ZTE developed its small-scale telephone switch mainly to serve China's rural market. In 1995, ZTE developed ZXJ10, one of the main domestic large-scale PDSSs (public digital switch systems), which led to its first financial success. As was the case with Huawei, ZTE developed key technologies and swiftly diversified into subsectors and products such as access, transmission, video conferencing, and power supply, after the saturation of the switch market. It also expanded rapidly into the global market in the late 1990s by entering the developing markets of Asia and Africa, and in the 2000s it has gained access to established markets in Europe

[2] DMTF is an industry organization that develops, maintains and promotes standards for systems management in enterprise IT environments.

and the United States (Fan 2010b). It became one of the top players in the global Digital Subscriber Line (DSL) and Code Division Multiple Access (CDMA) equipment markets in 2006. Its revenue from the overseas market increased significantly from 12.9 percent in 2003 to 54.18 percent in 2010. Like Huawei, ZTE has transformed itself into a significant global player, based not only on market share but on developing emerging technology using standards from advanced countries. Some examples of this are ZTE's multiple mode phones, including WCDMA/GSM (developed in 2004) and TDSCDMA/GSM U 350 (the first one in the world, developed in 2006). Both provided flexibility for subscribers to use phones in networks of different standards. Another example was its Gogo technology, which provides in-flight broadband internet and other connectivity services. Gogo is currently on 2,000 commercial and 6,000 business aircrafts of several US airlines, including American Airlines, Delta, and US Airways. Upon the invitation of AirCell, whose name changed to Gogo in 2011, ZTE jointly developed the technology with Qualcomm because of ZTE's unique strength in *CDMA EV-DO Rev. A* networking equipment, a type of wireless broadband equipment that transmits data at high data rates through radio signals. ZTE is also ranked high in innovation capability measured by number of patents: it applied for 2,826 patents through World Intellectual Property Organization (WIPO) in 2011,[3] listed as the number one patent applicant in the world, surpassing the long-term leader Matsushita. The innovation capability comes not only from its seven R&D units in China, but also its eight global R&D facilities. Nevertheless, ZTE is facing some challenges: currently, there is a slowdown in demand for telecom-infrastructure equipment as well as weakening economies in China and abroad in 2012–13. ZTE lost 2.84 billion yuan in 2012 due to delays in network projects, but was able to turn profitable (1.2–1.5 billion yuan ($198.3–$247.9 million)[4]) in 2013 through restructuring to increase efficiency and cut costs and the fulfillment of some low-margin contracts in Africa, South America, and Asia, despite the fall in revenue from its handset operations and the increased costs of building of the fourth-generation telecom networks.

DATANG

Founded in September 1998 in Beijing by the China Academy of Telecommunication Technology from its Science and Technology Research Institute, Datang Telecom is a much smaller, state-owned company and is not as

[3] See WIPO IP Facts and Figures (2012). <http://www.wipo.int/export/sites/www/freepublications/en/statistics/943/wipo_pub_943_2012.pdf>.

[4] See Reuters, "BRIEF-ZTE to swing to profit in 2013 after stringent control over signing low-margin contracts." <http://uk.mobile.reuters.com/article/telcommunicationsServicesSector/idUKL3N0KU2I120140120>, accessed Mar. 2014.

well-known as Huawei or ZTE outside of China. Within China, however, Datang has played a highly important role in the development of the telecom-equipment industry by developing TD-SCDMA, the home-grown international 3G mobile technology, through its subsidiary Datang Mobile. TD-SCDMA was accepted as a 3G international technology standard by the International Telecommunications Union (ITU) in May 2000.

TD-SCDMA has some advantages in technological improvement over the existing standards such as synchronized CDMA, intelligent antennas, software-based wireless, and high-speed data transmission. Intelligent antennas[5] and software-based wireless are unique to TD-SCDMA. In addition, TD-SCDMA is claimed to have high spectrum utilization because TD-SCDMA uses the Time Division Duplex (TDD) model and only needs a single channel for bi-directional communications[6] (*Communication World* 2002). Since the distribution of 3G licenses in 2009, China's TD-SCDMA market has grown rapidly, with 54 million subscribers in January 2012, compared with 38 million and 43 million subscribers for CDMA 2000 and WCDMA, respectively. In 2013, more than 100 million people became TD-SCDMA handset users, more than the sum of WCDMA and CDMA 2000 users. However, there have been persistent charges that the TD network is less reliable and slower than its competitors. It is suspected that the flawed network is behind the China mobile's eager switch to 4G networks, making TD a short-lived 3G network (Tan 2014). TD-LTE ADVANCED, the next generation technology of TD-SCDMA, became a 4G technology standard in 2012, and Datang and other local firms are actively providing TD-LTE equipment for both the domestic market and the global market. In fact, China Mobile has built the largest 4G network in the world, with more than 400,000 base stations, and it has attracted more than 20 million 4G users in China in 2013.

Regardless the debate, what is not disputable is the fact that, before TD-SCDMA, the Chinese telecom-equipment market, especially the wireless equipment market, was dominated by MNCs (Lou 2008). Leading local telecom-equipment firms such as Huawei, ZTE, and Datang were not yet the dominant players in China, and in fact only accounted for a small percentage of the market (for GSM, about 5 percent; for CDMA, about 10 percent) (Li 2006). In 2009, after TD-SCDMA, local firms' market share rose dramatically: in TD-SCDMA, they held more than 80 percent, and in the overall 3G (TD-SCDMA+WCDMA + CDMA2000) market, more than 75 percent.

[5] Intelligent antenna or smart antenna technology is a key technology of TD-SCDMA, it greatly reduces complexity and increase system capacity by using digital signal processing (DSP) algorithm and dynamically generate the beam pattern (Mitjana et al. 2000).

[6] For instance, TD-SCDMA only uses 1.6 MHz of spectrum for both directions whereas WCDMA and CDMA 2000 use 5 MHz and 1.25 MHz, respectively for each direction of communication.

As the leading developer of TD-SCDMA, Datang Mobile has focused on the commercialization of TD-SCDMA. Datang Mobile and its partners currently claims 30 percent of the China TD-SCDMA equipment market, providing TD-SCDMA network solutions and integrated services for mobile operators in more than 20 provinces. However, unlike Huawei and ZTE, Datang does not produce brand-name consumer products such as handsets.

It is interesting to note that all of the three leading Chinese telecom companies have emphasized R&D and innovation capability during the course of their rapid growth, although Huawei and ZTE have chosen a very different path in commercializing their R&D results from Datang. The theory of innovative enterprise can be applied to explain why these particular enterprises have succeeded in innovation for growth. First, the CEOs of the three companies, Ren Zhengfei (Huawei), Hou Weigui (ZTE), and Zhou Huan (Datang) all have solid strategic control and financial commitment to R&D. With their long-term control of the companies as founders, they have all allocated substantial amount of the firms' resources to R&D (about 10–20 percent of total revenue), even at the earliest stage of development. Having previously been engineers or scientists at state technological organizations, these leaders all have interest and conviction in developing China's indigenous technology. Although Datang did not have a strong revenue base for R&D, its key decision-maker was able to gain considerable government financial backing for its R&D in TD-SCDMA.

It is worth mentioning that in addition to the top decision-makers, many other upper and middle-level managers also have backgrounds in engineering and are keen to maintain the technological edge of the companies. Another organizational advantage is that these companies have strong management cultures. They use competitive salaries and employee ownership (in Huawei's case) to attract and retain high-quality R&D staff. Other measures, such as regular training by internal staff and outside experts and reforming the career path, also enhanced the organizational integration. For instance, Huawei designed a five-level career path; after passing the second level of technological certification, every employee can choose either the management path or the professional path. This not only ensures that managers have an excellent technological base—since only employees with the second-level certification may become the third-level managers—but it also means that technical professionals can be cultivated within the company (Fan 2010a). With these strategies of organizational integration, these companies effectively create financial incentives for employees to "apply their skills and efforts to strategic objective of innovation" (Fan 2010a, 2010b).

Finally, unlike most Chinese firms emphasizing technological transfer, these firms focused on internal development of technology. When compared with multinationals in China, they have particular advantages and motivation to ramp up their innovation capability and offer training in technological learning. The following sections elaborate on these areas.

Internal Development of Technology vs. Technology Transfer or Joint Development

Our research findings have indicated that the successful Chinese telecom companies have a history of focusing on internal development rather than external collaboration or technology transfer from MNCs because of their early successful experience. Internally developed technology was critical to boost the early growth of these companies. A technology like the aforementioned Public Digital Switch System (PDSS) served this purpose in the mid-1990s, such as C&C08 for Huawei, ZXJ-10 for ZTE, and SP-30 for Datang. Many Chinese companies have emphasized technological transfer through joint ventures, joint cooperation, or importations because it seems to offer a shortcut on catching up with the technological frontier. But these three companies believe that in-house R&D and internally developed technologies are the key to industrial leadership (Fan 2006a, 2006b; Gao 2003). Internal development in latecomer companies cultivates firms' technological capacities and grants them strategic control of the future development of the companies.

Continuous learning and innovation from these three companies eventually differentiated them from those who chose not to innovate at the earlier stages of development. The companies that grew primarily through technological transfer had to withdraw from the industry later once the market for small- or large-scale switch equipment became saturated (Fan 2006b). One of the authors' earlier studies revealed that innovation capability and self-developed technologies allow domestic companies to catch up with the MNCs later on. While the external alliance with MNCs is important, the effectiveness of such an alliance is contingent on the strength of the in-house R&D development. Without strong internal R&D strength, it is hard to convince others to become partners in technological collaboration. Thus, a strategy that focuses on internal development complemented with an external alliance is recommended (Fan 2006a).

The choice to develop technologies internally was in part forced upon these three firms by internal and external circumstances. When these companies were young and small, importing expensive cutting-edge technology was often beyond their resources. External factors, such as the export control imposed on China by foreign countries in the telecommunication sector, have also made it hard for Chinese firms to import the necessary telecommunication technology. Both of these factors strengthened the determination of the companies' efforts in developing proprietary technologies. After all, they believe that self-developed technologies are necessary for industrial leadership since they give the company strategic control over technological trajectories (Fan et al. 2007).

Advantages in and Motivations for Innovation: Chinese Firms vs. MNCs

The experience of the leading Chinese telecom companies demonstrates that home demand and a close and cooperative relationship with domestic customers can be transformed into motivation of—and pressure for—innovation. Porter has argued that the size of home demand is far less significant than the character of home demand for innovation. We argue, however, that both the size and the character of China's domestic market have been significant in driving domestic innovation. The character of demand can be significant in providing opportunities for domestic companies to innovate faster and achieve more sophisticated competitive advantages than their foreign rivals (Porter 1990). Large and diverse domestic demand, meanwhile, provides opportunities for product variations and allows domestic companies to concentrate their innovation efforts on specific areas. The size also ensures significant profit if the company is able to provide higher quality and/or lower cost products than foreign vendors in limited areas. For instance, Huawei and ZTE's initial profit came from meeting the substantial demand for small-scale public switch equipment from the countryside, a low-end market segment largely ignored by the MNCs. Later, Huawei, ZTE, and Datang succeeded in producing large-scale public digital switches in the 1990s. This product development benefitted from the seemingly insatiable home demand for small- and large-scale switch equipment as China quickly ramped up its telecom infrastructure, a demand that the existing foreign imports and joint venture production could not meet quickly enough (Fan et al. 2007). Of course, rapidly growing domestic demand may benefit all companies in the market for a given period, but only those that show long-term vision and strategic commitment to the improvement of innovation capacities can sustain success in the long run.

Since Chinese firms focus on the Chinese market, they have transformed their better understanding of home demand and the close relationship with customers into a better integration of market and R&D. This often can give them a local advantage over foreign-based MNCs. While MNCs in China usually staff R&D and marketing teams with local personnel for local customization, domestic companies enjoy close connections between their entire marketing and R&D teams. This integration allows them to react quickly to develop solutions and products in response to the market (Fan 2006a). This is particularly important in China because the rapid economic growth and diverse geographical conditions mean that the market can often change quickly; it needs close monitoring and fast reaction. Such home turf advantage was used by Lenovo in its dramatic rise to be the top personal computer seller in China in the mid-1990s. Lenovo was able to offer less expensive personal computers equipped with the latest operation system, while its foreign competitors were marketing computers that were one generation behind the global mainstream (Zhou 2008).

In contrast to this, foreign firms, separated from the core R&D and without localized autonomy in decision-making, generally have a difficult time reacting to clients' need as quickly as domestic companies (Fan 2006a; Zhou 2008). This has recently been changing, however, since many MNCs have established more substantial R&D centers in China as it grew to become a significant share of their global markets. However, various communication barriers and strategic differences between the Chinese subsidiaries and global headquarters cannot be easily overcome. For instance, it was the Chinese companies, not Samsung or Apple, who first developed low-cost smartphones that became popular in Chinese markets (see Chapter 10), swiftly responding to the demand for affordable smartphones in China. Clearly successful Chinese domestic firms still have local advantages over their foreign rivals.

From Imitation to Innovation

In terms of technological learning, the Chinese telecom companies have followed the general pathway of latecomers described by Kim (1997) with some adaptations. First, although they imitated mature foreign technologies such as those in switch equipment, they actually succeeded not by directly acquiring the technologies or learning from the MNCs as the Korean firms did, but through developing their switch technology in their own way. They then repeated the same process on higher-level technologies in the consolidation stage of advanced countries. Catching up in several key technologies of the telecom-equipment sector first, all three firms transformed themselves from imitators into innovators, eventually generating emerging technologies such as the third-generation and fourth-generation mobile communication technology (Fan 2011).

It should be mentioned that, unlike other latecomers from newly industrialized economies in the late twentieth century, Chinese companies had great difficulty in acquiring foreign technologies in the telecommunication sector in the 1980s and 1990s. They were forced to build their own innovation capability, mostly by using domestic resources, such as technical experts from domestic research institutes (e.g. MPT's Telecom S&T Research Institute and People's Liberation Army Information Engineering Institute) or higher education institutions, such as Beijing University of Posts and Telecommunications and Nanjing University of Posts and Telecommunications, as they developed the switch technology. Further, for these Chinese firms, collaboration with established industrial partners only came at a later stage, after they had developed a higher level of innovation capability. In the early stages of catching-up, they built their innovation capability almost exclusively through conducting in-house R&D on their core technologies, recruiting domestic talent, and working with domestic institutes.

After making a technological breakthrough by developing their own large-scale switch equipment, at first Huawei and ZTE shifted their focus to access system equipment,[7] a well-defined R&D segment that had the most potential in terms of market and innovation capability at the time. More recently, they have expanded into new fields for growth such as optical transport, data communication, and mobile communications. In comparison, Datang concentrated all its investment in TD-SCDMA. Currently, all three companies still conduct R&D on cutting-edge telecom technologies, and Huawei and ZTE now have similar strategies to other first-tier global players. For example, there are parallel research teams on CDMA and GSM systems in ZTE. Both Huawei and ZTE have multiple research teams on the third generation of mobile communication technology, CDMA2000, WCDMA, and TD-SCDMA (Fan 2011; Feng 2010; Gao and Liu 2012).

As these Chinese companies grew, they also started to collaborate with established firms, but only after they already processed many of their own technologies. For instance, as mentioned before, ZTE was sought out by Aircell to collaborate with Qualcomm in developing Gogo technology because of ZTE's unique strength in *CDMA EV-DO Rev. A* networking equipment. 3Com collaborated with Huawei because of Huawei's strength in data communication, which is comparable to 3Com's long-term rival Cisco (Fan 2011). These examples support the argument that internal innovative capacities are the necessary tickets for companies to be invited to sit at the global table of technological innovation.

Conclusion

We studied the development of China's telecommunication equipment industry since the 1980s by focusing on the technological learning of three leading companies, Huawei, ZTE, and Datang. This study is situated in the heated debate on whether or not innovation is important for latecomers, how innovation capability can be developed for latecomers, and what the

[7] The access network refers to the network which connects subscribers to their immediate service provider. Access system equipment are telecom equipment for the access network.

role of government should be in industrialization and innovation. China's telecom-equipment market is one of the most dynamic ones in the world due to its rapid growth and intense market competition. Since the late 2000s, Chinese telecom multinationals such as Huawei and ZTE have become global leaders side by side with Ericsson, Alcatel-Lucent, and Nokia Siemens Networks. Datang has become a global key technological provider due to its development of China's 3G standard TD-SCDMA and related technologies and products of TD-LTE.

State intervention has assisted the technological learning of the firms in the telecommunication equipment industry by carefully modulating its support in relation to the development of domestic firms. In particular, intervention in demand patterns, either through matchmaking between the market and domestic companies or restructuring of technological communication services to increase competition, has promoted innovation by stimulating demand for domestic companies. Government intervention is also reflected in its crucial support of local firms' participation in industrial standards, illustrated especially by the case of TD-SCDMA.

Further, innovative enterprises in China's telecom-equipment industry have focused on internal development as their primary approach for gaining innovation capability. They have utilized their advantages in understanding the domestic market and close relationships with customers to produce high-quality, low-cost products to serve the domestic market.

The theory of innovative enterprises emphasizes social conditions of strategic control, organizational integration, and financial commitment of the enterprises. Our cases show that these leading firms, under the leadership of their long-term founders and CEOs, have created such social conditions by remaining financially committed to and focused on internal technological development. This was in part a choice of necessity because of the external control of high-tech export by the United States and its allies. But in-house research has been critical in nurturing the technological capacity of the firms, and it also enhances the strategic control of technological trajectories which sets up companies for industrial leadership. Their accumulated in-house capacities earn them the respect to explore other technological options, such as transfer and R&D collaborations with established foreign parties. We found that the domestic firms have taken advantage of their understanding of the character and size of home demand and their close relationship with domestic telecommunication service providers. Such relationships motivate and incentivize domestic firms to innovate in the domestic market and create local advantages over MNCs. Overall, the size and quality of the Chinese market can and does support innovative firms—but only those with long-term visions that include the commitment to in-house technological development.

References

Amsden, A. (1989), *Asia's Next Giant: South Korea and Late Industrialization*. Oxford: Oxford University Press.

Austin, J. E. (1990), *Managing in Developing Countries: Strategic Analysis and Operating Techniques*. New York: Free Press.

Bernstein Research, (2010), *Chinese Telecommunications: The Three Kingdoms*. New York: Sanford C. Bernstein & Co.

Bell, B., M. Carpenter, H. Glimstedt, and W. Lazonick (2013), *From Innovation to Financialization: How Cisco Became Focused on its Stock Price and Lost its Way*. Working Paper. Cambridge, MA: The Academic-Industry Research Network.

Carpenter, M., W. Lazonick, and M. O'Sullivan (2003), "The Stock Market and Innovative Capability in the New Economy: The Optical Networking Industry," *Industrial and Corporate Change*, 12(5): 963–1034.

China Daily (2012), "China's Internet users hit 538 million," <http://www.chinadaily.com.cn/business/2012-07/19/content_15601041.htm> [accessed Jan. 2013].

Chu, W. W. (2011), "How the Chinese Government Promoted a Global Automobile Industry," *Industrial and Corporate Change*, 20(5): 1235–76.

Communications World (2002), "After Division of 3G Frequency Resource," Nov. 28 (Chinese Magazine).

Fan, P. (2006a), "Catching up through Developing Innovation Capability: Evidence from China's Telecom-Equipment Industry," *Technovation*, 26(3): 359–68.

Fan, P. (2006b), "Promoting Indigenous Capability: The Chinese Government and the Catching-up of Domestic Telecom-Equipment Firms," *China Review*, 6(1): 9–35.

Fan, P. (2010a), "From a Latecomer to a Global Telecom Giant: The Development Path of Huawei," *International Journal of Business System Research*, 4(5/6): 691–719.

Fan, P. (2010b), "Catching up through Staged Development and Innovation: The Case of Chinese Telecom Companies," *Journal of Science and Technology Policy of China*, 1(1): 64–91.

Fan, P. (2011), "Innovation, Globalization, and Catch-up of Latecomers: Cases of Chinese Telecom Firms," *Environment and Planning A*, 43(4): 830–49.

Fan, P., and C. Watanabe (2006), "Promoting Industrial Development through Technology Policy: Lessons from Japan and China," *Technology in Society*, 28(3): 303–20.

Fan, P., X. Gao, and K. Watanabe (2007), "Technology Strategies of Innovative Chinese Domestic Companies," *International Journal of Technology and Globalization*, 3(4): 344–63.

Feng, K. (2010), "Catching up or Being Dependent: The Growth of Capabilities among Indigenous Technological Integrators during Chinese Development." Doctoral dissertation, University of Sussex.

Gao, X. (2003), "Technological Capability Catching up: Follow the Normal Way or Deviate." Doctoral dissertation, Massachusetts Institute of Technology.

Gao, X., and J. Liu (2012), "Catching up through the Development of Technology Standard: The Case of TD-SCDMA in China," *Telecommunications Policy*, 36(7): 531–45.

Hobday, M. (1995), *Innovation in East Asia: The Challenge to Japan*. Cheltenham: Edward Elgar.

Huang, Y. (2003), *Selling China: Foreign Investment during the Reform Era*. New York: Cambridge University Press.

Johnson, C. (1982), *MITI and the Japanese Miracle*. Stanford, CA: Stanford University Press.

Kim, L. (1997), *From Imitation to Innovation: Dynamics of Korea's Technological Learning*. Cambridge, MA: Harvard Business School Press.

Kshetri, N., P. Palvia, and H. Dai (2011), "Chinese Institutions and Standardization: The Case of Government Support to Domestic Third Generation Cellular Standard," *Telecommunications Policy*, 35: 399–412.

Lazonick, W. (2003), "The Theory of the Market Economy and the Social Foundations of Innovative Enterprise," *Economic and Industrial Democracy*, 24(1): 9–44.

Lazonick, W. (2004), "Indigenous Innovation and Economic Development: Lessons from China's Leap into the Information Age," *Industry and Innovation*, 11(4): 273–97.

Lazonick, W. (2006), "The Innovative Firm," in J. Fagerberg (ed.), *The Oxford Handbook of Innovation*, 29–55. New York: Oxford University Press.

Lazonick, W. (2007), "Varieties of Capitalism and Innovative Enterprise," *Comparative Social Research*, 24(1): 21–69.

Lazonick, W. (2009), "The New Economy Business Model and the Crisis of US Capitalism," *Capitalism and Society*, 4(2): article 4.

Lazonick, W. (2010), "The Chandlerian Corporation and the Theory of Innovative Enterprise," *Industrial and Corporate Change*, 19(2): 317–49.

Lazonick, W. (2011), "The Innovative Enterprise and the Developmental State: Toward an Economics of 'Organizational Success'," in *Annual Conference of the Institute for New Economic Thinking*, 15–16. Bretton Woods, NH: Institute for New Economic Thinking.

Lazonick, W., and E. March (2012), "The Rise and Demise of Lucent Technologies," Munich Personal RePEe Archive paper 2012, <http://mpra.ub.uni-muenchen.de/22012>.

Leonard-Barton, D. (1995), *Wellspring of Knowledge: Building and Sustaining the Sources of Innovation*. Cambridge, MA: Harvard Business School Press.

Li, J. (2006), *Reflections on Innovation* (in Chinese). Beijing: Intellectual Property Right Press.

Liu, X., and S. White (2001), "Comparing Innovation Systems: A Framework and Application to China's Transition Context," *Research Policy*, 30: 1091–1114.

Lou, Q. (ed.) (2008), *A 10 Year History of TD-SCDMA*. Beijing: Electronics Industry Press.

Lu, Q. (2000), *China's Leap into the Information Age: Innovation and Organization in the Computer Industry*. Oxford: Oxford University Press.

Lu, Q., and W. Lazonick (2001), "The Organization of Innovation in a Transitional Economy: Business and Government in Chinese Electronic Publishing," *Research Policy*, 30(1): 55–77.

Motohashi, K., and Y. Xiao (2007), "China's Innovation System Reform and Growing Industry and Science Linkages," *Research Policy*, 36: 1431–42.

Naughton, B. (2007), *The Chinese Economy: Transitions and Growth*. Cambridge, MA: MIT Press.

Nelson, R. (1993), *National Systems of Innovation: A Comparative Study*. New York: Oxford University Press.

New Beijing News (2014), "TD-SCDMA Debate: Is 200 Billion RMB Investment Worthwhile?," *New Beijing News*, Dec. 25, <http://tech.163.com/14/1225/10/AEA7AN75000915BE.html> accessed Aug. 2015].

Organization for Economic Cooperation and Development (OECD) (1972), *The Industrial Policy of Japan*. Paris: OECD.

Porter, M. E. (1990), *The Competitive Advantage of Nations*. New York: Free Press.

Rosovsky, H. (1972), "What are the Lessons of Japanese Economic Development?" in A. Youngson (ed.), *Economic Development in the Long-Run*. London: Allen & Unwin.

Reuters (2012), "China's Mobile Subscribers up 1 pct in October at 1.09," <http://www.reuters.com/article/2012/11/20/china-mobilesubscribers-idUSL3E8M90X420121120> accessed Jan. 2013.

Sakakibara, M., and D. S. Cho (2002), "Cooperative R&D in Japan and Korea: A Comparison of Industrial Policy," *Research Policy*, 31(5): 673–92.

Seekingalpha.com (2012), "As Chinese Companies Conquer Global Telecom Equipment, Other Industries are Bound to Follow," http://seekingalpha.com/article/789791> accessed Jan. 2013.

Sumiya, M. (2000), *A History of Japanese Trade and Industry Policy*. Oxford: Oxford University Press.

Tan, M. (2014), "Innovation in TD style" (TD 式创新), *Caixin, New Century*, 47. Dec. 8, http://weekly.caixin.com/2014-12-05/100759605.html.

Wan, Y. (2001), "Sector Reform," in Jintong Lin, Xiongjian Liang, and Yan Wan (eds), *Telecommunications in China: Development and Prospects*, 159–80. Huntington, NY: Nova Science Publishers.

Wilson, R. W., P. K. Ashton, and T. P. Egan (1980), *Innovation, Competition, and Government Policy in the Semiconductor Industry*. Lexington, MA: Lexington Books.

World Bank (2014), *World Development Indicator*. Washington, DC: World Bank.

Zhan, A., and Z. Tan (2010), "Standardization and Innovation in China: TD-SCDMA Standard as a Case," *International Journal of Technology Management*, 51(2–4): 453–68.

Zhou, Y. (2008), *The Inside Story of China's High-Tech Industry: Making Silicon Valley in Beijing*. Lanham, MD: Rowman & Littlefield.

9

State, Multinational Corporations, Returnees, and Development of China's Integrated Circuit (IC) Design Industry

Yifei Sun, Zi Xue, and Debin Du

Introduction

The semiconductor is a critical component associated with the rise of the information technology industry during the last few decades. Popularly known as the brain of IT products, it is ubiquitously embedded in products ranging from ID cards to super computers, and it serves indispensable functions in industry and daily life. Given the strategic value of the semiconductor, it is not surprising that many countries have tried to establish this industry. However, as pointed out in the earlier chapter by Li, developing the semiconductor industry is very challenging since the industry demands highly experienced personnel, is extremely capital intensive, and makes very rapid technological progress. These conditions make it hard for late developers to unseat incumbents in a globally integrated and highly competitive industry.

The most successful catch-up cases in the world have been Japan and, more recently, Taiwan and South Korea (Chang and Hsu 1998; Chen and Sewell 1996; Cho et al. 1998; Choung et al. 2000; Lee and von Tunzelmann 2005; Liu 1993; Lu et al. 2004; Kim 1997, 1998; Kimura 1988; Mathews 1997; Mathews and Cho 1999, 2000). During the last three decades, a number of firms such as Taiwan Semiconductor Manufacturing Corporation (TSMC), Samsung, and Chartered Semiconductor in Singapore have also become significant competitors in the global market (Mathews 1999). Such examples have inspired other emerging economies, especially China, to develop their own semiconductor industry. With several false starts in the 1980s, China has achieved major progress in this industry. Revenue in the industry grew from less than $200 million in 1994 to $38.1 billion in 2011 (Hu 2006; Zhu 2006; Klaus 2003; PWC 2012). By 2014, China had developed a sizeable semiconductor industry value chain from assembly and testing to fabrication and IC design. In assembly and testing, leading multinationals including Intel have built numerous facilities in China. As detailed in the previous chapter by Li, manufacturing firms such as Semiconductor Manufacturing International (SMIC) in Shanghai and Shanghai Huahong NEC Electronics Co. Ltd (HHNEC) have become significant players in the world semiconductor foundry market. In the IC design segment, about 500 design firms have emerged since the late 1990s.

With progress in technology, the semiconductor global value chain (GVC) (Gereffi et al. 2005) has become increasingly disintegrated. The original integrated device manufacturing (IDM) model where one business includes all the functions—design, manufacturing, assembly, packaging, and testing—has shifted to the current state of coexistence of the IDM model and the disintegrated model in which each segment of the GVC is occupied by independent entities. The disintegration of the semiconductor value chain offers more opportunities for small and medium enterprises (SMEs) to engage in specified segments of the GVC, gain in-depth knowledge, and make technological progress. On the one hand, it opens up opportunities for enterprises in less developed countries to engage in segments of the whole value chain. On the other hand, such engagement is also very challenging since players in this industry have become increasingly specialized and thus possess in-depth knowledge, especially tacit knowledge that can only be learned by doing.

The disintegrated value chain for the semiconductor industry presents China with both opportunities and challenges. The chapter by Li focused on the manufacturing segment of China's semiconductor industry. In this chapter, we will focus on the design segment of the semiconductor value chain. The IC design industry differs from the manufacturing industry in several aspects. Manufacturing is much more capital-intensive and larger in scale than the IC design industry, but is relatively distant from the consumer. In contrast, IC design is heavily dependent on the close interactions between design firms and final consumer demands. The historical trajectory, industrial structure, and dynamics are different as well.

The development of China's IC design industry raises fascinating questions about the roles of the Chinese state, the globalization of IC manufacturing, the evolution of an industrial ecosystem, and the circulation of global talent. Those are the major questions we try to address in this chapter. We argue that China's measured success in the IC design industry in recent years is associated with its growing economic power in general, and the rapid expansion and diversification of domestic market for IC product in particular. The IC design industry in China has also benefitted from the maturing IC foundry industry and the large number of overseas Chinese who have returned to China, both of which have been supported by the Chinese government.

A Brief History of China's IC Industry between the 1960s and Late 1990s

China's IC industry was included in the governmental agenda soon after the establishment of People's Republic of China in the 1950s. Since 1956 when the Chinese government included semiconductors as a key industry in its "Twelve-Year Science and Technology Development Long-Term Program," the IC design industry has developed through three stages: (1) the beginning stage (1965–80), (2) falling behind—the slow growth stage (1980–2000), and (3) the catch-up stage (post-2000). Each stage shows a different trajectory.

As seen in the previous chapter, China was isolated from the West and the Soviet Union during the beginning stage (1965–80). It had no choice but to develop the integrated circuit (IC) technology through its own efforts. During this period, China's national innovation system (NIS) was organized in a hierarchical fashion where R&D was separated from manufacturing (Sun 2002); government-owned laboratories performed R&D while state owned-enterprises (SOEs) managed manufacturing. China's first IC was developed successfully in the mid-1960s by a couple of governmental R&D laboratories, including the Chinese Academy of Science, the Institute of Semiconductors, the Hebei Semiconductor Institute, and the Beijing Semiconductor Device Institute (Zhu 2006: 67). Such ICs were small and primarily used in computers for military purposes.

During this initial period when overall IC technology changed slowly, China made evident IC technology progress. Starting with small-scale integration (SSI) in 1965, it increased to medium-scale integration (MSI) in 1972 and further to large-scale integration (LSI) in 1972, only lagging behind the leading country by about six to seven years in terms of numbers of transistors on the IC. As was generally the case in the world market at the time, most IC factories were very small in production capability since the market was confined to computers with limited military applications, with less than 1 million units a year. By the end of the 1970s, only the No. 19 Shanghai Wireless Electronics Factory possessed the capability of more than 5 million units. The initial development of the IC design industry was limited and was an integral part of the production process. Such earlier efforts though did lay down foundational institutions, train a core group of experts, and plant the seeds for later growth.

The second period was from 1980 to 1995, corresponding to the early stage of China's open door policy. China's indigenous IC was stagnant, facing the competition from foreign sources during a time when the global progress on IC was growing by leaps and bounds. The gap between China's own technology and the global frontier widened considerably. Realizing the strategic role of the IC industry and the advancement of technologies in foreign countries, the state council decided to initiate a large-scale and internationally competitive IC industry in China (Chen 2005). The effort went through several ups and downs, as discussed in the previous chapter by Li, but more attention was devoted to manufacturing than to design capacity. The state council decided to establish the "Computer and Large-Scale IC Leading Group," led by then Vice-Premier Wan Li in 1982. The group laid out China's IC development plan during the 6th Five-Year Plan (1981–5). During this period, a number of production lines were imported first by SOEs and later by joint ventures, and IC design was primarily a function of many integrated device manufacturing facilities. Significant players included Wuxi Jiangnan Wireless Device Factory (No. 742 Factory), which imported the competed production line from Toshiba, Wuxi Huajing Microelectronics Corporation, Huayue Microelectronics Corporation in Shaoxin, Shanghai Belling Microelectronics Corporation Limited, Shanghai Philip Semiconductor Corporation, and Beijing Capital Steel-NEC Electronics Corporation. IC design was a minor function of the manufacturing enterprises; the primary focus was on raising production capability.

Rise of China's IC Design Industry since 2000

The third stage started in the late 1990s, and may be better termed business-led development. At this stage, IC development was characterized by the global shift of IC production to China and the aggressive expansion of indigenous manufacturing firms with transnational resources. Both set the stage for the rise in the IC design sector. By the late 1990s, IC technology and production had experienced great progress and expansion outside mainland China. Japan, South Korea, and Taiwan had become significant players in the global IC market, while China's IC production and sales made up less than 1 percent of the world market, despite the rapid growth of China's economy.

In 1996, sensing the failure of its semiconductor development strategy and the growing demand for IC chips, the Chinese government restarted its effort in the 9th Five-Year Plan by substantially increasing state investment through Project 909. Particularly important in this round of efforts was that the Chinese government realized that manufacturing was only part of the value chain. Simple expansion of manufacturing without improved design capabilities would not lead to comprehensive development. Consequently, China also started building eight IC design centers along with Project 909, in the hope that IC design capabilities would progress along with manufacturing, and that the two components would build a healthy ecosystem for the semiconductor industry.

China's IC progress since 2000 is the result of multiple forces: the rise of many indigenous multinational corporations in information telecommunication technology (ICT), China's entry into the WTO, and a large number of returnees who had formerly emigrated, among other reasons. A major structural change in the IC industry during this period is the emergence and significant growth of independent IC design houses. Before, IC design was part of the integrated device companies (IDC) and all the major IC enterprises had their own in-house IC design departments. In addition, some large assemblers and final electronics product companies also had an IC development department for application-specific IC (ASIC). IC design enterprises in China did not become a significant independent sector until after 2000 when IC manufacturing companies such as SMIC and Huarun (Wu Xi) started to specialize in foundry facilities that provide fabrication services to independent IC design firms. While China's Ministry of Electronics Industry built the first independent IC design company in 1986 (China Huada IC Design Company), by the end of 1999, China had 20 IC design houses (MOST 1999) and sales for all the IC design firms were about 200 million RMB. In comparison, hundreds of IC design firms have already populated the Hsinchu Science Park by the end of the 2000s (Zhou and Hsu 2010).

Since 2000, hundreds of independent IC design companies have emerged (Figures 9.1 and 9.2). Many of these were set up by returned engineers from overseas and researchers associated with former domestic IC research institutes. The IC design sector and fabrication have been the faster growing segments of China's semiconductor industry than the assembly and test segment, and IC design is the only segment of China's semiconductor industry "that achieved positive year-over-year (YoY) revenue growth for every year

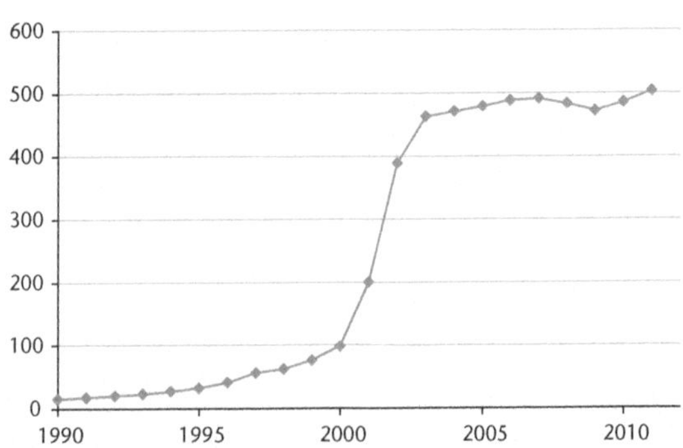

Figure 9.1. Number of IC Design Enterprises in China (1990–2011)

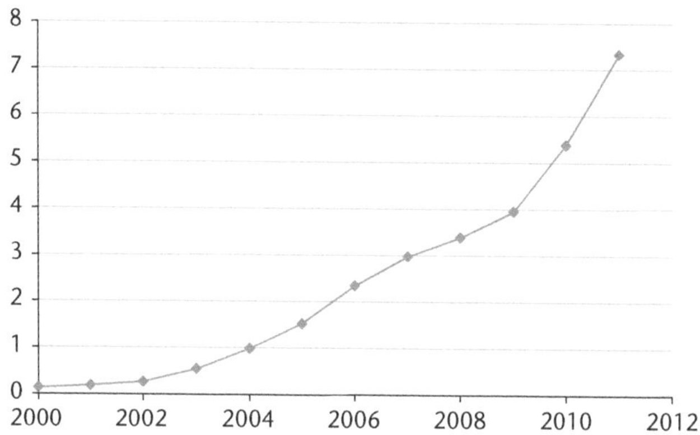

Figure 9.2. Revenues of China's IC Design Sector (2000–11)

Table 9.1. Growth of China's IC Industry (2000–11)

Year	Chinese IC Production and Sales (RMB bn)	IC Design (%)	Fabrication (%)	Assembly and Test (%)
2000	18.6			
2001	18.8			
2002	26.8	8.06	12.54	79.85
2003	35.1	12.79	17.24	70.09
2004	54.5	15.01	33.03	52.02
2005	70.2	17.66	33.19	49.15
2006	100.6	18.49	30.72	50.80
2007	125.1	17.99	31.81	50.20
2008	124.7	18.85	31.52	49.64
2009	110.9	24.35	30.75	44.91
2010	144	25.28	31.04	43.68
2011	157.2	30.15	30.98	38.93

since 2000" (PWC 2012: 3.1). In 2002, the assembly and test sector claimed about 80 percent of China's IC industry output (Table 9.1), while the shares for the design and fabrication sectors were only about 8 and 12.5 percent, respectively. In comparison, the shares for design, fabrication, and assembly and testing in China's IC industry have shifted to 30.15:30.98:38.93 in 2011. In particular, the IC design segment increased its share in the IC industry from 8.06 percent to 30.15 percent, or almost one-third of China's IC industry. More significantly, this proportion increased despite the fact that the packaging and assembly sector nearly tripled its sales from 21.4 billion RMB in 2002 to 61.2 billion RMB in 2011, while sales for the IC fabrication sector grew from 3.4 billion RMB in 2002 to 48.7 billion RMB 2011, with an annual growth rate of 29.7 percent. In comparison, the revenue of the IC design sector grew from US$178 million in 2001 to 7.3 billion in 2011, achieving a compounded annual growth rate of 45 percent. China's IC design sector even

Table 9.2. Top 10 Chinese IC Design Firms in 2011

Rank	Firm	2011 Sales (100m RMB)
1	Hisilicon Technologies C. Ltd (Shenzhen)	66.69
2	Spreadtrum Communications Inc.	42.88
3	RDA Microelectronics, Inc.	15.9
4	Hangzhou Silian Microelectronics C. Ltd.	13.3
5	Galaxycore Inc.	11.68
6	Shenzhen State Microelectronics Co. Ltd	11.2
7	Leadcore Technology Co. Ltd	9.44
8	CEC Huada Electronics Design Co. Ltd	8.24
9	Datang Microelectronics	6.24
10	Shanghai Huahong IC Co. Ltd	6.1

Source: *Shanghai IC Industry Development Report*, 2012: 50–1.

grew during the period between 2008 and 2009 when global semiconductor demand declined. In the year 2011, China's IC design sector grew by 36 percent compared to the worldwide market growth of 0.4 percent (PWC 2012: 31). In particular, most revenues of China's IC design sector came from fabless IC design companies which accounted for about 10 percent of the $74 billion worldwide fabless IC design industry, up from a mere 1 percent in 2001 and 4 percent in 2004 (Table 9.2).

The IC design industry has also shown significant progress in scale, technology, and production. Among the 503 IC design companies reported by China's Semiconductor Industry Association, PWC counted that about 240 were IC design houses for foreign companies and "no more than 100, possibly less, of the local indigenous IC design enterprises... are truly viable fabless semiconductor companies" (PWC 2012: 3.3). Meanwhile, China's Semiconductor Industry Association reported that 99 IC design enterprises had sales above 100 million RMB in 2011. Indeed, 32 IC design enterprises had more than 500 employees in 2011, while only 26 IC design firms had more than 500 employees a year earlier. Clearly, select IC design houses have expanded significantly with the overall growth of Chinese IC market.

In addition, Chinese IC design houses have made significant technical progress, which can partly be traced by looking at the size of the design line. The finer the line width in design, the more detailed and precise the work it can accomplish. More than 50 IC design enterprises in China have adopted process technology with a design line width less than 90 nm. An additional 163 enterprises are able to design with a line width of less than 0.25 micron. In comparison, the number of enterprises with less sophisticated technology (with a line width bigger than 1.0 micro) were about 67 in 2011. Further technical evidence of Chinese IC design houses can be demonstrated by their licensing of advanced technologies. ARM has reached licensing deals with more than 34 Chinese companies for its Cortex processor and Mali graphics processor cores, which indicate that such IC design houses are engaged in advanced design activities. Indeed, many of the Chinese IC design houses (63 percent of EE Time Survey respondents) are using foundry services from Taiwan instead of SMIC and other Chinese domestic foundry service providers, because the more advanced processing capabilities were not available at domestic foundries.

Chinese IC design houses are engaged in various markets. About one-third of them specialize in design, development, and sales of chips for telecommunication, computers, and multimedia applications. In particular, a number of IC design enterprises such as Datang, HiSilicon, and Spreadtrum have made significant progress in network communication, mobile devices, and multimedia chip design. However, about two-thirds of IC design firms still focus on low-end applications such as display-drivers, power management, and identification cards (SHIC and SHICA 2012). A number of IC design firms have become significant players in China and even throughout the world market. Datang, with help from Siemens, developed the Chinese 3G standard, Time Division-Synchronous Code Division Multiple Access (TD-SCDMA). The standard was accepted by the Institute of Electrical and Electronic Engineers (IEEE) in May 2000 as one of the global third-generation mobile communications standards (see Chapters 2, 7, and 8). Datang, as the developer, has since accumulated significant capabilities in IC design with applications along the wireless communication value chain. In 2011, HiSilicon Technologies became China's first billion dollar IC design company with strong capabilities in network equipment as well as mobile phone chip design. Spreadtrum and RDA Microelectronics have become two of the world's top five fastest growing IC design companies, though both were acquired by Tsinghua Unigroup in 2013, making Unigroup a potential major player in the IC design industry. Spreadtrum specializes in cellular phone chip design and provides 2G and 3G chip designs based on TD-SCDMA and WCDMA to leading global brands such as Samsung and HTC. RDA Microelectronics designs and develops wireless systems-on-chip and radio-frequency semiconductors for cellular, connectivity, and broadcast applications. It was the number one supplier of power amplifiers and Bluetooth systems-on-chip and the number two supplier of 2G baseband chips to Chinese manufacturers in terms of units shipped in the fourth quarter of 2012. While the Chinese IC design sector reported a productivity with average sales of $69,000 per employee, much lower than the world average $525,000, Spreadtrum Communications and RDA Microelectronics reported average sales of $1,006,000 and $825,000 respectively, far exceeding the worldwide average. The comparison between the number of IC design firms and total sales demonstrates that average size of IC design firms has gone up significantly. All in all, the upgrading of China's IC design capacity is beyond dispute.

In short, China's IC industry has experienced "parallel upgrading" (Chen 2005; Chen and Xue 2010) in all the three major segments along the IC value chains, including production, design, assembly, and testing. However, China's IC industry is still lagging behind the world-frontier technologies in terms of the sophistication of design and fabrication technology.

State, Market, and Development of China's IC Design Industry

The development of the IC design industry after 2000 deserves careful examination. IC growth escalated after the Chinese state changed its strategy from a hands-on approach of directly building state-owned large IC businesses to a more flexible and decentralized approach of encouraging development of non-state firms, and the whole industrial chain, wherein IC design is an essential component. We argue that the state was not the determining factor in the growth of the design industry, which resulted from multiple forces: growing domestic demand for information technology products, the rise of domestic electronics and home appliance manufacturers, the disintegration of the global IC value chain, and the expanding numbers of experienced overseas Chinese engineers who have decided to return (returnees). These forces together has constructed an industrial ecosystem that both attracted the establishment of MNCs and encouraged the growth of China's indigenous IC design firms. In particular, the establishment of foundry firms in China provided a critical infrastructure for the ecosystem.

To begin with, the rise of China's electronics and information technology industry since the mid-1990s has been extremely rapid. The industry output for the electronics and telecommunication industry grew more than 10 times from 583 billion in 1999 to 6,380 trillion RMB in 2011. A number of domestic giants have become world players. This includes such companies as Lenovo in personal computers; Haier, Hisense, Changhong, and Konka in home appliances; and Huawei and ZTE in telecommunication equipment and recently in mobile devices. For example, Lenovo's worldwide revenue grew more than ten times, from US$1.9 billion in 2004 to US$21.6 billion in 2011. Huawei has become the second largest telecom equipment provider; its revenue grew by eight times from US$4.5 billion in 2004 to $35 billion in 2012. Haier has become the top white home appliance manufacturer and its revenue grew from US$13 billion in 2004 to over $20 billion in 2011. In addition, multinational corporations such as Nokia, Motorola, Samsung, Apple, and others either have established subsidiaries or have subcontracted their production to original equipment manufacturers in China. From 1999 to 2011, output of foreign-invested enterprises in China's electronics and information technology industry grew from 402 billion to 4.9 trillion billion RMB. Their share in

Table 9.3. China's Growth in Electronic and Information Technology Products

Year	Output (100m RMB)	ByFDI	Mobile Phones (m)	Computer (m)	Color TV (m)	Cameras (m)
2000	7550	5403	52.5	6.72	39.4	55
2001	8990	6631	80.3	8.78	40.9	59
2002	11288	8281	121.5	14.64	51.6	54
2003	15839	12209	182.3	32.17	65.4	61
2004	21463	17580	237.5	59.75	74.3	62
2005	26994	22712	303.5	80.85	82.8	82
2006	33077	27172	480.1	93.36	83.8	85.5
2007	39223	32967	548.6	144.4	84.8	86.9
2008	43902	34684	559.6	133.6	90.3	89
2009	44563	34713	681.9	182.2	98	84.6
2010	54970	42539	998.3	245.8	118	93.3
2011	63795	48549	1132.6	320.4	122	82.4

Sources: Compiled by the authors from different sources.

China's total output in the industry increased from 69 percent in 1999 to 76 percent in 2011. China has become the leading source country for many products, from personal computers and laptops to cell phones and color TVs, all of which have experienced dramatic growth since the late 1990s (Table 9.3). For example, China manufactured 52.5 million mobile phones in 2000 and the production rose to 1.1 billion units in 2011. During the same period, China's personal computer production grew from 0.84 million units to more than 320 million units.

All such electronics and telecommunication products contain ICs. According to some estimates, ICs account for nearly 27 percent of the final product value (PWC 2006: 17). This expansion in output created a huge market for IC products, some of which were imported and others of which were produced domestically. Having major appliance makers in China facilitated the growth of design houses who need to have intensive interactions with the clients. In some cases, spin-offs from the appliance makers are common. HiSilicon, the largest IC design firm in China, for example, was a spin-off company from what was originally Huawei's internal design department. Even now, more than 90 percent of HiSilicon's revenue still comes from Huawei. This is very different from China's earlier experience before 2000 in the IC industry, when limited domestic demand hindered growth.

China's development of the IC industry in general, and IC design in particular, has also benefitted from the changing strategies of multinational corporations. Until the 1990s, multinational corporations were not interested in making major investments in China's IC industry, due to the lack of market distinction and concerns with China's uncertain directions during the post-1989 period. In the early 1990s, the only foreign project was Shanghai Belling Semiconductor invested by Philips. When China started Project 909, it had great difficulty attracting multinationals to join its efforts to develop the first 8-inch fabrication facility (Hu 2006). The new Project 909 aimed to build the whole value chain, not just manufacturing capability or advanced design technologies. In addition to advanced manufacturing capability, the Chinese government aimed to build a competent technology and management team, develop strong design capability, and have access to global capital. The successful completion of Project 909 in 1999 laid down the infrastructure for the IC industry growth after 2000 from building construction, equipment transport and installation, power supply and fire protection, among others. It also helped raise the confidence of multinationals and overseas Chinese in China's IC development capabilities. Since then, many multinational corporations have started to invest in China's IC industry. Most of them started with establishing assembly and testing facilities to take advantage of the growing market and cheap labor before making an investment in fabrication.

With China's growing capabilities in the IC industry and the rapidly rising Chinese manufacturing capabilities in information and communications technology, Intel Hynix-Numonyx, Taiwan Semiconductor Manufacturing Corporation (TSMC), and many others have started IC fabrication in China. The expansion of MNCs' fabrication, particularly of foundry services, has also contributed to the rise of China's fabless IC design houses in the post-2000 period; investment for starting an IC design house is relatively small, and having foundry services nearby makes starting fabless IC design houses relatively convenient. Proximity between foundries and IC design houses facilitates their interactions for solving technical problems in processing technologies, since foundry service providers do not necessarily have the capabilities to solve all the technical issues by themselves. This is particularly important in explaining the concentration of IC design houses in Shanghai and the nearby Yangtze River Delta. During our interviews, executives of many IC design houses in Shanghai cited the importance and advantage of locating next to the major foundry service providers. In-house know-how about the processing technologies and close interactions with these foundry service providers give such IC design houses an advantage for competing in the global market.

One important development after 2000 was the changing role of the Chinese state in the development of the IC industry. As we mentioned before, the central government has identified semiconductors as a strategic area for development since the 1950s. However, over the next several decades, the major applications for semiconductor were primarily military, while the civilian market for ICs remained small. During the reform era, China's shortage in capital and limited foreign investment in the 1980s and the earlier 1990s meant that China had to start with low-tech sectors and prioritize employment. Under the centrally planned regime where the state was directly involved in planning, investment, and implementation of the various projects, the slow policymaking process made the system incompatible with the fast-moving global IC industry, which requires quick decision-making and action from all parties involved. Without simultaneous investment in IC design, and manufacturing, China did not achieve much success in its earlier efforts of growing its semiconductor industry. The Chinese government learned its lesson and made the decision regarding Project 909 swiftly with the goal of improving the competence of firms along the whole IC value chain in China.

After 2000, the Chinese central government decided to cease its direct involvement in setting up businesses in the IC industry, including IC design, and instead changed its policy to encourage the development of the whole industry. In 2000, the State Council announced various policies to promote the IC design industry along with the software industry (the No. 18 Document). The No. 18 Document included policies on investments, taxes, tariffs, and intellectual protection, among other things. In 2001, 2005, and 2011, the state council announced further policies to promote the development of the IC design industry. Some of these policies included funding up to 50 percent of a firm's R&D spending, waiving the income tax for the first two years, and cutting the income tax by half for the following three years.[1] IC firms with more advanced technology under 0.25 micron or with investment more than 8 billion RMB could enjoy even more extended benefits: a waived income tax for the first five years and income tax reduced by half for the sixth to the tenth year. Such policies minimized risk and uncertainties for IC design firms and helped domestic as well as foreign firms. Businesses responded favorably and registered fast growth shortly after the announcement of the policies.

The Chinese government also decided to set industrial technology standards as part of the intellectual property rights (IPR) laws in order to promote indigenous technology. In particular, promotion of the third-generation TD-SCDMA in wireless communication[2] has played a critical role in assisting many domestic IC design firms to survive and upgrade their technologies in the face of intense international competition (Fan 2006; Gao 2014; Gao and Liu 2012). Wideband Code Division Multiple Access (WCDMA) and Code Division Multiple Access (CDMA) 2000 already had the advantages of brand recognition and early adoption by major domestic telecom service providers such as China Mobile and China Telecom. Network equipment and mobile device manufacturers, such as Huawei and ZTE, were also strong supporters

[1] The government employed similar promotional policies in the auto industry in the early 1990s. See Ch. 5 for more information on this.

[2] As already mentioned, TD-SCDMA was developed by Datang, a Chinese firm.

of these foreign standards. In comparison, TD-SCDMA was developed by a domestic company, a latecomer in the industry with fewer resources and less market recognition. All the major companies—as well as government officials—were concerned about the technological and market uncertainties associated with TD-SCDMA (Fan 2006; Gao and Liu 2012; Gao 2014). Cell phone brand names, and especially foreign brand names, were reluctant to invest in TD-SCDMA-related technologies. This became a great obstacle for domestic chip design houses, since they could not find customers. Datang, the developer of the standard, worked closely with a variety of stakeholders, including prominent scholars and scientists, government officials, and businesses, and became the driver for TD-SCDMA's development and deployment (Gao and Liu 2012; Gao 2013). China Mobile, the leading mobile service provider in China, eventually made the official announcement of adopting TD-SCDMA in January 2009, at the request of the central government. Recently, China issued the 4G license. With the experience of TD-SCDMA, Datang has successfully developed the TD-LTE 4G standard, which is one of the two worldwide 4G standards. The deployment of the TDs by China Mobile give an opportunities for many Chinese design houses because they do not have to face major competition from existing foreign corporations. Leadcore and Spreadtrum, the two major IC design houses benefitted greatly from the standard.

With the help of China's 3G standard, some of these design houses have even become major global players, such as Spreadtrum and RDA. A brief history of Spreadtrum can illustrate the influence of a domestically developed standard on China's domestic IC design industry. Spreadtrum was founded in 2001 by a group of returnees from the US and funded by a number of VCs. At the very beginning, it developed mobile phone chips based on the GSM/GPRS standards. It successfully developed the GSM/GPRS chip in 2002, becoming a competitor in the Chinese mobile phone chip market. However, the founders were very technologically ambitious and driven to become a competitor in the 3G market. Initially in 2001, Spreadtrum had planned to start with WCDMA due to its more mature market. Soon after, Spreadtrum decided to focus instead on TD-SCDMA in 2003. By 2004, it had successfully developed the cellular chip based on TD. Despite the fact that it was losing money for many years, it had a successful IPO with NASDAQ in 2007, given its unique position in China's chip design industry. Despite this technological accomplishment, Spreadtrum faced great difficulty due to uncertain prospects of TD before 2009. Only after China's official adoption of TD as one of its 3G standards in 2009 did Spreadtrum's situation begin to improve. Its sales grew rapidly from $100 million in 2009 to more than $1 billion in 2013, since it was the pioneer in the TD market. It has become the third largest cellular phone chip provider in the world, right after Qualcomm and MediaTek. Though it still experienced difficulties, Spreadtrum would never have achieved its current success without the official adoption of TD in China. Since few foreign companies worked on this domestic wireless standard, the complex work to deployment ended up boosting the domestic IC chip design industry.

Returnees and China's IC Design Industry Growth

The pre-2000 Chinese IC design industry was powered by Chinese domestic trained experts. One critical development after 2000 has been the growing contribution of returnees to China. The roles of returnees in upgrading latecomer companies has been widely recognized (Fuller 2010). Between 1978 and 2012, more than one million Chinese students and scholars went abroad for study and research, and the number has been exploding in recent years. The annual count of students who go abroad has risen almost 20 times from 20,000 in the late 1990s to nearly 400,000 in recent years (Kenney et al. 2013; Saxenian 2006, 2007; Wang and Lu, 2012; Wang et al. 2009). Learning from the successful experience of Taiwan across the strait, various levels of government in China have initiated large-scale efforts to attract overseas talent. These have ranged from the "1000 Talents Program" and the "1000 Young Talents Program (the so-called "Little 1000 Talents Program") at the national level to various programs initiated by local governments, such as Nanjing's "321 Program," Wuxi's "530 Program," and Shenzhen's "Peacock Program." They usually aim to attract talent in the areas of technology, especially experienced engineers and entrepreneurs (Wang et al. 2009). These engineers have often worked in multinational corporations for many years and accumulated rich experience. With China's fast-growing economy and the maturing opportunities or business development as well as the slowing down of Western economies, many overseas Chinese have chosen to return and started new businesses or work for other ventures (Kenney et al. 2013).

The IC design industry has been a major magnet for former emigrants, and this new workforce has helped transform the industry (Fuller 2010). The roles of returnees in China's development of the IC industry have been manifold. To begin with, returnees have played significant parts in developing China's IC manufacturing industry in the foundry segment, as demonstrated by Li in the previous chapter. In addition, in 1996, a group of returnees planned to start the first fabless model in China, New Wave (Xintao) Science and Technology. With VC support from the US, Taiwan, and China's Huahong NEC, New Wave started operation as China's first fabless IC design firm in 1997. In 2001, New Wave was acquired by IDT (Integrated Device Technology) for the price of US$85 million in 2001. This was among the top 10 acquisition cases in

China in 2001 and was considered a milestone in China's IC development, encouraging more returnees to consider establishing fabless IC firms in China. A few significant IC fabless firms have emerged since, including Beijing Vimicro for image processors, Spreadtrum for wireless communications, and GalaxyCore Inc for image sensors. Meanwhile, many domestic IC firms also learned to establish their own fabless companies such as Zhuhai Actions, Hangzhou Silan, and Leadcore (formerly known as Datang Microelectronics), among others. Without chip design houses founded with the expertise of such returnees, China's effort to build the TD-SCDMA wireless network would have been slowed down.

Furthermore, returnees have helped connect China's IC firms to the global capital market. For example, Beijing Vimicro, founded by Dr Deng Zhonghan, a returnee from Silicon Valley, was a leader in developing mixed-signal multimedia chips. In 2005, Beijing Vimicro became the first fabless chip company from China to be listed on NASDAQ. After Vimicro, Zhuhai Actions (a company focusing on portable multimedia chips) and Spreadtrum were also listed on NASDAQ. Returnee-founded IC design houses are more easily recognized by the foreign capital market, probably because of trust built through the returnees' prior connections (Wang et al. 2009).

Returnees gravitate to the IC industry because of the growing Chinese market for ICs, improved IC production capabilities through foundry services, and preferential governmental treatment of the IC industry. Most former Chinese students who went overseas also concentrated on science and engineering and ended up working in technological sectors in the US (Wang et al. 2009); many of them worked in the semiconductor field. With their personal experience in foreign MNCs as well as their global connections, returnee entrepreneurs and engineers have found a thriving field in China's catch-up in the IC design industry, which also in turn helped fulfill their personal aspirations (Wang et al. 2009). The general migration of semiconductor global production system toward East Asia, including Japan, South Korea, and Taiwan also made the trend self-evident. Many returnees who worked in semiconductor in the United States had traveled frequently to East Asia, and it was reasonable for them to realize the promise of mainland China as a major semiconductor industrial site, given evident growth in this sector.

Returnees' major competitive advantages over domestic Chinese enterprises are their global connections and experience with advanced technology. Meanwhile, their advantage against global IC giants is their familiarity with the Chinese market and culture—as well as relatively low costs for talent and operation in China. As such, many of them have adopted a strategy that relies on "high quality-cost ratio." First, enterprises specialize in designing and manufacturing "good-enough" products, similar to the mobile phone industry as documented by Chapter 10 of this book. "Good-enough" products are considered products that are not at the frontier of world technology but are good enough to provide technologies and products for the second or lower tier markets. These lower tier consumer bases still make up a significant portion of the market in China and other emerging economies, and they provide business opportunities for the growth of such IC design houses. According to our interviews with executives in many returnee IC design houses, many of China's IC design houses can achieve 70–80 percent of the performance that has been reached by leading IC brands. Such chips have been widely used in other appliances and devices that serve the second- and third-tier market in China, which is more sensitive to price. For example, image sensor chips designed by GalaxyCore have been widely used by mobile device manufacturers in China. Additionally, many chip design houses have developed chips that are based on TD-SCDMA, in which MNCs did not show interest until recently, when the market for TD-SCDMA-based products become increasingly mature.

The success of such Chinese IC design houses relies heavily on cost control. Due to the fact that costs for Chinese labor and other operations are still significantly lower than in advanced economies, Chinese firms' costs can be 50–70 percent lower, according to our interviews. In particular, regular IC design enterprises' costs are much lower than those in advanced economies. Based on personal interviews, it is expected for regular IC hardware and software engineers to make \$100,000–\$150,000 a year in the US, while typical IC design engineers in China make \$30,000–\$80,000 a year depending on their field and experience. Income at such levels allows returnee design engineers to maintain their high quality of life in China. For many IC design houses in China, their costs are 50 percent lower than major foreign competitors while the performance of their products could be as good as 70 percent of what their foreign competitors have achieved. The relatively lower cost and acceptable quality performance offer the opportunity for China's IC design houses. Such a strategy will have to continue for the foreseeable future, since it would be very challenging for Chinese firms to catch up to the world IC technological frontier in a short time, and China's cost advantage will also remain for the coming years. Consequently, China's IC design industry will remain in the catch-up phrase for a while as it accumulates technological capabilities.

Clearly, the progress of China's IC industry in recent decades is the result of a number of intersecting forces, both domestic and global. The most salient forces include the vertical disintegration of the global IC production chains, the evolution of government support, the growing Chinese market demand, broader manufacturing capability in the information and electronics industry, and the increasing number of overseas Chinese who are willing to return.

Discussion and Conclusions

China's IC industry in general, IC design in particular, has experienced drastic growth since the late 1990s, despite its rocky trajectory in the preceding decades. Such growth has been accompanied by significant upgrading in its technological capabilities. The early failure of China's IC industry had to do its compartmentalized national innovation systems. In particular, the separation between R&D and manufacturing as well as the unresponsiveness of its SOEs, which were the backbone of China's economy, prevented the IC industry from advancing. The state-centered inflexible approach to the IC industry was particularly fatal due to the nature of the industry itself: technology was progressing at such a rapid pace that it was impossible for a command economy to keep up. The financial and talent shortfall also contributed to the failures at the beginning stages of reform in the early 1980s. Yet, it would be wrong to completely dismiss these unsuccessful attempts. The rudimentary but persistent IC R&D and manufacturing effort provided the earlier seeds of IC design and manufacturing capacity and talents that laid the ground for the later development.

The significant growth of China's IC industry occurred at a different stage of China's integration with the global economy in the 2000s. Several major factors contributed to the acceleration. First, the entry into the WTO in 2001 fortified the economic opening to the global market and investment, which has been a boost to many industries. China's growth in domestic markets, expanded manufacturing capabilities in ICT industries due to both foreign investments and China's indigenous enterprises have created world largest and most diverse market for IC chips, setting up the stage for IC design houses.

Secondly, China's IC industry and IC design in particular, emerged from the trend of vertical disintegration of the semiconductor global value chain over the last few decades. Design, manufacturing, assembly and testing, and final IC consumers are now much more separated in space and in organization. This rearrangement of global semiconductor chains has afforded China the opportunity to make progress in selected segments of the industry as the both the costumers and IC assembly operations have gravitated to China. IC design houses emerged amid the growth of the IC value chain.

Thirdly, the China's developmental/entrepreneur state has gained experience and learned from its failed experiments of earlier decades, employed more flexibility and diverse approaches. The Chinese government largely gave up initiating semiconductor manufacturing industry through a centrally controlled state-owned company. Instead, it deployed more decentralized and diversified industrial policies that incorporate non-state actors, and target the development of the entire IC industrial through globalization. The case of the

IC design industry demonstrates that the state—as powerful and authoritarian as it is in China—cannot single-handedly create and maintain a dynamic high-tech industry. The limited success of China's IC design industry so far had to be achieved through cooperation with other parties such as multinational corporations (MNCs), the market, and returnee entrepreneurs.

Finally, China's development of its IC design industry has benefitted from the increasing number of overseas Chinese engineers and people with strong technical capabilities and global business experience. While the returnees contributed substantially to its "908" and "909" projects in the 1980s and 1990s, their numbers have increased substantially and they have taken a leadership role in its recently burgeoning IC foundry services and independent IC design houses. These returnee funded or staffed firms have taken advantage of China's many environmental assets, including cheaper labor, favorable government policies, governmental flexibility in attracting foreign capital and talents, China's growing power in the world market, and the growing presence of domestic multinational corporations in the IC value chain. The returnees' leadership suggests a deep integration with the global technological trends and progress China has made in technological capabilities.

China's growth in the IC design industry does exhibit some unique patterns in comparison to the experiences of Singapore, South Korea, and Taiwan. The IC design industry in these three economies has followed different paths (Chang and Hsu 1998; Chen and Sewell 1996; Cho et al. 1998; Choung et al. 2000; Lee and von Tunzelmann 2005; Liu 1993; Lu et al. 2004; Kim 1997, 1998; Kimura 1988; Mathews 1997; Mathews and Cho 1999, 2000). Singapore, as the smallest economy of the three, has adopted a strategy that focuses on attracting multinational corporations in semiconductors, taking advantage of its relatively open business environment. South Korea has relied on nationally owned large conglomerates such as Samsung, utilizing their diversified businesses and deep pockets to specialize in integrated device manufacturers (IDMs). In comparison, Taiwan's IC design industry has initiated model of made-to-order foundry firms and development of its small and medium enterprises (SMEs) of fabless houses, constituting an industrial ecosystem. China's IC design industry has learned from Taiwan's experience and exhibits a similar trajectory to that of Taiwan. The increasing economic and technological interactions across the Taiwan Strait have also clearly benefitted mainland China's IC design industry, from capital and technology to business model. Taiwan's contribution to China's IC design industry cannot be overstated.

Nevertheless, mainland China has unique features in its IC design industry due to its large domestic market, more powerful governments, and stronger attraction to multinational corporations. China has attempted to achieve measured success in the entire the ICT industry value chain, from telecommunication equipment to mobile devices. In each segment, domestic IC design firms have emerged to offer valued products and have accumulated significant technological and market capabilities. The advance is particularly noticeable in mobile phone chip design based on China's TD-SCDMA and TD-LTE standards.

Overall, China is still lagging behind the leading countries in the IC design industry. China's IC design sector also faces many challenges in the years ahead. In particular, China has continued to rely on leading countries in integrated processor units (IPU) and key IC design technologies, particularly in the mobile device market. Intel and Advanced Micro Devices (AMD) still dominate the personal computer and laptop markets, while IPU technologies, Apple, and Qualcomm control smartphone IC technologies. But there is a chance that some Chinese IC design firms will narrow the technological gaps.

For example, IC design firms such as HiSilicon have accumulated significant technological competence and are very aggressive in investing in new technologies. Indeed, some analysts even speculate that Huawei will become a powerful competitor in the 4G mobile phone market, given its technological competence in the field. If this becomes true, HiSilicon, as Huawei's spin-off, will make further progress in the high-end market. China's development of its TD standards in the telecommunication fields also affords IC design houses such as Spreadtrum and Datang unique opportunities for future progress. As a part of Tsinghua Unigroup, Speadtrum is likely to broaden its market beyond TD-based technologies, as it had become the third largest cellular phone chip provider in the world market in 2013. Should China's 4G standard become adopted by other leading markets, Chinese IC design firms will have a brighter future. With accumulated technological capabilities and the dynamics of the technical changes in the telecommunication industries, it is possible for Chinese IC design firms to further narrow their gaps with leading global players.

The lessons on innovation in the IC design houses have to do with the maturation of an industrial ecosystem, including foundry, and diverse customers and their needs of valued products. The role of the state is important in providing a supportive and stable macro-environment, but is not the determining factor. Global collaboration and globalized talents have been crucial in the industry.

References

Chang, P. L., and C. W. Hsu (1998), "The Development Strategies for Taiwan's Semiconductor Industry," *IEEE Transactions on Engineering Management*, 45(4): 349–56.

Chen, L., and L. Xue (2010), "Global Production Network and the Upgrading of China's Integrated Circuit Industry," *China and World Economy*, 18(6): 109–26.

Chen, L. (2005), *Institution, Elites and Consensus: The Policy Process of China's Semiconductor Industrial Policies*. Ph.D. dissertation, Tsinghua University.

Chen, C. F., and G. Sewel (1996), "Strategies for Technological Development in South Korea and Taiwan: The Case of Semiconductors," *Research Policy*, 25(5): 759–83.

Cho, D. S., D. J. Kim, and D. K. Rhee (1998), "Latecomer Strategies: Evidence from the Semiconductor Industry in Japan and Korea," *Organization Science*, 9(4): 489–505.

Choung, J. Y., H. R. Hwang, J. H. Choi, and M. H. Rim (2000), "Transition of Latecomer Firms from Technology Users to Technology Generators: Korean Semiconductor Firms," *World Development*, 28(5): 969–82.

Fan, P. (2006), "Catching up through Developing Innovation Capability: Evidence from China's Telecom-Equipment Industry," *Technovation*, 26(3): 359–68.

Fuller, D. B. (2010), "Networks and Nations: The Interplay of Transnational Networks and Domestic Institutions in China's Chip Design Industry," *International Journal of Technology Management*, 51(2/3/4): 239–55.

Gao, X. (2014), "A Latecomer's Strategy to Promote a Technology Standard: The Case of Datang and TD-SCDMA," *Research Policy*, 43(3): 597–607.

Gao, X., and J. Liu (2012), "Catching up through the Development of Technology Standard: The Case of TD-SCDMA in China," *Telecommunications Policy*, 36(10): 817–31.

Gereffi, G., J. Humphrey, and T. Sturgeon (2005), "The Governance of Global Value Chains," *Review of International Political Economy*, 12(1): 78–104.

Hu, Q. (2006), *China's Chip Path (Xinlu Lichen): China's 909 Project*, Beijing: Publishing House of Electronics Industry.

Klaus, M. (2003), "Red Chips: Implications of the Semiconductor Industry's Relocation to China," *Asian Affairs: An American Review*, 29(4): 237.

Kenney, M., D. Breznitz, and M. Murphree (2013), "Coming Back Home After the Sun Rises: Returnee Entrepreneurs and Growth of High Tech Industries," *Research Policy*, 42(2): 391–407.

Kim, S. R. (1997), "The Dynamics of Samsung's Technological Learning in Semiconductors," *California Management Review*, 393: 86–99.

Kim, S. R. (1998), "The Korean System of Innovation and the Semiconductor Industry: A Governance Perspective," *Industrial and Corporate Change*, 7(2): 275–309.

Kimura, Y. (1988), *The Japanese Semiconductor Industry: Structure, Competitive Strategies, and Performance*. Greenwich, CT: JAI Press.

Lee, T. L., and N. von Tunzelmann (2005), "A Dynamic Analytic Approach to National Innovation Systems: The IC Industry in Taiwan," *Research Policy*, 34(4): 425–40.

Liu, C. Y. (1993), "Government's Role in Developing a High-Tech Industry: The Case of Taiwan's Semiconductor Industry." *Technovation*, 13(5): 299–309.

Lu, L. Y. Y., S. W. Hung, and C. Yang (2004), "Successful Factors of the Fabless IC Industry in Taiwan," *International Journal of Manufacturing Technology and Management*, 6(1): 98–111.

Mathews, J. A. (1997), "A Silicon Valley of the East: Creating Taiwan's Semiconductor Industry," *California Management Review*, 39(4): 26–54.

Mathews, J. A. (1999), "A Silicon Island of the East: Creating a Semiconductor Industry in Singapore," *California Management Review* 41(2): 55–78.

Mathews, J. A., and D. S. Cho (1999), "Combinative Capabilities and Organizational Learning in Latecomer Firms: The Case of the Korean Semiconductor Industry," *Journal of World Business*, 34(2): 139–56.

Mathews, J. A., and D. S. Cho (2000), *Tiger Technology: The Creation of a Semiconductor Industry in East Asia*. Cambridge: Cambridge University Press.

Ministry of Science and Technology (MOST) (1999), *China New High-Tech Industrialization Development Report*. Beijing: Scientific Press.

Price Waterhouse Coopers (PWC) (2006), *Continued Growth: China's Impact on Semiconductor Industry*, 2006 Update. Los Angeles: Price Waterhouse Coopers.

Price Waterhouse Coopers (PWC) (2012), *Continued Growth: China's Impact on Semiconductor Industry*, 2012 Update. Los Angeles: Price Waterhouse Coopers.

Saxenian, A. (2006), *The New Argonauts: Regional Advantage in a Global Economy*. Cambridge, MA: Harvard University Press.

Saxenian, A. (2007), "Brain Circulation and Regional Innovation: The Silicon Valley-Hsinchu-Shanghai Triangle," in Karen R. Polenske (ed.), *The Economic Geography of Innovation*, 190–209. Cambridge and New York: Cambridge University Press.

Shanghai Information Industry Commission (SHIC) and Shanghai IC Industry Association (SHICA) (2012), *Shanghai IC Industry Development Report* (in Chinese). Shanghai: Shanghai Educational Publishing House

Sun, Y. (2002), "China's National Innovation System in Transition," *Eurasia Geography and Economics*, 43(6): 476–92.

Wang, H. Y., and J. Y. Lu (2012), *Annual Report on Chinese Returnee Entrepreneurship*. Beijing: Social Sciences Academic Press China.

Wang, H. Y., D. G. Miao, and X. Cheng (2009), *The Report on the Development of Chinese Overseas Educated Talents* (in Chinese). Beijing: China Machine Industry Press.

Zhou, Y., and J. Y. Hsu, (2010), "Divergent Engagements: Roles and Strategies of Taiwanes and Mainland Chinese Returnee Entrepreneurs in the IT Industry," *Global Networks*, 10(4): 398–419.

Zhu, Y. (2006), *Collections of Research on China's IC Industry* (in Chinese). Beijing: China New Times Publisher.

10

The Evolution of China's Mobile Phone Industry and Good-Enough Innovation

Shin-Horng Chen and Pei-Chang Wen

Introduction

The communications industry is a landmark sector in China's pursuit of indigenous innovation. Not only have two Chinese firms, Huawei and ZTE, become the leading international communication equipment manufacturers, but China has also established its own flagship international industrial standard, TD-SCDMA (Time Division-Synchronous Code Division Multiple Access) for third-generation mobile phones (3G). As late as 2008, China's market for second-generation (2G) mobile communications devices was dominated by many "Shanzhai" (also called guerrilla) handsets,[1] the manufacturers of which were often accused of brand imitation in terms of product appearance, design, and name. While controversial, the proliferation of Shanzhai handsets brought about the progressive expansion of China's mobile communications market. It enabled an army of local, non-branded handset producers to gain a strong foothold in the market, exceeding the market shares of leading international brands and legal local brands.

In essence, the development of China's mobile phone sector exemplifies two distinctive but connected approaches to innovation. The development

[1] Shanzhai handset makers were mostly clustered in Shenzhen, Guangdong Province. There, thousands of small-sized "guerrilla" (at least initially) phone workshops used to form a comprehensive supply chain, ranging from project designing, software development, assembling, printing, packaging, logistic distribution to sales and after-sales service. However, the ecosystem of the Shanzhai handset sector had a much bigger and more meaningful impact on local clustering than the terms, bandit, copy cats, and piracy can imply. For detailed discussions on the Chinese way of innovation by Shanzhai handset makers, see Chen et al. (2013).

of the Shanzhai handset sector demonstrated a bottom-up approach. This is parallel to the governmental top-down approach, which is represented by the establishment of industrial standards, state-owned companies, and provision of generous state support and regulation to a few national champions, including Datang, Huawei, and China Mobile (Liu and Zhou 2013). The bottom-up innovation model was influenced by diverse grassroots demands, local development policies, restructuring of the production system, and the cross-strait innovation network, which involves Taiwanese firms such as Mediatek, a leading IC (Integrated Circuit) design house (Chen et al. 2013; Liu and Chao 2009; Sheng and Shi 2010; Tse et al. 2009). Shanzhai handsets can be considered a result of indigenous innovation that was stimulated by strong local demands, particularly from the market's lower tiers, and products of the entrepreneurial spirit of cell phone makers. Elsewhere, we have argued that out of the Shanzhai handset phenomenon might emerge a broadly defined "Shanzhai Economy" with Chinese flavor of innovation (Chen et al. 2013). We call it "good-enough innovation (economy)" (see also Brandt and Thun 2010; Gadiesh et al. 2007).

But the situation changed dramatically as consumers began to migrate towards 3G carriers and smartphones. Home-grown brands such as Xiaomi, Lenovo, Coolpad, Huawei, and ZTE have gained footholds in the Chinese market, outperforming Shanzhai handset makers as well as international rivals such as Apple, Samsung, and Nokia in 2014. Most remarkable here is the proliferation of low-cost smartphones with price tags below CNY 2,000 (sometimes as low as CNY 1,000), intended for the vast market of lower income users in China. On the surface, it seems that the emergence of these Chinese home-grown brands, like the earlier dominance of Shanzhai handset makers, is a result of the market demand patterns. But we argue that the consolidation of handset brands was also driven by technological innovation, particularly platform-based development, or "platformization," during the migration towards smartphones and mobile digital services (Eaton et al. 2011; Feijóo et al. 2009; Kenney and Pon 2011). Such innovation trends were created by both international and Chinese domestic firms, but Chinese companies have taken good advantage of them.

This chapter is structured as follows. First we highlight factors that, based on the theoretical framework of the co-evolution of technology and market, are particularly relevant to China's migration towards 3G services and smartphones. Then we describe the rise of a few indigenous brands through the adoption of 3G mobile communications and smartphones in China. We then portray the key factors in the new industrial ecosystem and transformation. The next section focuses on the role of industrial standards in facilitating the Chinese migration to smartphones, before the conclusion.

The Changing Ecosystem and Chinese Market for the Smartphone Industry

In the field of technology management and innovation, researchers have long explored the interactions between the dynamics of technology and factors such as society (Dosi 2000; Geels 2005), industrial structure (Nelson 1994), and the market (Struben 2008). Freeman and Perez (1988) examined how technology shapes and co-evolves with the broadly defined production system over time, describing this relationship with the term "techno-economic paradigm." Underlying this process is the feedback loop between new technologies and socio-institutional systems (Perez 1985: 445). Against the theoretical backdrop of the co-evolution of technology and the market, we examine the evolution of China's mobile phone industry, with a focus on the effect of the migration to smartphones on the industrial ecosystem and the upgrading of the mobile phone industry.

Remarkable economic growth over the past three decades has turned China into a manufacturing powerhouse (Holz 2008). China's quest for technological leadership through the promotion of indigenous innovation represents a new stage of development (Sigurdson 2005; Suttmeier and Yao 2004; Rowen and Hancock 2008). Indeed, a number of indigenous firms, most notably Huawei and ZTE, have managed to climb the technological ladder from the switching system by taking advantage of the large domestic market and particular features of the technological regime such as the dominance of state-owned telecommunication service companies (Mu and Lee 2005). China's quest for technological leadership in 3G wireless communication is epitomized by its development of an indigenous industrial standard (TD-SCDMA) as an alternative to the two competing global standards, CDMA2000 and W-CDMA (Liu and Zhou 2013; Yan 2007).[2]

The R&D for TD-SCDMA started in the mid-1990s, when few companies understood either what specific services 3G networks could offer or how they would impact the mobile communications ecosystem. It has now become clear that the physical network standard is only one element needed for broadband mobile communications services to prevail and prosper. In China, particular features of the market and the evolution of its industrial system have also critically shaped innovation in this sector.

Recently, academics have been paying more attention to how domestic demand and entrepreneurship have affected China's economic and industrial development (Minagawa et al. 2007; Yueh 2009, Christensen 2003; Brandt

[2] For example, Liu and Zhou (2013: 163) argue that in the IT industry, "It seems that the climate has helped TD-SCDMA to gain favour over existing multinationals in China in the 3G market, but the technology still faces many uncertainties."

and Thun 2010; Gadiesh et al. 2007; Zhou 2008). For example, Brandt and Thun (2010) showed that, since China's accession to the World Trade Organization (WTO), domestic firms (mostly private ones) in a few sectors (automotive, construction, and machine tools) have been able to compete with well-established foreign firms, even leading to the upgrading of local industries. Local competitors achieved such success by taking advantage of the sheer size of low-end market segments and exploiting strong, pre-existing capacity in those sectors. In competing for the low to middle market, these companies were able to produce "reliable-enough products at low-enough prices to attract the cream of China's fast-growing cohort of midlevel consumers" (Gadiesh et al. 2007: 82). Such "good enough" innovation had been the primary mode of operating in the proliferation of "Shanzhai" phones.

Indeed, China—along with Russia, India, and Brazil—has been marked as an "emerging market." This distinction is given to countries with rapid paces of economic development, high income inequality, market diversity, and high price elasticity of demand (Dawar and Chattopadhyay 2000; Walters and Samiee 2003). These characteristics imply that China's massive market cannot be sufficiently served by products developed in rich countries, despite the existence of wealthy communities in China that are able to consume the state-of-the-art, premium products marketed to the developed world. Prahalad (2005) and Christensen et al. (2001) were among the leading authors to draw our attention to innovation—and the creation of an entirely new business model—that revolved around this underserved segment in emerging markets (in Prahalad's words: the bottom of the pyramid, or "BOP"). Christensen et al. (2001: 94) in particular argue that "[exactly] what kinds of disruptive technologies might emerge within countries such as India and China cannot be easily extrapolated from the market needs and success stories of developed economies . . . technologies emerging from these countries may have profound but unpredictable implications for the rich world's markets." Christensen (2003) further distinguishes disruptive technologies from sustaining technologies in the context of BOP innovations. Disruptive technologies are "innovations that result in worse product performance, at least in the near term, but are generally cheaper, simpler, smaller, and frequently, more convenient to use." They are also often the most effective means for new entrants to serve and expand the lower tiers of the market overlooked by incumbent firms.

In other words, China's uneven spatial and social development, especially in terms of large and diverse demands at lower tiers of the market, affects the development of indigenous innovation. Local firms in China (and other emerging markets) are setting up their own playing field by taking advantage of specific and underserved local demands through "BOP" and "good-enough" innovation such as Shanzhai phones (Chen et al. 2013). Such phones cost a fraction of what mainstream phones cost, but were able to outperform local and international brands by taking advantage of Mediatek's turnkey solutions, which incorporated OSs and application software into a single chip, making it much easier for Shanzhai manufacturers to focus on external design and the production of a wide variety of mobile phones (Chen et al. 2013; Rong and Shi 2009). For example, some of the more exotic Shanzhai handsets looked like watches or a packet of premium cigarettes, and were intended to show off; others provided striking new features such as solar chargers, telephoto lenses, super-loud speakers, or ultraviolet lights to detect forged currency (Chen et al. 2013). Some adaptations are particularly significant in China, for example having multiple SIM card slots to accommodate different standards. In their own way, Chinese Shanzhai handset makers quickly produced a variety of trendy handsets that were affordable, fashionable, and even tailor-made for migrant workers, rural farmers, or urban young white-collared workers. A few of them, most notably Tianyu, even developed a strong brand presence before the arrival of smartphones.

But as Chinese industries upgraded and adapted to the migration of Chinese customers to 3G services and smartphones, the Shanzhai golden age quickly drew to a close. Figure 10.1 shows key milestones in the evolution of China's mobile phone and mobile communications sectors in recent years.

After the Chinese government liberalized license regulations on the manufacturing and sales of handsets in 2007, Shanzhai cell phones grew rapidly and captured a large segment of the low-end 2G market. Since 2010, however, the

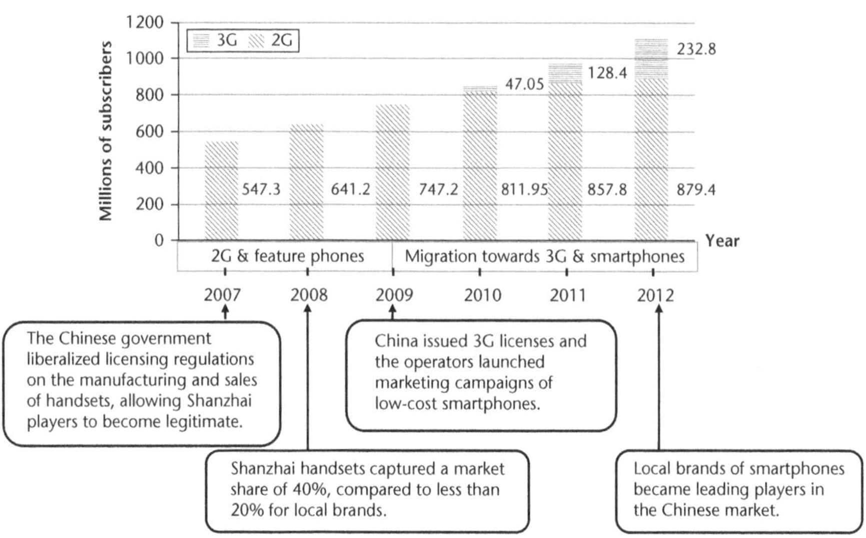

Figure 10.1. Key Milestones in the Evolution of China's Mobile Phone and Mobile Communications Sectors

2G mobile phones have been losing popularity with the rise of smartphones. In the migration to 3G, both services and devices have become more driven by applications, which has brought about platform-based development and competition such as mobile operating system (OS) platforms and service platforms (Ballon 2009; Ballon and Walravens 2009; Feijóo et al. 2009; Hammershøj et al. 2009; Tilson et al. 2012). The source of competition for players in the mobile phone industry has moved from the sphere of architectural design (the infrastructure level) to the functions of middleware/service platforms (such as Apple's App Store) and applications, as shown by the role of apps in the rise of iPhone and iPad. The transformed ecosystem underscores the importance of service platforms. Mobile internet applications grew exponentially, which worked to the advantage of a few home-grown brands at the expense of Shanzhai handset makers.

It has become clear that smartphones are not simply technological upgrades. Rather, they represent a profound change in telecommunication, from mainly voice telephony terminals to multimedia data communications and mobile internet devices. Researchers working for the China-European Union Standards project propose a "three level model for standards and innovation in ICT." The model distinguishes between the three layers of smartphone industry, namely the infrastructure, middleware, and applications, as shown in Figure 10.2. Middleware refers to a software platform between a smartphone OS and a third-party application: Sun's J2ME and Qualcomm's BREW are typical examples (Lin and Ye 2009: 620). We define the middleware level more broadly to include service platforms, as shown in Figure 10.2. In addition, applications have become essential parts of any mobile communications services.

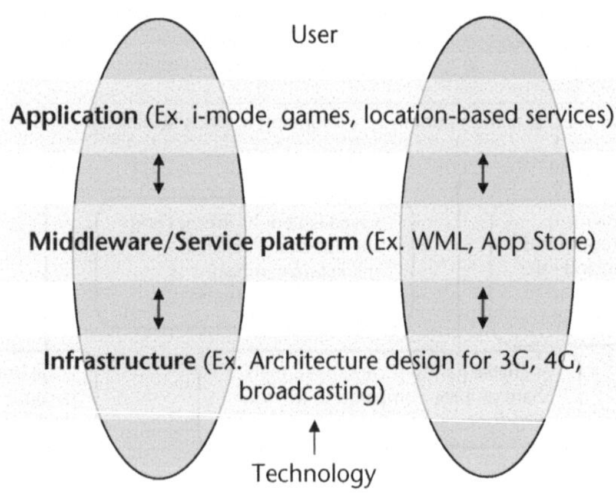

Figure 10.2. The Three-Level Model for Standards and Innovation in ICT

Thus, a 3G industrial ecosystem is far more complex than one from the 2G period. Yoo et al. (2010: 725) argue that the advent of digital innovation, such as the mobile internet and the e-book, has brought about a new type of product architecture. Called by them "the layered modular architecture," this is a hybrid of the modular architecture of a physical product and the layered architecture of digital technology. Modular architecture provides a scheme by which a physical product is broken up into its components, attributed functionality, and interconnected through pre-specified interfaces. The layered architecture of digital technology is embedded into physical products, enhancing product functionality with software-based capabilities. In the professional terms of the mobile communications industry, Yoo et al. (2010) dissect layered architecture into four layers: devices, networks, services, and contents. It is worth mentioning that the revised model shown in Figure 10.2 can be considered as a simplified version of the conceptual framework of Yoo et al. (2010) for the new industrial ecosystem of mobile digital services.

Mobile OS platforms such as Google's Android and Apple's iOS—now pivotal in the creation of mobile service industry ecological systems and proliferation of new mobile internet services—are typical examples of features in this new technological landscape (Ballon 2009; Ballon and Walravens 2009; Hammershøj et al. 2009; Tilson et al. 2012). The developers of mobile OS's and service platforms (e.g. Apple for App Store, Google for Play, and China Mobile for Mobile Market) differ in their treatment of mobile service development, provisions, and network control. This difference reflects their varying strategies, core competencies, and organizational boundaries within the evolving ecosystems (Ballon 2009; Ballon and Walravens 2009; Hammershøj et al. 2009, Eaton et al. 2011; Tilson et al. 2012). Apple's strategy features a proprietary platform—iTunes and App Store—and tight control over its relatively closed ecosystem. Google adopts a hand-off approach with a free and open-source OS. Once, Google even publicly announced that it would welcome Shanzhai handset makers to take advantage of Android (Chen et al. 2013). These differences affect China's 3G mobile frameworks: currently an overwhelming share (86.4 percent) of smartphones produced by domestic firms is Android-based (China Academy of Telecommunication Research 2013). The implication of such ecosystem transformation is discussed later in the chapter.

The Adoption of 3G Mobile Communications: A New Ecosystem

The migration from 2G to 3G marked a turning point for the mobile communication service sector. In the 2G period, most contents (mainly voice communications and short messages) delivered over the mobile communications network were produced by the user. The service operators did not have to pay much attention to applications or even the capabilities of the handset. In fact, none of the three mobile carriers in China offered subsidized service plans to their customers until the launch of their 3G services. With the shift toward 3G, Apple pioneered app services (known as the App Economy) that have not only boosted the popularity and proliferation of smartphones but also cemented the significance of applications and user experiences to mobile communications. In China, home-grown internet and e-commerce services have already developed rapidly within a "walled garden," built both by the government's internet censorship and the unique preferences of Chinese internet users. Some of the most popular international applications and websites, including Facebook and Twitter, are blocked from the Chinese market. This has required smartphone vendors and service providers to develop localized user interfaces for the China, so that domestic internet companies, such as Baidu, Tencent, and Alibaba dominate the market. Because of such censorship, the Chinese internet ecosystem has a strong domestic flavor.

China launched its 3G mobile communications services only after 2009, when its indigenous standard of TD-SCDMA finally became technologically ready. By the end of 2012, 232.8 million people used 3G mobile phones out of 1.11 billion mobile phone users in China.[3] Despite a low 3G penetration of around 29 percent for 2013, China has by now overtaken the USA to become the largest smartphone market in the world.[4]

The massive adoption of smartphones in China was made possible by the advent of low-cost smartphones for the masses, with price tags below CNY 2,000 and even around CNY 1,000. For example, according to estimates by Canalysys, a local analyst house, these cheap entry-level smartphones made up 25 percent of smartphones sold in the Chinese market in 2012. By 2015, they are expected to reach 40 percent.[5] In a same vein, a report by Eguan, another local marketing consultant, suggests that the average price for Android-based smartphones in China had declined from CNY 2,020 in the second quarter of 2011 to CNY 1,560 in the second quarter of 2012.[6]

Concomitant with this trend, the Chinese market has recently witnessed the surge of home-grown smartphone manufacturers. Figure 10.3 shows a market-share comparison of leading smartphone brands in the global and Chinese markets in 2013.

[3] <www.miit.gov.cn/n11293472/n11293832/n11294132/n12858447/15036653.html>, accessed Jan. 2013.

[4] http://allthingsd.com/20130529/mary-meekers-internet-trends-report-is-back-at-d11-slides, accessed June 2013.

[5] <http://micgadget.com/27390/sub-200-smartphones-expecting-to-proliferate-in-china-by-2015>, accessed Jan. 2013.

[6] <http://lazure2.wordpress.com/2012/09/10/the-low-priced-android-based-smartphones-of-china-will-change-the-global-market>, accessed Jan. 2013.

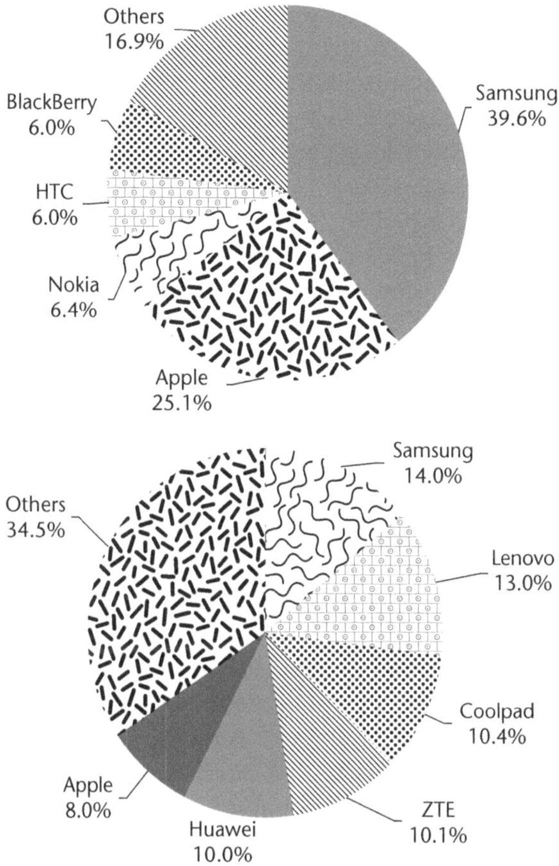

Figure 10.3. A Market-Share Comparison of Leading Smartphone Brands in the Global and Chinese Markets, 2012

While the global market is dominated by internationally premium brands such as Samsung (39.6 percent), Apple (25.1 percent), BlackBerry (6.0 percent), and HTC (6.0 percent), the Chinese market is commanded by a few local brands such as Lenovo (13.0 percent), Coolpad (10.4 percent), ZTE (10.1 percent), and Huawei (10.0 percent), with some international competition from Samsung (14 percent), according to IDC. In particular, although Apple is the pioneer of the "App Economy" and stands as a pop culture icon in the USA as well as in China, it has been outperformed in market share in China by the domestic brands mentioned. The local brands have various specializations: ZTE and Huawei have had a strong market foothold in the communications equipment industry. Lenovo was a domestic player for 2G handsets as well as personal computers, while Coolpad and Xiaomi are newcomers to the mobile device industry. In 2014 Lenovo acquired the Motorola Mobility smartphone business from Google, which signified the company's strategic intent to strengthen its ability to compete in the mobile communications industry.

Assorted "others" claim about 34 percent of the market share in China. This category includes some new local players, still largely unknown in the developed world. Handset vendors like K-Touch, Gionee, Meizu, Tianyu, Oppo, and Bubugao, some of which had Shanzhai origins, have built up their brands and served as white-box handset OEM (Original Equipment Manufacturing) producers. Additionally, some Chinese internet companies, such as Baidu (known as "the Google of China"), Qihoo, and Alibaba, have started to enter the marketplace with customized smartphones to take advantage of their popular contents. In other words, no longer "guerrillas," there are many potentially promising home-grown smartphone producers in China. In addition to the established brands such as Lenovo, Huawei, and ZTE, the rise of Xiaomi handset is particularly dramatic. Founded only in 2010 by veteran Beijing IT entrepreneur Lei Jun, it offers a clean design similar to iPhone. But it has developed customized software on the Android platform, called MIUI. Avid fans are invited to help the design by giving feedback online and the company releases a software update every Friday. Powerful, inexpensive, and with a cult-like following, Xiaomi became the top smartphone seller in China in 2014, and ranked the third in the global market.[7]

It should be noted that the Chinese firms mentioned are just a small sampling of the Chinese smartphone and 3G mobile communications sectors. The China Academy of Telecommunication Research (2013: 13)[8] has reported that about 73 percent (more than 380 firms) of the 529 handset makers in China are engaged in the manufacturing of smartphones. These manufacturers are mainly clustered in Guangdong, though there are secondary concentrations in Beijing, Tianjin and Fujian (China Academy of Telecommunication Research 2013: 40). The concentration in Guangdong and especially in Shenzhen, the first and arguably most successful of China's Special Economic Zones, suggests substantial continuity in the supply chain, since this area has been the hotbed of the *Shanzhai* handset sector.

Figure 10.4 identifies the key factors and facilitators in the new ecosystem and industrial transformation resulting from the migration to 3G in China. In this transition, firms with different domains have sought to establish versatile platforms as the gatekeepers and value capturers for the development and provision of mobile digital services. This has resulted in a typology of

[7] <http://bits.blogs.nytimes.com/2014/08/04/xiaomi-tops-chinese-smartphone-market>, accessed Dec. 2014. See also Mozur and Wang (2014).

[8] The China Academy of Telecommunications Research is an important institute under the Ministry of Industry and Information Technology (MIIT).

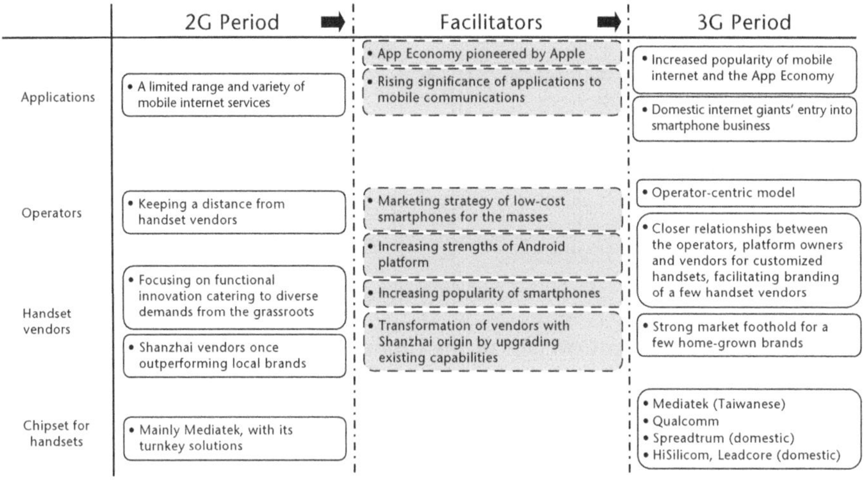

Figure 10.4. The New Ecosystem and Industrial Transformation of Smartphones and Services in China

platforms, as suggested by Ballon (2009), including the teleco-centric or operator-centric platform model (i.e. Vodafone Live!), the device-centric platform model (i.e. the iPhone), and the aggregator-centric platform model (i.e. Google). In China, the operator-centric model has prevailed (Huang 2011), and because the three operators use different 3G technologies, they must be actively involved in smartphone sourcing and distribution. China Mobile has attempted to mobilize external suppliers and developers by launching a program of entrepreneurship development of apps. In this way it can solicit the development of smartphones and mobile digital services that are compatible with its unique TD-SCDMA standard, its OPhone OS platform, and the Mobile Market service platform. This leads operators to pay more attention to the development and marketing of customized smartphones. The operators then will promote low-cost and/or entry-level smartphones for the broad base of low-income Chinese consumers in order to increase the penetration rate of 3G services. This has resulted in a closer relationship between operators and domestic smartphone vendors for customized handsets.

In addition, as in other countries, there are more than just operator-centric models in China. Major domestic internet firms, such as Baidu, Tencent, Qihoo, and Alibaba have also come out with customized smartphones via outsourcing to expand the reach of their popular social network services (e.g. QQ services) and e-commerce services (e.g. Taobao Mall). This has given rise to the aggregator-centric platform model in China. Xiaomi, initially an internet firm, follows Apple's device-centric platform model to promote its premium smartphones and its MIUI OS platform. Figure 10.5 illustrates the value chain coverage of the key Chinese players under these three different models.

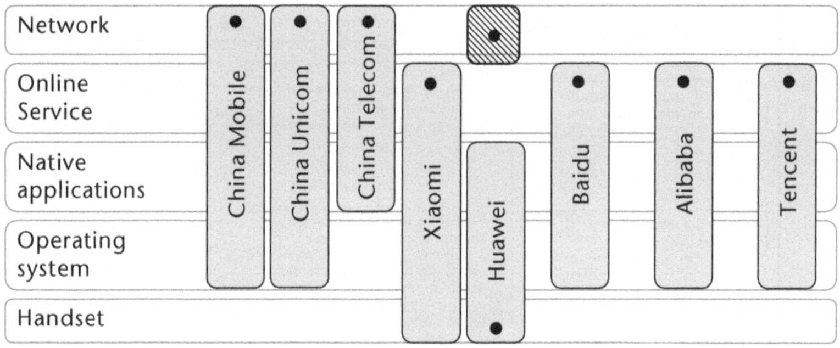

Figure 10.5. Value Chain Coverage of the Chinese Key Players under the Different Models in China

All of the models mentioned require the smartphone vendors to establish strong relationships with either the operators or the content aggregators throughout the design process. It is also necessary for the operators and content aggregators to get involved in the sourcing and marketing of customized smartphones in order to capture value from their OS platforms and/or service platforms (Eaton et al. 2011; Kenney and Pon 2011). In theory, both the legitimate Chinese handset vendors and the numerous *Shanzhai* vendors could use their experience with the low-end GSM handset business to supply low-cost designs and manufacturing knowledge. In reality, though, the shift in 3G services and smartphones to platform-based development tends to work in favor of larger smartphone vendors.

As a result, only a few home-grown brands have flourished in China's smartphone industry. Huawei and ZTE have long-established and strong relationships with Chinese and international operators from their existing telecom equipment business and their operator-branded handsets (with Vodafone and Orange). Other companies have succeeded in shedding their *Shanzhai* origins and becoming household name brands in China. For example, in 2012, China Wireless (under Coolpad) established a major R&D center with more than 1,700 R&D engineers in Shenzhen. The company's dual-mode dual-working handset (or four-channel handset under China Telecom's brand name) is a distinctive type of handset for the China market, which supports simultaneous call waiting and phone calls on both the GSM and CDMA networks. Coolpad has developed close relationships with all three Chinese mobile carriers in order to tap the growing 3G boom in China. In short, the availability and popularity of affordable, capable entry-level smartphones has led to consolidation in the domestic handset industry.[9]

[9] In the Chinese market, there are still Shanzhai smartphones, such as MEOX 1 and Android iPhone 4S (fake iPhone 4S), HDC Galaxy S3 (fake Samsung Galaxy S3) and HDC One X (fake HTC

It should also be noted that such a turnkey solution as Mediatek's chipset—the essential hardware required for compatibility with 3G—remains indispensable to the transformation. It is estimated that Mediatek has captured about 50 percent of the smartphone chipset market share in China. In part, this is a result of the company's shift in customers; originally serving *Shanzhai* handset makers, it now caters to more well-established Chinese smartphone vendors and/or brands. There are two existing players in the chipset maker market in China, Mediatek and Spreadtrum (based in the mainland, purchased by Tsinghua Unigroup in 2013). In addition to these, the US-based Qualcomm has penetrated the market with the Qualcomm Reference Design (QRD), with the belief that entry-level smartphones will drive future mobile growth.[10] The company has also invested in Xiaomi, a Chinese internet firm, with a famous iPhone-like model. A few other local players have also grown in this market segment, including HiSilicon, a local IC design house and a spin-off of Huawei and Leadcore, the chip design arm of Chinese communications equipment company Datang Group. Compared to the 2G era, there have been more routes for China's IC designers to be involved in the smartphone market.

Since Apple initiated the "App Economy," there have been increasing contributions of Chinese-influenced applications from Chinese firms. For example, Tencent's Wechat (Weixin) pioneered and popularized voice-recorded message services, which are nowadays widely copied by other companies with social media services. Taobao Alipay also dominates local online payment service, partly because of a low penetration rate of credit cards in China. The local key players, such as the mobile carriers and content aggregators, have put substantial effort into entrepreneurial programs to cultivate app developers. For example, China Mobile's program to incubate one million app entrepreneurs was designed to cultivate the speedy proliferation of home-grown applications.[11]

However, all this rapid indigenous development does not mean that China is no longer dependent on foreign firms for communication architecture. The communication ecosystem in China has benefitted from the overwhelming dominance of Google's Android OS in the Chinese market, with a market share of about 86 percent. In particular, the Android OS platform has lowered technology barriers between handset brand vendors capable of high-end

One X), which imitate premium models from Apple, Samsung, and HTC. For more information, see <http://www.shanzhaiji.com>.

[10] See Qualcomm's PowerPoint presentation by Yan Zhuang, at Upling 2012 Conference. https://www.uplinq.com/.../The_QC_Reference_Design_App_Ecosyst.

[11] According to an unofficial source, there are actually more than 200 Android app stores because Google Play is not widely available in China. <http://bambooinnovator.com/2013/04/06/china-is-finally-becoming-a-lucrative-market-for-app-makers>, accessed May 2013.

product development and handset OEMs capable only of handset manufacturing and low-end handset design. Taking advantage of this Android OS, firms like China Wireless have succeeded in shedding their *Shanzhai* origins and becoming leading branded players.

Discussion

The 3G transformation of China goes beyond the advent and popularity of low-cost smartphones. It has given China the footing to interact and compete with foreign players in the global industrial mobile communications ecosystem. Chinese customers have vastly varying income levels and unique user preference with regards to handsets and mobile applications. This means that their 3G subscribers need widely varying prices for handsets compared with developed countries (Huang 2011). Leading international smartphone brands tend to follow a global model that targets consumers in developed countries and the elite in Tier 1 cities in China. As a consequence, their products usually turn out to be unnecessarily technologically advanced, or at least simply unaffordable for the less wealthy and rural populations in China (the sluggish tiers of the market; Christensen et al. 2001: 81–2). The burgeoning local brands have managed to build their portfolio with feature-rich, multi-SIM handsets[12] that span the ultra-low and entry-tier segments. Increasingly the startups such as Xiaomi, with a clever design and avid use of social media, are providing quality smartphones substantially cheaper than the compatible foreign brands. It is by catering to the growing number of low to middle class, and rural subscribers, that local brands will bring smartphones to the majority of Chinese.

By comparing China's handset market to that of Japan, we can see clearly how the system in China is strongly shaped by vast variation in income and needs among Chinese users. In a study on Japan's mobile phone market, Chen et al. (2007: 17) found that the prevailing customized handset model, which is a collaboration between handset vendors and operators, has resulted in a high concentration of high-end handsets in the Japanese market. They attribute this to "the demanding nature of Japan's customers for well-functioned mobile phones." However, though a similar customized handset collaboration model also dominates in China, the income inequality, diverse demand from different sectors, and the existence of the multi-3G standard have created opportunities in China for lower-cost smartphones. These opportunities have pushed the home-grown brands and the operators to collaborate in

[12] With multi-SIM handsets, Chinese mobile phone subscribers can save costs while roaming within China.

producing a variety of trendy smartphones in the lower to mid-price range that are affordable, fashionable, and useful for such enormous market segments as migrant workers, rural farmers, and urban young white-collar workers. It is the ongoing demand from these segments that has led to the creation and expansion of "good-enough" innovation.

In addition, as mentioned earlier, since there are more than 300 handset makers in China manufacturing smartphones, most of them form a camp of white-box producers. They do not have close relationships with the operators or the major content-aggregators investing in mobile service platforms. To fight for their market share, they have to design and manufacture a variety of smartphones with trendy features and micro innovations. By benchmarking their legitimate counterparts, they cater for diverse demands from the grass-roots sectors in the less wealthy parts of China. In this way, the 3G ecosystem shares certain elements of continuity with the *Shanzhai* economy in its low price, rapid shift between models, and micro innovations, but by now the industry has been consolidated in favor of a few home-grown brands. While both *Shanzhai* brands and these few home-grown brands have benefitted from the "lower tiers of the market," it is the "platformization" of smartphones and 3G services (Feijóo et al. 2009) that requires close relationships between smartphone makers and the operators and the service platform owners (Eaton et al. 2011; Kenney and Pon 2011). Thus, the established players are ultimately at an advantage.

Will Chinese smartphones be confined to the domestic market and the low-end market segments? Early signs indicate that this is unlikely. For most rising Chinese brands, there are actually several significant areas for growth in countries less developed than China, in particular India and Africa. To break into such markets, Coolpad has teamed up with Reliance Webstore, a subsidiary of Reliance Communications, to market its dual-mode smartphones in India. In fact, Give-Me-Five, a Chinese handset maker, grew from scratch to become a well-known household name in India. Coolpad has even cooperated with American operator MetroPCS to launch its first LTE (Long Term Evolution) 4G mobile phone, named Quattro, in the US. Xiaomi has also moved into internationally to India and Southeast Asia (Mozur and Wang 2014).

Many observers regard China's TD-SCDMA as a signature achievement, in its role as industrial standard in innovation. However, we have shown that analysts need to consider the broader meanings of industrial standards for 3G communication services. Based on the revised version of the "three level model for standards and innovation in ICT" (Figure 10.2), it can be argued that for the "App Economy" and broadband mobile communications services to succeed, it takes much more than the technologies and standards of infrastructure. It requires other (de facto) standards at the middleware/service platform and application levels (Ballon 2009; Ballon and Walravens 2009;

Hammershøj et al. 2009; Eaton et al. 2011; Tilson et al. 2012; Yoo et al. 2010). Apple's iOS and App Store are obvious examples of such a de facto standard. The sharp contrast in the global revenue of mobile phone business between Apple and Nokia shows a shift of the primary sites of competitiveness from architectural design to middleware/service platforms and applications. Apple follows other players' (e.g. Nokia's 3G) industrial standards at the architecture level. But the company's success with the iPod, the iPhone, and the iPad lies in its proprietary platforms, such as iTunes and the App Store, as well as its practice of providing software design kits to numerous external developers for designing applications for customer experiences. This implies that the mobile communications sector, in terms of both service and device, has become more application-driven than ever, even in the context of "the layered modular architecture" (Yoo et al. 2010). In other words, TD-SCDMA alone, though important to China, cannot fully support China's migration toward indigenous smartphones and broadband mobile communications services. There are other standards that may become more influential.

As we have mentioned, Google's Android, a free and open-source OS, has become the dominant platform upon which to develop smartphones and mobile internet services in China. As shown in the right part of Figure 10.6, there are a few Chinese versions of OSs for smartphones, based on Google's Android platform, including Baidu's Yi OS and Xiaomi's MIUI. Baidu's Yi OS is essentially a forked version of Android: in addition to much of the same functionality and services that are available from Google, Baidu throws in its own bundle of apps, such as native maps, reader, music, web apps, and even a program similar to Google Places. Alibaba's Aliyun is Linux-based, enabling

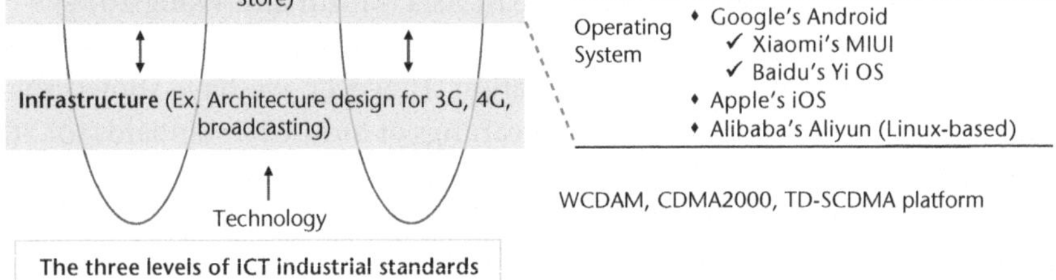

Figure 10.6. Industrial Standards and the New Ecosystem of Smartphones and Services in China

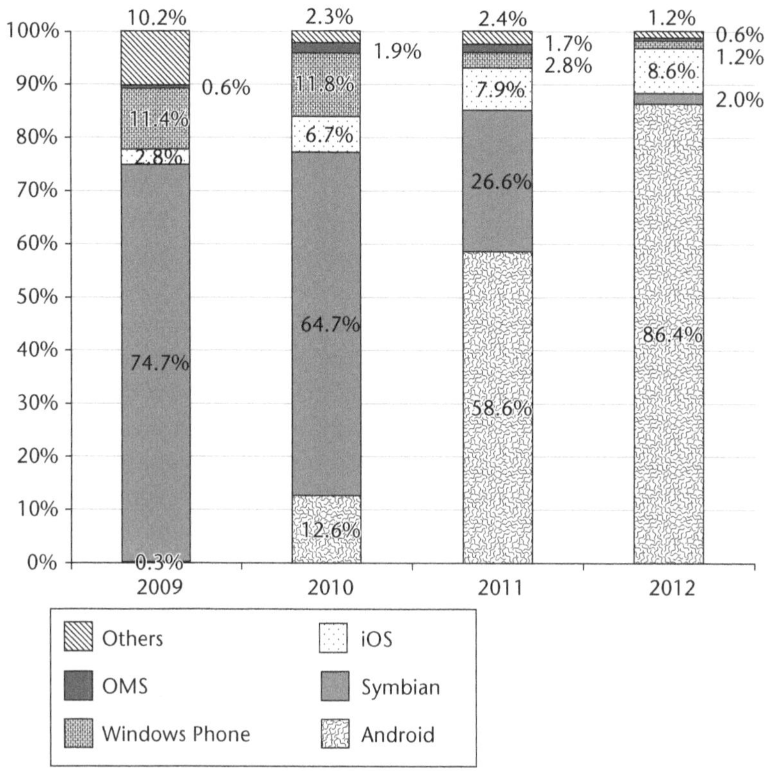

Figure 10.7. The Market Share of Mobile Operating Systems in China (2009–12)

Alibaba to take advantage of its popular e-commerce services. The overwhelming reliance on Google's OS platform of Android has caused some concerns about the lack of independence in China's smartphone developments (China Academy of Telecommunication Research 2013: 46). As shown in Figure 10.7, Nokia's Symbian, which once dominated the Chinese market as well as the global market, has lost much of its market share within just few years. In contrast, Android accounted for 86.4 percent of the market share for OS platform in China in 2012, while the Chinese home-developed OSs like Aliyun and Yi OS had market shares of less than 1 percent.

These Chinese internet companies may even benefit from China's latest policy to issue licenses for Mobile Virtual Network Operators (MVNO), allowing them to launch their own mobile communications services by leasing airtime capacity from the incumbent operators. Overall, China's mobile phone industry is still highly dependent on the foreign-provided Android platform despite the TD-SCDMA standard. At the same time, Chinese firms are making additional and possibly useful developments based on their existing strength and localized strategic development.

Chinese firms have used the fact that China's cyber world is a "walled garden" that excludes many popular internet applications and services from the developed world. But even if the proliferation of home-grown app services owes its success to this wall of censorship around the Chinese cyber world, their entrepreneurial and innovative work is still remarkable. But it remains to be seen how well the domestic innovation will prosper and reach outside China in the future. Tencent's Wechat (Weixin) with its voice-recorded message services is one example of a success story, since the program boasted more than 300 million subscribers as of July 2013.

The migration toward smartphones and broadband mobile communications services has brought tremendous changes in the Chinese innovation system in the 3G era. The indigenous 3G standard, TD-SCDMA, enables China Mobile to bargain with Apple for the network access of iPhones and facilitates the rise of domestic chip makers, such as Spreatrum, HiSilicon, and Leadcore. Since the mobile communications sector has become more application-driven than ever, and also since layered platforms have become the key to the ecosystem, TD-SCDMA's infrastructure standard is no longer as important in Chinese indigenous innovations.

Underlying Chinese efforts to develop and promote its indigenous TD-SCDMA standard are the government's intentions to reduce dependence on foreign technologies, including royalties paid to the foreign architecture providers and standard setters, and to establish its own playing field (Liu and Zhou 2013; Yan 2007). Through its accumulated strengths in mobile communications services, China Mobile has done relatively well in promoting its TD-SCDMA 3G services.

The advent of 3G services and smartphones has also altered the ecosystem in China with the rise of home-grown brands for smartphones, at the expense of *Shanzhai* handset vendors. While both have benefitted more from the "lower tiers of the market," the "platformization" of smartphones and 3G services (Feijóo et al. 2009) requires close relationships between smartphone makers and the operators and the service platform owners (Eaton et al. 2011; Kenney and Pon 2011), thus favoring the established players.

At the application layers, the proliferation of home-grown app services popular for smartphones has benefitted from the "walled garden" of the Chinese cyber world, but their entrepreneurial and innovative works are also remarkable. The evolution of China's mobile phone industry and its currently prevailing good-enough innovations can be better comprehended by referring to the idea of the co-evolution of technology, industrial structure, and supporting institutions (Nelson 1994).

Conclusions

The Chinese government has put great effort and resources into promoting indigenous innovations and industrial standards in the mobile communications sector. How well has this lived up to their high expectations? TD-SCDMA has been endorsed by ITU (the International Telecommunication Union) and actively promoted by its business champion, China Mobile. A few home-grown brands emerged to become the leading suppliers of smartphones in the Chinese market, outperforming the once-dominant players of *Shanzhai* handsets and even a few international premium brands. The top-down and bottom-up innovation in 3G has merged into a complex industrial ecosystem, with continued strong capabilities and multiple Chinese players specialized in different areas.

The rise of Chinese home-grown brands has been possible because of the growing availability of low-cost smartphones in supplying affordable, mid-tier, and even high-end smartphones. Such development shows the potential of "good-enough innovation" under the new ecosystem through the industrial transformation taking place in China. There has been some skepticism about the sustainability of the *Shanzhai* model in high-tech. The migration from 2G to 3G in China has brought about platform-based development and new business models, rendering new organizing logic between the smartphone vendors and other important stakeholders. The prevailing operator-centric model has forced operators to pay more attention to the development, sourcing, and marketing of customized smartphones. This has resulted in a closer relationship between the operators and the domestic smartphone vendors for customized handsets, leading to the consolidation of the Chinese smartphone industry. Many *Shanzhai* players disappeared, while a few others have grown into leading brands. In addition, major internet firms, such as Baidu, Tencent, Qihoo, and Alibaba have come out with customized smart phones to take advantage of their popular social network services and e-commerce services. The grassroots demands in the Chinese market and the growing popularity of mobile internet services within the "walled garden" have brought indigenous innovation in different layers in the technological ecosystem.

As for the roles of industrial standards in the Chinese integration of smartphones, we argue that TD-SCDMA, the indigenous industrial standard at the infrastructure level, can only be part of Chinese solutions in the mobile phone sector. At the middleware and service platform level, there are a number of Chinese versions of mobile OSs based on Google's Android platform, such as Baidu's Yi OS and Xiaomi's MIUI. Since the mobile communications sector has become more application-driven than ever and layered platforms have become the key to the ecosystem, China's quest for indigenous innovations and industrial standards must be viewed in a broader context than just that of infrastructure.

Given the strong hand the Chinese state has in innovation and development, it is useful to evaluate how state policies have affected the mobile phone industry. It seems that TD-SCDMA has played only a limited role, but China's regulation and censorship have effectively created an environment conducive to the rapid growth of domestic internet service companies, some of which have become important mobile service providers. Grassroots creativity, represented by *Shanzhai* makers in the 2G era, has morphed into more regular brand-name competition. But the low-cost approach, production networks centered in Guangdong, and collaboration between handset makers and Taiwan's MediaTek continue to show considerable creativity and energy in manufacturing affordable, trendy, and rapidly changing handsets. At a more general level, China's mobile phone industry has developed a distinct business model from developed countries, popularized by different internet and handset producers. It is not a segregated model, however, since it follows the trend of platformization and depends overwhelmingly on the Android system. China's experience seems to show that indigenous innovation does *not* simply follow the state plan, nor is it exclusive to the Chinese players. The top-down and bottom-up forms of innovation, Chinese and foreign players are interacting to create a vibrant and rapidly evolving Chinese market for mobile phone services.

References

Ballon, P., and N. Walravens (2009), "Towards a New Typology for Mobile Platforms: Validation through Case Study Analysis," 1st Europe, Middle East, North Africa Regional ITS conference (20th European Regional ITS Conference), Manama, Kingdom of Bahrain, Oct. 26–28.

Ballon, P. (2009), "Platform Types and Gatekeeper Roles: The Case of the Mobile Communications Industry," Paper presented at the Summer Conference 2009 of Copenhagen Business School, Frederiksberg, Denmark, June 17–19.

Brandt, L., and E. Thun (2010), "The Fight for the Middle: Upgrading, Competition, and Industrial Development in China," *World Development*, 38: 1555–74.

Chen, C., C. Watanabe, and C. Griffy-Brown (2007), "The Co-evolution Process of Technological Innovation: An Empirical Study of Mobile Phone Vendors and Telecommunication Service Operators," *Technology in Society*, 29: 1–22.

Chen, S. H., P. C. Wen, and C. Y. Tai (2013), "Shanzhai Handsets and China's Bottom of the Pyramid Innovation," in Phil Cooke, Glen Searle, and Kevin O'Connor (eds), *The Economic Geography of the IT Industry in the Asia Pacific Region*, 169–88. London and New York: Routledge.

China Academy of Telecommunication Research (2013), *White Paper on Mobile Internet*. Beijing: China Academy of Telecommunication Research of MIIT (in Chinese).

Christensen, C. M. (2003), *The Innovator's Dilemma*. Cambridge, MA: Harvard University Press.

Christensen, C., T. Craig, and S. Hart (2001), "The Great Disruption," *Foreign Affairs*, 80(2): 80–95.

Dawar, N., and A. Chattopadhyay (2000), *Rethinking Marketing Programs for Emerging Markets*. Ann Arbor, MI: William Davidson Institute Working Paper, 320.

Dosi, G. (2000), *Innovation, Organization and Economic Dynamics*. Cheltenham: Edward Elgar.

Eaton, B., S. Elaluf-Calderwood, C. Sørensen, and Y. Yoo (2011), *Dynamic Structures of Control and Generativity in Digital Ecosystem Service Innovation: The Cases of the Apple and Google Mobile App Stores*. Working Paper Series, 183, Information Systems and Innovation Group, London School of Economics and Political Science, Apr.

Feijóo, C., C. Pascu, G. Misuraca, and W. Lusoli (2009), "The Next Paradigm Shift in the Mobile Ecosystem: Mobile Social Computing and the Increasing Relevance of Users," *Communications and Strategies*, 75(3): 57–76.

Freeman, C., and C. Perez (1988), "Structural Crisis of Adjustment, Business Cycle and Investment Behavior," in G. Dosi, C. Freeman, R. Nelson, G. Silverberg, and L. Soete (eds), *Technical Change and Economic Theory*, 38–66. London: Pinter.

Gadiesh, O., P. Leung, and T. Vestring (2007), "The Battle for China's Good-Enough Market," *Harvard Business Review*, 85(9): 81–9.

Geels, F. (2005), "Co-evolution of Technology and Society: The Transition in Water Supply and Personal Hygiene in the Netherlands (1850–1930)—a Case Study in Multi-Level Perspective," *Technology in Society*, 27: 363–97.

Hammershøj, A., A. Sapuppo, and R. Tadayoni (2009), "Mobile Platforms: An Analysis of Mobile Operating Systems and Software Development Platforms," Paper presented at CMI international conference on social networking and communities, Nov. 25–26, Copenhagen.

Holz, C. (2008), "China's Economic Growth 1978–2025: What we Know Today about China's Economic Growth Tomorrow," *World Development*, 36(10): 1665–91.

Huang, L. (2011), "The RMB 1,000 Revolution," in *Smartphones: China Telecoms and Technology*. Tokyo: Nomura.

Kenney, M., and B. Pon (2011), *Structuring the Smartphone Industry: Is the Mobile Internet OS Platform the Key?* Keskusteluaiheita Discussion Papers, 1238. Helsinki: Research Institute of the Finnish Economy, Feb. 10.

Lin, F., and W. Ye (2009), "Operating System Battle in the Ecosystem of Smartphone Industry," Paper in 2009 International Symposium on Information Engineering and Electronic Commerce, 617–21.

Liu, X. F., and Z. X. Chao (2009), "The Strategic Upgrading and Restructuring of China's Shanzhai Cellphone Industry under Financial Crisis," *International Conference on Information Management, Innovation Management and Industrial Engineering*, i. 562–5. Xi'an: IEEE.

Liu, X., and J. Zhou (2013), "China's Catch-up and Innovation Model in IT," in Phil Cooke, Glen Searle, and Kevin O'Connor (eds.), *The Economic Geography of the IT Industry in the Asia Pacific Region*, 144–68. London and New York: Routledge.

Minagawa, T. Jr., P. Trott, and A. Hoecht (2007), "Counterfeit, Imitation, Reverse Engineering and Learning: Reflections from Chinese Manufacturing Firms," *R&D Management*, 37(5): 455–67.

Mu, Q., and K. Lee (2005), "Knowledge Diffusion, Market Segmentation and Technological Catch-up: The Case of Telecommunication Industry in China," *Research Policy*, 34: 759–83.

Nelson, R. (1994), "The Co-evolution of Technology, Industrial Structure, and Supporting Institutions," *Industrial and Corporate Change*, 3(1): 47–63.

Perez, C. (1985), "Microelectronics, Long Waves and World Structural Change: New Perspective for Developing Countries," *World Development*, 13(3): 441–63.

Prahalad, C. K. (2005), *The Fortune at the Bottom of the Pyramid*. Philadelphia: Wharton School Publishing.

Rong, K., and Y. Shi (2009), "Constructing Business Ecosystem from Firm Perspective: Cases in High-Tech Industry," Proceedings of the International ACM Conference on Management of Emergent Digital EcoSystems (MEDES) in Lyon, Oct. 27–30.

Rowen, H., and M. Hancock (2008), *Greater China's Quest for Innovation*. Baltimore: Brookings Institution Press.

Sheng, Z., and Y. Shi (2010), "Shanzhai Manufacturing: An Alternative Innovation Phenomenon in China: Its Value Chain and Implications for Chinese Science and Technology Policies," *Journal of Science and Technology Policy in China*, 1(1): 29–49.

Sigurdson, J. (2005), *Technological Superpower China*. Northampton, MA: Edward Elgar Publishing.

Struben, J. (2008), "Technological Coevolution: Critical Dynamics in Industry Transformation," Paper presented at the 25th Celebration Conference on Entrepreneurship and Innovation—Organizations, Institutions, Systems, and Regions. Copenhagen, June 17–20.

Suttmeier, R. P., and X. Yao (2004), *China's Post WTO Technology Policy: Standards, Software, and the Changing Nature of Techno-Nationalism*. Washington, DC: NBER Special Report 7.

Tilson, D., C. Sørensen, and K. Lyytinen (2012), "Change and Control Paradoxes in Mobile Infrastructure Innovation: The Android and iOS Mobile Operating Systems Cases," Paper presented at 2012 45th Hawaii International Conference on System Sciences. IEEE Computer Society, 1324–33.

Tse, E., K. Ma, and Y. Huang (2009), *Shan Zhai: A Chinese Phenomenon*. Beijing: Booz & Co.

Walters, P., and S. Samiee (2003), "Marketing Strategy in Emerging Markets: The Case of China," *Journal of International Marketing*, 11(1): 97–106.

Yoo, Y., O. Henfridsson, and K. Lyytinen (2010), "The New Organizing Logic of Digital Innovation: An Agenda for Information Systems Research," *Information Systems Research*, 21(4): 724–35.

Yan, H. (2007), *The 3G Standard Setting Strategy and Indigenous Innovation Policy in China: Is TD-SCDMA a Flagship?* DRUID Working Paper, 07-01, Copenhagen, Danish Research Unit for Industrial Dynamics.

Yueh, L. (2009), "China's Entrepreneurs," *World Development*, 37(4): 778–86.

Zhou, Y. (2008), *The Inside Story of China's High-Tech Industry: Making Silicon Valley in Beijing*. Lanham, MD: Rowman & Littlefield.

11

The Development of China's Wind Power Technology Sector

Characterizing National Policy Support, Technology Acquisition, and Technological Learning

Joanna I. Lewis

Introduction

In assessing China's innovative capacity, its energy sector is especially worthy of attention. Innovation in energy technologies, particularly for cleaner and low-emission energy sources, is crucial to China's ability to continue to grow its economy without leaving a massive environmental and health toll on its population. China's success in adopting the technologies it so critically needs will be determined in part by its ability to become an innovator and global leader in the clean energy sector. Its growth in this sector also has important implications for the global diffusion of these technologies.

Wind power holds particular promise both worldwide and in China for providing relatively low-cost, low-emission electricity. A late adopter of wind power technology, China has quickly risen to become the largest wind power market in the world, while Chinese firms are now among the leading manufacturers of wind turbine technology globally. Providing a 2 percent share of China's electricity in 2012, wind power is still a very small part of China's electricity mix. However, wind was also the third largest source of electricity that year, just behind coal and hydropower. Realistically, wind power does not have great potential to displace large amounts of coal electricity in China, at least in the short term. But it is a useful example of a low-carbon energy technology sector in which China grew from having no experience to being a major global producer and innovator in a very short period of time. As such,

China's experience with wind power technology provides an excellent case to understand how foreign firms transferred technology to China and how China supported and indigenized wind power technology.

This chapter examines various aspects of wind power technology development in China. It assesses the status of the Chinese wind power industry in an international context, then studies China's technological achievements and limitations in wind power technology over the past decade and the key policy decisions that shaped them. Next it turns to China's current leading technology firms in the wind power sector, focusing on their technology acquisition and innovation strategies in order to understand how firms were able to rapidly build up their knowledge through using foreign-sourced technologies. A few metrics are then used to evaluate technological learning in China, and the chapter concludes with a few reflections about the implications of the analysis. This chapter presents the case of China's entry into the wind power technology sector to further our understanding of China's future capabilities in low-carbon energy technology innovation.

China's Wind Industry in Global Context

China is the biggest wind power market in the world and now builds almost all of its wind turbines at home. China's wind power capacity has increased over 100-fold in the past decade from 344 MW in 2000 to 44,733 MW in 2010, and estimates for 2012 put installed wind capacity at just under 80 GW (Figure 11.1). Just a decade ago, the country had only a handful of wind turbines in operation, all of which were imported from Europe and the United States.

In 2012, the top five wind turbine manufacturing firms in the world were GE from the United States, Vestas from Denmark, Siemens from Germany,[1] Enercon from Germany, and Suzlon from India. The next tier of wind turbine manufacturers making up the top 10 includes Gamesa from Spain and the four top Chinese firms: Goldwind, United Power, Sinovel, and Minyang (Figure 11.2). Firms 10 through 15 include Nordex of Germany, Chinese firms XEMC and Sewind, Wind World India (previously Enercon India), and Alstom Wind, a relative newcomer to the wind power industry and the only firm from France among the top 15.

The vast majority of wind turbine manufacturers got their start in their home markets, benefitting from frequently preferential domestic policy support (Lewis and Wiser 2007). However, the wind power industry now has a

[1] Siemens entered the wind power industry when it purchased Bonus, a former leading Danish wind power firm, in 2004.

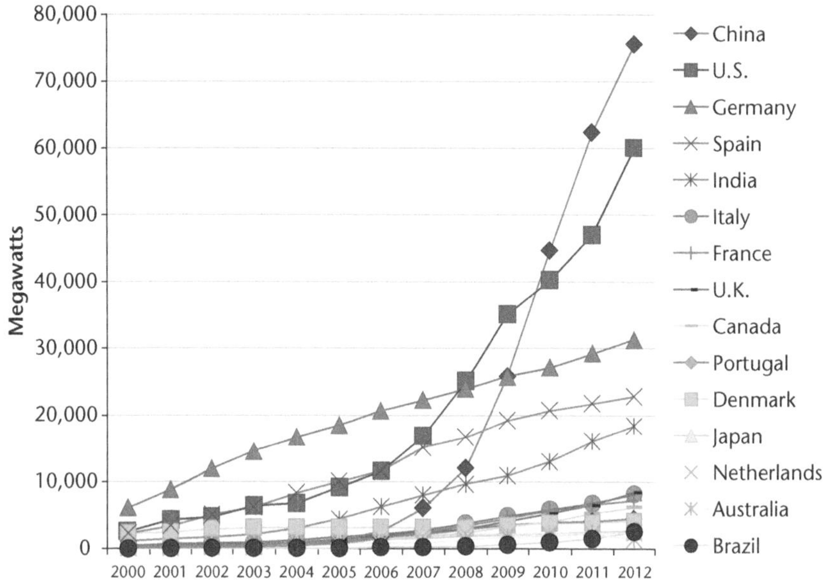

Figure 11.1. Global Wind Power Installations: Leading Countries
Source: Author's database.

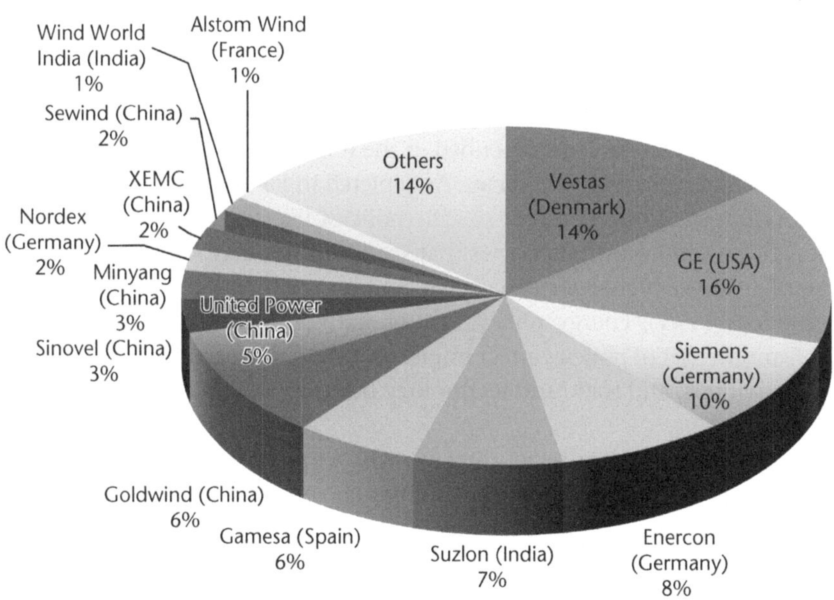

Figure 11.2. Top 10 Wind Turbine Manufacturers in 2012
Source: BTM and Navigant Research (2013).

technically advanced and mature global supply chain (Global Wind Energy Council and International Renewable Energy Agency 2012). As a result, the national ownership of wind power technology firms is increasingly less relevant. Even GE, perceived as a national company, in reality draws on a complex global supply chain to build its US wind turbines. Danish wind power champion Vestas, one of the first wind power technology firms in the world, is currently active in 28 countries (Vestas 2013). While it had its origins in Denmark, its current market is almost entirely abroad (Lewis and Wiser 2007).

The National Wind Power Technology Support System: Policies and Institutions

China has politically pursued the development of an indigenous wind power industry almost from the very beginning of utilizing wind power. A core national innovation strategy in China has been one that targets domestic development of technologies even if they were initially based on foreign-innovated designs. Given this priority, the Chinese state opted to support the development of wind power technology with a strategy similar to what it used in other industries. China's wind power industry has benefitted from various forms of government policy support; some policies have specifically targeted industrial development for the wind power industry, while others have indirectly supported industrial development by establishing a local market for wind power. The government has used policies to emphasize different areas of support at different times, and policy structure has undoubtedly influenced firm strategies for technology development in wind power.

The support mechanisms described in the following sections can be viewed as an interacting system of policies. As depicted in Figure 11.3, broader framework policies, including sector-specific policies like the National Renewable Energy Law and industrial policies that span multiple sectors, operate under a system of even higher level national policies such as general science and technology (S&T), energy policy, and climate policy strategies. At a lower level, another set of policies affect implementation, including pricing policies, R&D programs, and trade and technology transfer policies that affect specific firms.

All of these policies interact directly with wind technology companies at various stages of the technology development process. In some cases, companies are vertically integrated actors, but in this sector it is more common to combine different actors at different stages of the technology development process with varying levels of interactions. The levels of interaction greatly depend on the technology development strategy of the firm in question, as well as the technology acquisition, transfer, or innovation model used.

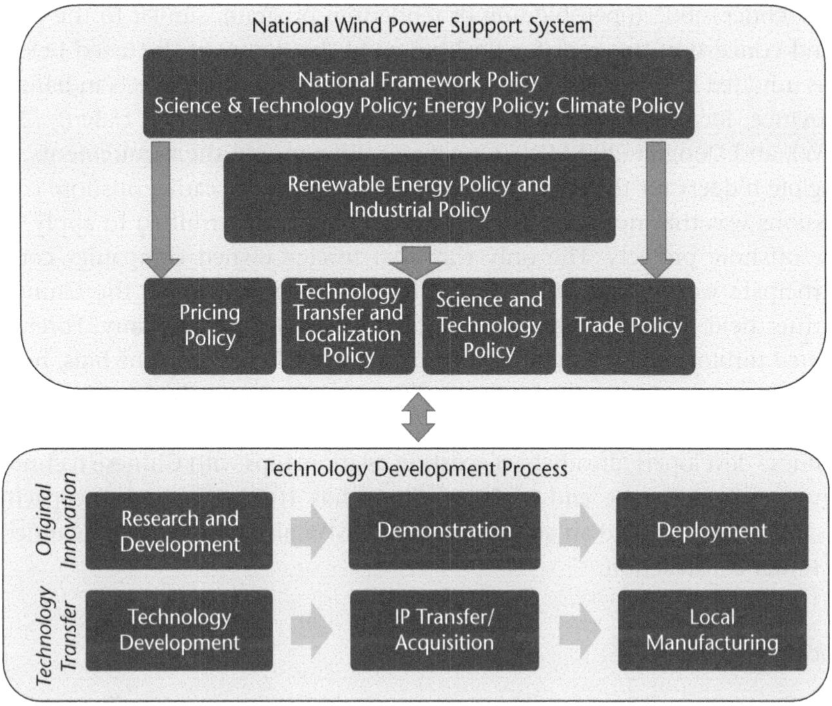

Figure 11.3. China's National Wind Power Support System and Conventional Technology Development Processes

Source: Created by author, based on framework developed by Dai and Xue (2014).

Because of this variation, these specific firm-level strategies are explored in more detail in the next section.

National Framework Policy

A new era of policies supporting renewable energy development in China began in 2006 with the launch of the Renewable Energy Law of the People's Republic of China (National People's Congress 2005). The Renewable Energy Law, while not exclusive to wind energy, directly benefitted wind power development in China by establishing a framework for regulating renewable energy. This law established a basis for setting national renewable energy targets informed by provincial energy plans, enacted a mandatory connection and purchase policy and authorized the establishment of feed-in tariffs for renewable electricity. The law also set the groundwork for a cost-sharing mechanism for renewables by creating a special fund for renewable energy development and called for much-needed national surveys of available renewable energy resources.

A "concession" (open bidding or tendering) program, similar to the early wind concession program for onshore wind development discussed below, was initiated in May 2010. It presented four offshore wind projects in Jiangsu Province, located in Binhai (300 MW), Shenyang (300 MW), Dafeng (200 MW), and Dongtai (200 MW). One major difference in the requirements for eligible bidders for these offshore concessions from the earlier onshore concessions was that no foreign-owned companies were permitted to apply for the offshore projects. The only way that foreign-owned companies could participate was as part of a Sino-foreign joint venture where the Chinese partner held over a 50 percent controlling share in the company. Foreign-owned turbine technology was technically not excluded from the bids, however it is proving increasingly rare in China for Chinese-owned developers to partner with foreign-owned turbine manufacturers, particularly since many Chinese developers already have existing relationships with Chinese technology suppliers. By the end of 2010, China has 100 MW of wind capacity installed offshore, and an additional 15,100 MW already proposed, planned, or under construction.

Pricing Policies

The first major Chinese policy to support wind power specifically came in 1994 when the government, with help from what was then the Ministry of Electric Power, released the *Provisions for Grid-Connected Wind Farm Management.* Between 2003 and 2007, onshore wind resource concessions were awarded to developers through a competitive bidding process for government-selected sites. These concessions became key drivers of wind development at this time. The government awarded winning bidders with approval to develop the selected project site, a power purchase agreement for the first 30,000 hours of the project, guaranteed grid interconnection, financial support for grid extension and access roads, and preferential tax and loan conditions. Five rounds of such wind concessions yielded 18 wind projects ranging from 100 MW to 300 MW in size, totaling 3,350 MW of new wind installations. While the wind concession projects got off to a bumpy start due to reported gaming with the bidding system, they succeeded in helping the government to determine the current price for wind power in China and set the groundwork for the establishment of feed-in tariffs (fixed, subsidized wind power prices). Tariffs issued under the wind concession program between 2003 and 2007 ranged from 0.42 to 0.551 RMB/kWh.

In August 2009, the National Development and Reform Commission (NDRC), released the *Notice on Policy to Improve Grid-Connected Power Pricing for Wind Power Generation,* which established a unified nationwide pricing standard and return on investment, thereby standardizing the development process for wind farms in China (NDRC Pricing Department 2009). Having a unified pricing policy makes it easier for project developers to predict the market and for manufacturers to plan their technology production. The policy set four feed-in tariff levels across the country, varying by region based on wind resource class. Category I resource areas had the best wind resources and therefore received the lowest tariff, while Category IV areas had the poorest wind resources and therefore received the highest tariff. Category I resource areas received 0.51 yuan/kWh, Category II resource areas, 0.54 yuan/kWh; Category III resource areas were to receive 0.58 yuan/kWh, and Category IV resource areas were to receive 0.61 yuan/kWh. Most of southern and southwestern China have lower wind speeds and are designated Category IV resource areas, while Inner Mongolia and Northern Xinjiang have the highest wind speeds and are designated Category I. Setting a higher tariff in low wind resource regions encourages wind power development despite lesser opportunities for electricity production.

Technology Transfer and Localization Policies

While framework policies set the national stage for the promotion of renewable energy and pricing policies promoted its deployment, another set of policies aimed at promoting the technology transfer and then the localization of wind power technology. This section reviews some of the key policies that were implemented at the national level, though many provinces implemented their own technology promotion programs as well. While the official technology transfer policies are reviewed here, there are also numerous reports of "unofficial" practices that either encourage or explicitly require technology transfer in firm-to-firm interactions that are not enshrined in official government policies.

In 1997, the SETC launched the *Double Increase Program* that aimed to double the 80 MW of wind capacity that were then installed in China to date. It also encouraged—but did not mandate—that a larger share of local content be incorporated into the turbines used. However, the future outlook for wind power utilization in China was probably too uncertain and 80 MW too small a quantity to encourage local manufacturing by turbine suppliers at this stage. Additionally, local content requirements conflicted with the requirement of most foreign government loans, which were already being used to support many wind farm ventures in China. These loans were typically in the form of tied-aid that came from various foreign governments (including Denmark, Germany, and the US) to support the sales of their own domestic wind farm technology to China. The tied-aid from foreign governments helped to subsidize the cost of early wind power development in China. About 74 MW of wind power was successfully installed under this program, essentially meeting the program target (Ministry of Science and Technology et al. 2002).

In 1997 the SDPC began its *Ride the Wind Program* in order to promote a model of "demand created by the government, production by joint venture enterprise, and ordered competition in the market." Two joint venture enterprises were established to manufacture wind turbines domestically. One was formed between the Spanish company Made and Chinese company Yituo, part of China's Luoyang First Tractor Factory, a commercial wing of the Chinese Ministry of Machinery. The technology transfers carried out through this program started with a 20 percent local content requirement and a planned increase to 80 percent as learning on the Chinese side progressed (Lew 2000). The Made-Yituo joint venture focused on a 660 kW turbine transferred by Made. The other joint venture was between the German company Nordex and the Chinese company Xi'an, which focused on a 600 kW turbine transferred by Nordex. The *Ride the Wind Program* had limited success, which many foreign companies attributed to the fact that they were not able to choose their Chinese partners; rather, the Chinese government selected them. The government selected companies from industries that they believed to be appropriate to wind technology—primarily the aerospace industry—but in reality had little experience or interest in manufacturing wind turbines. This problem was not unlike what occurred in the early years of the wind industry in the United States, and China fell far short of its target of 1,000 MW of wind by the year 2000. Members of the wind industry blamed this shortcoming on murky approval procedures and unrealistic local content requirements (Feifel 2001).

The wind concession projects were the first meaningful instances in which the use of locally made wind turbines was requested and rewarded. While earlier guidelines required that all wind farm projects approved by the NDRC during the 9th Five-Year Plan (1996–2000) included wind turbine equipment containing at least 40 percent locally made components, by the 2003 wind concession program that percentage had increased to first 50 and then 70 percent. Since there were very few Chinese turbine manufacturers at this time, these local content requirements mainly affected the foreign wind turbine manufacturers, causing most of them to establish manufacturing facilities in China.

China's local content requirement for wind turbines was further institutionalized in the 2005 NDRC *Notice on the Relevant Requirements for the Administration of the Construction of Wind Farms* (NDRC, 2005). This Notice clarified the basis upon which wind projects would be approved, with the major criteria being the project's proximity to the power grid to facilitate the dispatch of electricity, and the rate of using domestically manufactured equipment. If the localization rate for the project was less than 70 percent, it would not be allowed to be built. While some components were still expected to be imported, the customs administration applied import duties on any wind equipment brought into China from abroad (NDRC 2005). The 2005 *Requirements* also clarified that wind farm projects with outputs greater than 50 MW must be approved by the NDRC, while projects under 50 MW would be approved by the provincial or local Development and Reform Commission authorities.

Various preferential tax policies have also been directed at wind technology equipment manufacturers over the past decade. In 2001, the value-added tax was reduced by half on wind electricity (Ministry of Finance and State Administration of Taxation 2001). In 2005, the Renewable Energy Law called for new tax benefits to be put into place to promote industrial development in renewable energy. This led to the Renewable Energy Industrial Development Guidance Catalog, which gave special tax status to wind power generation projects and equipment manufacturers (NDRC 2005) and to the Enterprise Income Tax Law, which levied reduced income tax rates on wind manufacturers.

Science and Technology (S&T) Policies

The Ministry of Science and Technology (MOST) has subsidized wind energy R&D expenditures at varied levels over time. In an effort to help Chinese turbine manufacturers develop new products and technologies, MOST funded research to develop technologies for 600 kW machines as part of the 9th Five-Year Plan (1996–2000) (Ministry of Science and Technology et al. 2002). MOST is now supporting the development of megawatt-size wind turbines, including technologies for variable pitch rotors and variable speed generators, as part of the 863 National High Tech R&D Program. The 11th Five Year Development Plan of Science and Technology (2006–10) included support for the commercialization of 2 to 3 MW-sized wind turbines.

In April 2008, the Chinese Ministry of Finance issued a new regulation stating that the tax revenue for the key components and raw materials for large turbines (2.5 MW and above) would be channeled back into technology innovation and capacity building in the wind industry. In the same year, the Ministry of Finance announced the *Interim Measures on Management of Special Project Funds for the Industrialization of Wind Power Generation Equipment,* which provided funding support for the commercialization of wind power generation equipment (Ministry of Finance 2008). It specified that for all "domestic brand" wind turbines (with over 51 percent Chinese investment), the first 50 wind turbines over 1 MW produced would be rewarded with RMB 600/kW (€60) from the government. The measures further required that the wind turbines be tested and certified by China General Certification (CGC), enter the market, and be put into operation and connected to the grid. The National

Energy Bureau has reportedly granted licenses to 16 national energy research and development centers to research topics such as blade R&D, large-scale grid connected wind power systems, and offshore wind power equipment (Li et al. 2010).

Industrial and Trade Policies

In addition to explicit technology transfer policies, trade policies have been used in a variety of ways over time to try to encourage different modes of local manufacturing and industry development. For example, in 2001, the Ministry of Finance and State Administration of Taxation guidelines on the taxation of wind turbine imports stated that wind turbine components for turbines larger than 1.5 MW were exempt from customs duties and import sector value-added tax, while complete wind turbines less than 3 MW were subject to normal taxation. These guidelines discouraged the import of complete wind turbines and supported local manufacturers needing access to foreign components (Ministry of Finance and State Administration of Taxation 2001.

In 2009, US Department of Commerce Secretary Gary Locke traveled to China to ask for the removal of the local content requirement, arguing that it was a trade barrier for foreign firms. China agreed and recalled the requirement in the November 2009 (NDRC 2009) *Notice on Abolishing the Localization Rate Requirement for Equipment Procurement in Wind Power Projects*. While this was viewed as an achievement for foreign manufacturers, it was likely to have little impact in the Chinese wind sector where foreign firms had already established in-country manufacturing facilities. At this same time, there were concerns at the highest levels of government about the health of the Chinese wind sector due to reports of substantial overcapacity. In August 2009 the State Council listed wind turbine production as an "excess capacity sector," causing the Ministry of Land and Resources reportedly to deny all applications for new wind turbine manufacturing facilities.[2]

In early 2010, the Chinese Ministry of Industry and Information Technology released the *Access Conditions for Wind Power Equipment Industry* which aimed to "promote the optimization and upgrading of the industrial structure of the wind power equipment manufacturing industry, enhance enterprises' technical innovation, improve product quality, [and] restrict the introduction of redundant technology" to "guide the industry's healthy development." This was to be accomplished by restricting the operation of wind turbine manufacturers that did not have the capability to produce a 2.5 MW or larger

[2] In 2009 there were reportedly 83 wind producers with annual production capacity exceeding 50 GW. This was only a national capacity based on company estimates, not actual production. Annual demand that year was closer to only 10 GW.

turbine, could not claim at least five years of experience in a related industry, or could not meet various other financial, R&D, and quality control requirements (Baker Botts 2010). The necessity and efficacy of such regulations were widely disputed in the industry. Some stakeholders claimed that the regulations were not really about improving the "health" of the sector but were rather a case of the government showing preference for the top three Chinese wind turbine manufacturers in an attempt to thwart competition from other smaller manufacturers. All three of the leading Chinese wind turbine manufacturers are at least partially state-owned.

Firm-Level Models of Technology Acquisition

As Chinese firms entered the wind power industry over the past couple of decades, they frequently opted to acquire technology that was initially developed by firms in other countries. Their first decision was choosing among manufacturing complete wind turbine systems, manufacturing certain components and importing others, or even just serving as an assembly base for wind turbines imported from abroad. Each of these approaches represented different goals for manufacturing, different degrees of localization and technology ownership, and different policy incentives at work. In addition, each model required a different degree of localization ranging from partial to full (the turbine is only fully localized when it is completely manufactured in the home country).

Countries with lower wage rates, such as India and China, expect to be able to realize cost savings through domestic manufacturing of wind turbines compared to their European and American counterparts. This cost reduction is potentially significant for those turbine components that are particularly labor intensive.

Firm-Level Technology Transfer Models

A common strategy among latecomer firms has been technology transfer, or obtaining technology from a company that has already developed advanced wind turbine technology. Technology transfers can occur in several different modes. One mode is through a licensing agreement that gives the licensing firm access to a certain wind turbine model, often with some restrictions on where it can be sold. Another model is establishing joint-venture partnerships between two companies either to share a license or for collaborative research and development. Firms also can opt to collaborate to jointly develop a new technology design (joint development) and then share the associated intellectual property, which can be done without forming a new company or a joint- venture enterprise. If a latecomer firm has the capacity and means, it can also obtain access to technology through the purchase of ownership rights in a company with the desired technology or other forms of mergers and acquisitions (M&A). Many Chinese wind turbine manufacturers have used some of this set of technology transfer modes, while others have developed original designs, often in conjunction with research units at universities.

European and American wind turbine manufacturers were demonstrating their technology in China as early as the mid-1980s. These demonstrations created opportunities for learning, led to local partnerships, and eventually resulted in a shift from technology imports to local manufacturing. Along the way, technology transfers from these overseas companies to local Chinese companies, whether in the form of intellectual property, skilled personnel, or other informal means of knowledge transfer, helped lay the groundwork for the Chinese wind industry. The close proximity and small number of players in the Chinese wind industry encourage many informal learning networks that are influencing firm experience and strategy.

When the first utility-scale wind turbine was installed in China in 1985, it was imported from Denmark. At that point, the pioneering wind company Vestas largely had the Chinese wind market to itself. Over the next decade, a handful of other foreign wind turbine manufacturers imported turbines to China. The mid-1990s saw the establishment of the first Sino-foreign joint ventures in wind turbine manufacturing, and the first Chinese-owned wind turbine manufacturers were established in the late 1990s. By the mid-2000s, many new Chinese manufacturers had entered the Chinese market. Between 1999 and 2009, the number of Chinese-owned companies installing at least one wind turbine in China annually increased from 2 to 34. The actual number of existing Chinese wind companies is estimated to be much larger, probably in the range of 80 or more companies, although they are yet to participate in a commercial wind farm development in China. The emergence of the many new Chinese wind power firms has come at the expense of the foreign firms that had already entered the Chinese market, as illustrated in Figure 11.4.

Although the acquisition of technology from overseas companies is one of the easiest ways for a new wind company quickly to obtain advanced technology and begin manufacturing turbines, there is, of course, a disincentive for leading wind turbine manufacturers to license proprietary information to firms that could become competitors. In the past decade, the loosening of Chinese government restrictions on the ownership of foreign firms has resulted in far fewer joint-ventures. Instead, technology transfers are increasingly occurring between companies in different countries with synergies along the technology development continuum.

For example, German engineering design firms with little manufacturing ability are transferring technology to Chinese manufacturing dynamos with little innovative ability. But China is increasingly not only a recipient of

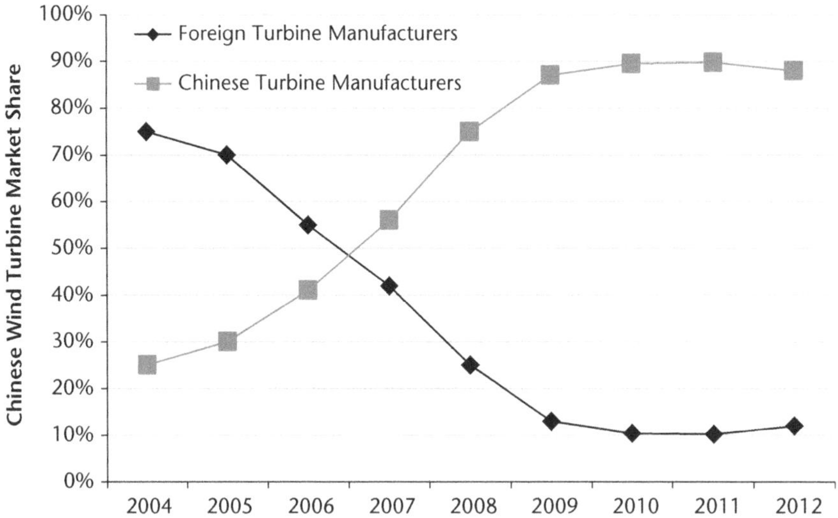

Figure 11.4. Chinese Market Shares of Chinese and Foreign Wind Turbine Manufacturers (2004–12)
Source: Lewis (2013).

technology from industrialized nations but also a source of technology being transferred to other developing nations. While such technology transfers are commonly facilitated via licensing agreements, models of technology transfer in which the recipient of the transfer plays a far more active role, such as mergers and acquisitions (M&A) and joint development, are prevalent.

Firm-Level Innovation Models

The largest market share in China in 2012 was held by Chinese firm Goldwind, which was China's first leading wind turbine manufacturer. The company has benefitted from a combination of sustained government support and an effective technology acquisition and development strategy. As the first Chinese-owned wind turbine manufacturer to produce a successful wind turbine design, it has developed a reputation for independent technology innovation. It began by licensing technology from Jacobs (which later became Repower) and later from Vensys Energiesysteme GmbH (both are German firms). In early 2008, when several other firms made a bid to purchase Vensys, Goldwind opted to purchase a 70 percent stake in the company outright so that it could continue its partnership. After its acquisition of Vensys, Goldwind began to jointly develop several new wind turbine designs in partnership with the company (Lewis 2013).

The 2012 market shares of other leading Chinese firms are shown in Figure 11.5. Sinovel obtained its 1.5 MW wind turbine technology through a

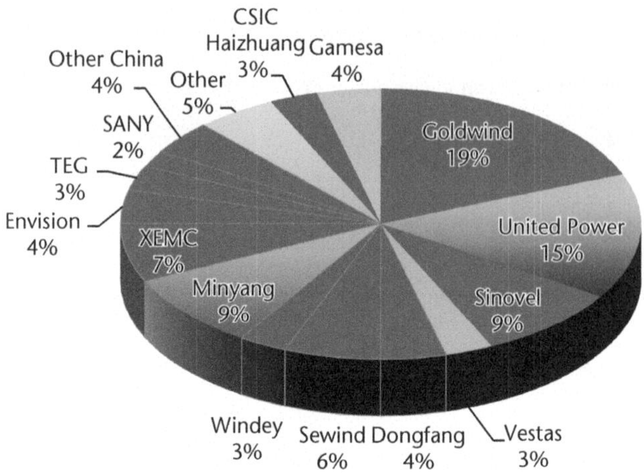

Figure 11.5. Wind Turbine Market Shares in the Chinese Market (2012)
Source: BTM and Navigant Research (2013).

licensing agreement with German firm Fuhrlander. It later partnered with American Superconductor and its wholly owned, Austrian-based subsidiary Windtec to jointly develop 3 MW and 5 MW turbines (May and Weinhold 2009). Dongfang Electric Corporation (DEC), based in Sichuan province, is part of a large, state-owned enterprise managed directly by the Chinese central government and is one of the largest power plant construction firms in China. It obtained its wind power technology through a licensing agreement with REpower for its 1.5 MW wind turbine. Another emerging Chinese firm, A-Power, obtained its technology through a licensing agreement with Fuhrlander for a 2.5 MW turbine, and with Danish firm Norwin for 225 kW and 750 kW wind turbine models, which also included the establishment of a joint-venture company (Electric Energy 2008).

Numerous other Chinese wind turbine manufacturers have relied on licenses as well. Sinovel and Beijing Beizhong have benefitted from licenses acquired from German firms Fuhrlander and DeWind. Dongfang Electric Corporation (DEC) and Windey, like Goldwind, have both licensed turbine designs from REpower. China Shipbuilding Industry Corporation Haizhuang Windpower Equipment Co., Ltd (CSIC) and Guodian United both obtained licenses from the German firm Aerodyn. AMSC-Windtec, an American owned firm with roots in Austria, has licensed wind turbine technology to Chinese firms Shenyang Blower Works (Group) Co., Ltd (SBW), XJ Group, Sinovel, and CSR Zhuzhou, as well as to Korea's Hyundai and several smaller Indian wind turbine manufacturers.

The sources and models of technology development used by Chinese wind companies are detailed in Table 11.1.

Table 11.1. Technology Transfer Models and Sources of Leading Chinese Wind Turbine Manufacturers

Chinese Company	Model of Technology Transfer	Source of Technology Transfer
A-Power (GaoKe)	License	Fuhrlander (Germany)
	License/Joint development	Norwin (Denmark)
Beijing Beizhong	License	DeWind (Germany/UK/US/Korea)
Changzing	Self-developed	Developed with Shanxi Science and Technology University (China)
CSIC Haizhuang	License	Frisia (Germany)
	Joint development	Aerodyn (Germany)
CSR Zhuzhou	License	AMSC-Windtec (USA/Austria)
DEC	License	REpower (Germany)
	Joint development	Aerodyn (Germany)
	Joint development	AMSC-Windtec (USA/Austria)
Engga	Self-developed	Developed with the Tsinghua Industrial Academy (China)
Envision	Joint development	Supported by the European Clean Energy Fund (EU)
Goldwind	License	Jacobs/REpower (Germany)
	Joint development	Vensys (Germany)
Guodian United	License	Aerodyn (Germany)
Hadian	Self-developed	Developed with Harbin Power Planet Equipment Corporation (China)
Hafei	Joint venture	WinWind (Finland)
Harbin Steam Turbine Co.	License	DeWind/EU Energy (Germany/UK)
Hewind	Joint development	Aerodyn (Germany)
Huachuang	Self-developed	Developed with the Shenyang University of Technology (China)
Huide	License	Fuhrlander (Germany)
Jiuhe	License	Windrad Engineering (Germany)
Minyang	License	Aerodyn (Germany)
	Joint development	Aerodyn (Germany)
New United	Self-developed	Developed with the Shenyang University of Technology (China)
REpower North	Joint venture	REpower (Germany)
SBW	Joint development	AMSC-Windtec (USA/Austria)
Sewind	License	DeWind/EU Energy (Germany/UK)
	Joint development	Aerodyn (Germany)
Shanghai Electric	License	DeWind (Germany/UK/US/Korea)
	Joint development	Aerodyn (Germany)
Sinovel	License	Fuhrlander (Germany)
	License	Windtec/AMSC (USA/Austria)
	Joint development	Windtec/AMSC (USA/Austria)
Tianwei	Joint development	Garrad Hassan (UK)
Windey	License	REpower (Germany)
	Joint development	REpower (Germany)
	Self-developed	Developed with Zhejiang Institute of Mechanical and Electrical Engineering (China)
Wuhan Guoce Nordic New Energy	License	Deltawind/Nordic Windpower (Sweden)
XEMC	License	Zephyros/Lagerwey (Netherlands)
XJ Group	Joint development	AMSC-Windtec (USA/Austria)
Yinhe Avantis	Joint development	Avantis Energy (Germany)
Yinxing	License	Mitsubishi (Japan)

Sources: Author's own database and Paul Recknagel, *Mapping WTG Manufacturers in China*.

As companies develop their own designs and manufacturing expertise, they may be more interested in co-developing wind turbine technology with firms that bring a different set of experiences to the partnership. One advantage of joint development is that there is no initial concern about market competition; when multiple manufacturers are involved, arrangements for the sharing of any resulting IPR are almost always made prior to the start of the joint work. These arrangements can be more straightforward when joint development involves a firm that primarily focuses on design working with a firm that primarily focuses on manufacturing. The risk with this model, however, is that if the design firm has no manufacturing experience and if manufacturers have no design experience, the resulting product may look great on paper but fail in the factory or in the field. This form of technology acquisition is becoming increasingly common in China, particularly among the larger firms. Examples include Sinovel's joint development with AMSC-Windtec, DEC's with AMSC-Windtec and with Aerodyn, Goldwind's with Vensys, A-Power's with Norwin, and Hewind and Sewind's with Aerodyn.

Several Chinese wind turbine firms have relied on government support for R&D, often in conjunction with a consortium of research institutes or universities. While this is a less common model than joint ventures or licensing, it is being used by several smaller Chinese manufacturers such as Windey, which originated at China's Zhejiang Institute of Mechanical and Electrical Engineering. As firms enhance their presence around the world by expanding manufacturing bases or R&D facilities, they are also increasingly able to tap into an expanded global knowledge base. Just a few years ago, Goldwind was principally a Chinese wind turbine company, primarily operating its manufacturing and R&D facilities in China. This domestic focus changed with the acquisition of Vensys in 2008, when it began to increase its R&D activities in Germany.

Many foreign-owned wind turbine manufacturers involved in the Chinese market have shifted larger shares of their total R&D expenditures into China. For example, in October 2010, Vestas announced that it had established a new wind power R&D Center in Beijing. Vestas committed an investment of $50 million within five years and a staff of 200 by 2012. Suzlon also announced that they would be opening a new R&D center in China in 2011. Although few Chinese-owned wind turbine manufacturers provide transparent reports of their R&D expenditures, there is substantial anecdotal evidence that investments in R&D are increasing as a percentage of total revenue. For example, Minyang increased its R&D expenditures by 31 percent from 2009 to 2010 as they worked to develop a new 3 MW wind turbine design (PR Newswire 2010).

Technological innovation in the wind industry is currently being driven to develop larger onshore and offshore wind turbine technology, reduce costs, increase efficiency, and improve grid interactions. These continuous advancements create a barrier to new entrants that may struggle to catch up with the best available technology. Firms intending to enter have to decide whether to compete with the currently popular turbine model and risk it being outdated in the near future, or to develop a larger size turbine that does not yet have a commercial application in the hope that it soon will. Otherwise, an entry-level firm must find another competitive edge such as producing a popular turbine type at a lower cost.

Certain wind turbine components, including blades and gearboxes, are technically sophisticated and must last for years with little maintenance. Quality control is therefore of primary importance in the wind industry. Many technologically advanced countries have been able to enter the wind market at a late stage without much prior experience in wind turbine manufacturing due to their relatively developed technical knowledge base. Countries with more general indigenous technical capacity are readier to develop new technologies, particularly wind turbines, as experience in other industries has been shown to translate as an asset to wind technology development (Kamp et al. 2004). Even the perception of poor quality can severely limit market growth.

Evidence of Technological Learning

Technology Size

One way to assess technological progress in wind power technology is by the average size of the wind turbines being installed annually. Since the size of individual wind turbine technology has increased over time, and since the majority of China's wind power installed in recent years has come from Chinese technology manufacturers (over 80 percent in 2009), the size of the turbines installed is an approximate measure of the technology level of Chinese wind technology providers. Overall, China is still installing smaller wind turbines on average than other countries (Figure 11.6), even those that have fallen behind China in numbers of annual installations. These countries, including Denmark, Germany, the United States, and Spain, were all earlier innovators in this industry. The local manufacturing of wind turbines in China began around 1996, about two decades after it began in Denmark. While China has clearly made strides, its companies are still working to catch up in terms of the level of technology they are manufacturing.

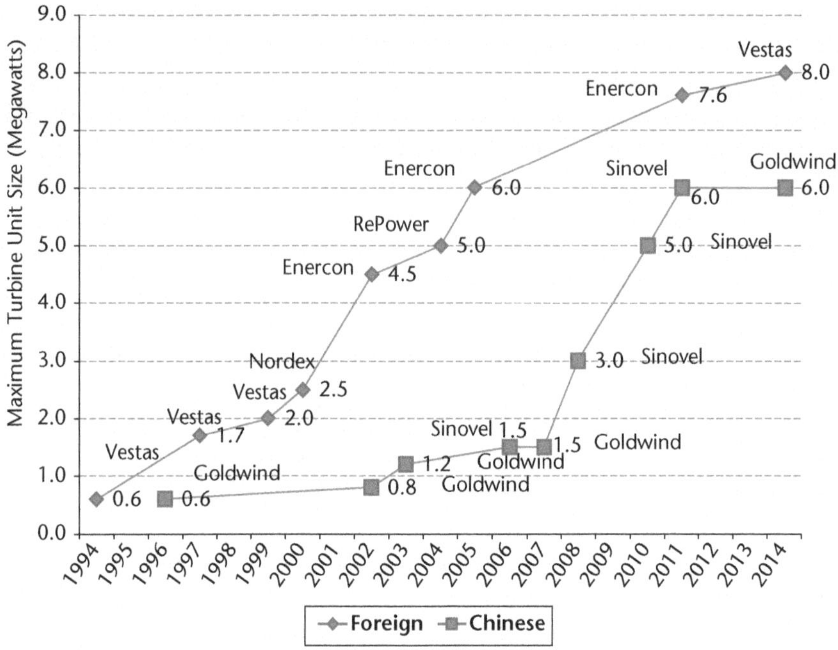

Figure 11.6. Maximum Wind Turbine Size of Foreign and Domestic Chinese Firms Compared

Source: Adapted from Ru et al. (2012) and updated.

Patents

China is sixth globally in terms of patenting activity in the wind power sector, as illustrated in Figure 11.7. However, we will see that number of patents is not always a reliable indicator of indigenous innovation.

To get a more global perspective on clean energy patenting trends, we can use the database of global patenting activity maintained by the World Intellectual Property Organization, or WIPO. Wind power is the largest non-fossil, non-nuclear, clean energy technology on an investment basis, and China is the location of the largest investment in wind energy. But a closer look at wind power patents shows that while the most patents are being filed in China, the leading patenters (even filing patents in China) are firms from the United States, Denmark, Germany, and Japan (Figure 11.8). This tells us that, although China has at least become the main location for manufacture and patent filing for wind power technology, Chinese firms are not yet the ones developing innovative ideas.

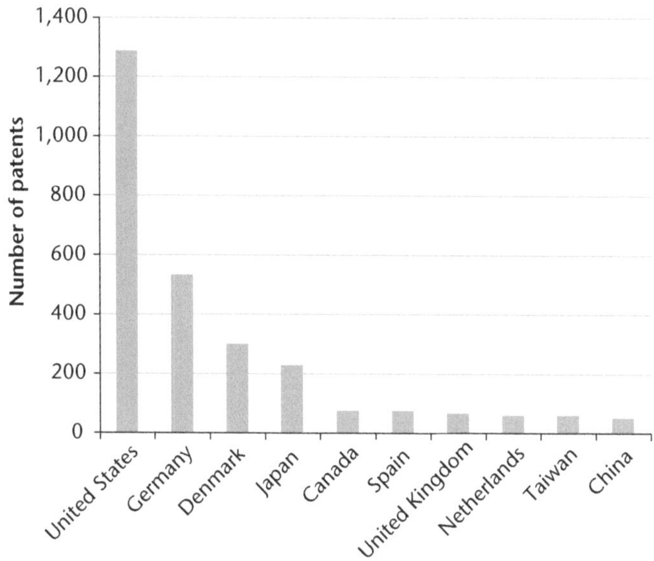

Figure 11.7. Total Patent Cooperation Treaty Wind Energy Patents (1999–2011)
Source: OECD (2013).

Learning and Technology Cost

The production of wind turbine technology has steadily increased globally, but it has not grown as quickly anywhere as much as it has in China. Therefore, if wind power technology costs decrease over time with manufacturing experience, China no doubt is playing an important role in that cost decrease.

With experience comes learning, which can be measured in the form of technology cost reductions. Many studies have calculated learning rates for wind power from 1980 to 2000, suggesting that historical cost reductions have been significant (though there is relatively little agreement on the magnitude of those reductions). For example, in the United States, the other major site of wind power development outside of China in the past decade, installed costs for wind power fell by about $2,700 between 1980 and 2000 (Figure 11.9), reaching a low point of about $700/kW between 2000 and 2002. Between 2002 and 2009, however, costs in the United States have increased by about $800/kW or over 100 percent, despite over 30 GW of new US wind power installations representing one-quarter of capacity installed globally over that period.

The challenge in trying to assess China's role in these cost reductions is that China has only become a major source of wind power deployment and manufacturing over the last five or so years, the same time period over

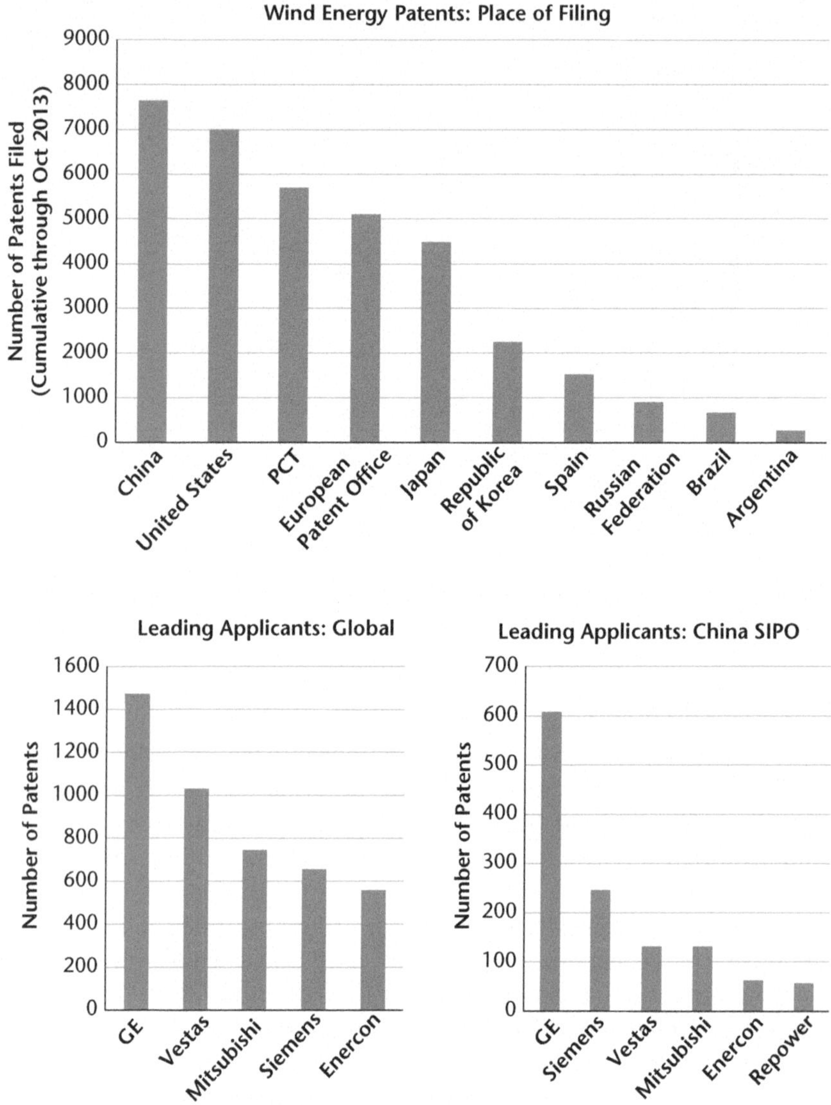

Figure 11.8. Wind Energy Patenting: Global Trends
Source: WIPO (2013).

which installed wind power costs have actually been increasing in many countries, making it difficult to assess recent learning rates.

Taken together, there are several clear indicators that learning is taking place in China's wind industry. While China is still installing smaller wind turbines on average than other countries, including the early wind power technology innovators that have fallen behind China in terms of annual installations,

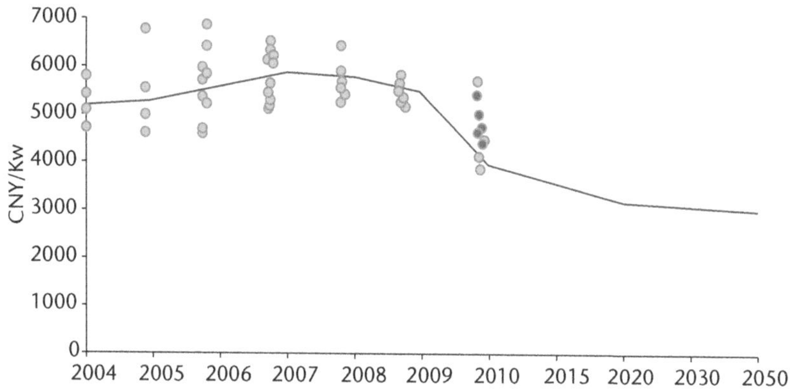

Figure 11.9. Wind Turbine Prices and Future Predictions
Sources: IEA and ERI (2011).

the average size of wind turbines installed in China have increased by almost 1 MW over the past decade. Over the same time period the number of Chinese-owned companies installing commercially viable wind turbines has increased 17-fold. The global reach of Chinese firms is also expanding, allowing for additional access to global learning networks.

Conclusions

China's indigenous development of wind power technology provides an excellent case to examine how foreign firms have transferred technology to China, how the Chinese state has supported a new technology, and how China's technological indigenization process unfolded.

The national innovation strategy in China tends to be one that targets domestic development of technologies even if they were initially based on foreign-innovated design. As a result, China has pursued an indigenous wind power technology industry almost from the very beginning of utilizing wind power in China.

China's wind power industry has benefitted from various forms of government policy support. Some policies have specifically targeted the wind power industry by directly supporting industrial development, while other polices have indirectly supported industrial development by establishing a local market for wind power. The government has used policies to emphasize different areas of support over time, and policy structure has undoubtedly influenced firm strategies. These support mechanisms can be viewed as an interacting system of policy types that includes targeted framework policies such as the

National Renewable Energy Law as well as broader industrial policies that span multiple sectors. The interacting system also operates in conjunction with even higher level national policies such as general S&T, energy policy, and climate policy strategies.

The consistent policy support specifically targeting Chinese firms has led to the emergence of a Chinese wind technology manufacturing industry built upon foreign technology transfers. While many Chinese wind turbine manufacturers have used a similar set of technology transfer mode, including licenses, mergers and acquisitions, and partnerships with other firms to conduct joint technology development, the trend has been to move toward models which allow for more technological autonomy as well as ownership to be in the hand of the Chinese firms. The wind industry relies far less on joint-venture manufacturing facilities in partnerships with foreign firms than other industries. Lower entry costs and widely accessible IP, primarily from smaller American and European engineering and design firms, permitted Chinese turbine manufacturers to rapidly catch up with foreign competence in this sector. While Chinese firms still lag in novel and frontier innovations in this sector, they are producing world-class technology that has dominated within the Chinese market and is increasingly being sold outside the Chinese market.

References

Baker Botts (2010), *Entry Standards for Wind Power Equipment Manufacturing Industry, Draft for Solicitation of Opinions* (tr. from Chinese). Beijing: Ministry of Industry and Information Technology, Mar. 31.

BTM and Navigant Research (2013), *International Wind Energy Development: World Market Update 2012*. Copenhagen: Navigant Consulting.

Dai, Y., and L. Xue (2014), "China's Policy Initiatives for the Development of Wind Energy Technology," *Climate Policy*, 0(0): 1–28. doi:10.1080/14693062.2014.863549.

Electric Energy (2008), "A-Power Signs Second Wind Turbine Deal, Enters Exclusive Agreement with Norwin A/S to Manufacture and Sell 750 kW and 225 kW Wind Turbines in China." *Electric Energy*, Jan. 20, <http://www.electricenergyonline.com/?page=show_news&id=82409>.

Feifel, K. (2001), "Wind Power in China, a German Company's Experience," *Industry and Environment*.

Global Wind Energy Council (GWEC), and International Renewable Energy Agency (IRENA) (2012), *30 Years of Policies for Wind Energy: Lessons from 12 Wind Energy Markets*. Bonn: IRENA.

International Energy Agency (IEA) and China Energy Research Institute (ERI) (2011), *China Wind Energy Development Roadmap 2050*. Paris: OECD/IEA. <http://www.iea.org/publications/freepublications/publication/china_wind.pdf>.

Kamp, L. M., R. E. H. M. Smits, and C. D. Andriesse (2004), "Notions on Learning Applied to Wind Turbine Development in the Netherlands and Denmark," *Energy Policy*, 32(14): 1625–37.

Lew, D. J. (2000), "Alternatives to Coal and Candles: Wind Power in China." *Energy Policy*, 28(4): 271–86. <http://dx.doi.org/10.1016/S0301-4215(99)00077-4>.

Lewis, J. I. (2013), *Green Innovation in China: China's Wind Power Industry and the Global Transition to a Low-Carbon Economy*. New York: Columbia University Press.

Lewis, J. I., and R. H. Wiser (2007), "Fostering a Renewable Energy Technology Industry: An International Comparison of Wind Industry Policy Support Mechanisms," *Energy Policy*, 35(3): 1844–57.

Li, J., P. Shi, and H. Gao (2010), *2010 China Wind Power Outlook*. Beijing: GWEC and Greenpeace. http://www.greenpeace.org/eastasia/press/reports/wind-power-report-english-2010>.

May, H., and N. Weinhold (2009), "New Turbines on Offer," *New Energy*, <http://www.newenergy.info/index.php?id=1542>.

Ministry of Finance (2008), *Regulation No. 476*. Beijing: Ministry of Finance.

Ministry of Finance and State Administration of Taxation (2001), *Notice on the List Related to the Interim Provision on Import Tax Policies for Major Technology Equipment*. Beijing: Ministry of Finance and State Administration of Taxation.

Ministry of Science and Technology, State Development Planning Commission, and State Economic and Trade Commission (2002), *Evaluation of Policies Designed to Promote the Commercialization of Wind Power Technology in China*. Beijing: Energy Foundation China Sustainable Energy Program, May 15.

National Development and Reform Commission (2005), *Regulation No. 2517*. Beijing: National Development and Reform Commission.

National People's Congress (2005), *The Renewable Energy Law of the People's Republic of China*. Beijing: National People's Congress.

NDRC Pricing Department (2009), *Regulation No. 1906*. Beijing: NDRC.

Office of the United States Trade Representative (2010), "WTO Dispute Settlement Proceedings: Subsidies on Wind Power Equipment, China." *Federal Register*, 75(249) (Dec. 29): 82130–2.

OECD (2013), "OECD Patent Database," <http://stats.oecd.org>.

PR Newswire (2010), "China Ming Yang Wind Power Group Limited Reports Third Quarter 2010 Results," Nov. 15, <http://www.prnewswire.com/news-releases/china-ming-yang-wind-power-group-limited-reports-third-quarter-2010-results-108141664.html>.

Recknagel, P. (2010), *Mapping WTG Manufacturers in China*. Bonn: GTZ Renewable Energy Program.

Ru, P., Q. Zhi, F. Zhang, X. Zhong, J. Li, and J. Su (2012), "Behind the Development of Technology: The Transition of Innovation Modes in China's Wind Turbine Manufacturing Industry," *Energy Policy*, 43: 58–69.

Vestas (2013), "Find Vestas," <http://www.vestas.com/en/about-vestas/find-vestas.aspx> accessed Apr. 2013.

WIPO (2013), "WIPO Patentscope Database," <http://patentscope.wipo.int>.

12

The Rise of the Chinese Solar Photovoltaic Industry

Firms, Governments, and Global Competition

Matthew Hopkins and Yin Li

Introduction

In the first decade of the twenty-first century, a group of Chinese manufacturers including Suntech Power, Yingli Green Energy, Trina Solar, and LDK Solar quickly established themselves as global leaders in the solar photovoltaic (PV) manufacturing industry. These young firms have scaled rapidly, reduced costs faster than foreign competitors, and replaced American, European, and Japanese industry leaders as dominant suppliers to the world solar PV panel market. By the early 2010s, seven of the global top 10 solar firms were Chinese (Lian 2014).

The rise of China's solar PV industry has profoundly reshaped the global landscape of solar energy production, evidenced not only by the tariff wars and industrial reorganization occurring as a result in advanced economies, but also in the increasing use of solar panels worldwide. We would note also that China's success is an important milestone in the broader global effort to promote clean technologies. China is demonstrating that the large-scale manufacture of high-performance, low-cost solar PV technologies is possible. The innovations behind lower cost, higher performance solar panels increase the likelihood that solar PV technology will become a major component of low-carbon economies in the future.

The rapid ascension of the Chinese solar industry raises important questions for the study of industrial development and public policy: How did Chinese solar PV manufacturers succeed, and how will that success endure given intense foreign competition? In this chapter, we answer these questions through an in-depth examination of the industry's development trajectory since the late 1990s. We argue that a number of factors have driven the growth of the Chinese solar PV industry, including: export-led growth, particularly to new markets created by European states; continuous improvement of crystalline-silicon PV technology that leads to lower cost, higher performance solar cells and panels; development of indigenous upstream production capabilities that enhanced Chinese firms' competitiveness; the success of entrepreneurs and Chinese cities in bidding on manufacturing sites in the early stage of industry formulation; and substantial quantities of public finance acting as patient capital for firms during their early stage growth and later when challenged by depressed global solar panel prices.

In the following sections, we document the origin of China's major solar companies. We then describe their supply chain and technological development. We explain the transition of Chinese firms away from roles as major exporters to roles based upon supplying a growing domestic market and corresponding policy changes driving that process. In the end, we engage in a discussion about what we view as the main causes of China's success in the solar PV industry.

Industry Formation and Local Initiative

China's historical research and development (R&D) in solar PV technology began in 1958, just four years after the 1954 discovery of the solar PV cell by Bell Labs in the United States (Marigo 2007: 145; Yang et al. 2003). But China's solar PV manufacturing industry has roots in its semiconductor development initiatives of the late 1970s. The first generation of Chinese solar PV manufacturers emerged when three state-owned semiconductor enterprises were converted to produce crystalline-silicon (C-Si) solar PV cells and modules (Marigo 2007; Yang et al. 2003). As in the United States, solar PV technology was initially deployed as a power source for space satellites (such as for the Dongfanghong 2 carrying electronic sensors). Later, researchers and businesses found terrestrial applications such as in supplying power to remote rural parts of the country without access to standard electric grid power (Perlin 1999; Liu 2009; Liu and Shiroyama 2013). But other than a few government demonstration projects for rural electrification, the early solar PV industry in China was small, with a sum total of just 5 MW of production capacity in 1995 (Liu 2009; Marigo 2007).[1]

[1] 1 Megawatt = 1 million watts. According to Earth Policy Institute (2011), world production was approximately 78 MWs that same year, with the US supplying approximately half that amount.

By 2000, non-state firms began to emerge in the Chinese PV industry. A draw was the 1996 Brightness Program, designed to electrify 20 million Chinese with solar power in rural western provinces. The program was given 3–5 billion yuan from national and local governments (Huo and Zhang 2012: 40; Yang et al. 2003: 706). Rural electrification projects were the primary source of domestic market demand until 2008, when strong subsidy and quota policies established China as a leading market for solar power.

The critical opportunity for new Chinese firms was the development of growing markets for solar power in other countries. For example, the German Renewable Energy Sources Act of 2000 subsidized solar PV energy and established Germany as the center of solar PV energy development for many years.[2] While technological barriers to entry were relatively low given standardized C-Si solar PV production lines (Tour et al. 2011: 764), a major challenge was to finance the scaling of operations, as neither the export nor the domestic market was yet established. China's emerging solar PV industry began with the success of Yingli Green Energy and Suntech Power, which presented two different approaches to how firms could form in China.

Yingli Green Energy: Indigenous Firm Formation

A pioneer of the Chinese solar PV industry, Yingli Green Energy (Chinese characters: 英利) has been described as one of the most important companies in China's "takeover" of the global solar industry. About one decade after its founding, it was able to disrupt one of the biggest global markets for solar PV in the world: "by the [end of 2009, Yingli] held 27 per cent of the California market" (Sanderson and Forsythe 2013: 150). A darling of the stock market in 2007, Yingli enjoyed a market capitalization of approximately $5 billion and earned its founder Miao Lingsheng,[3] then 51, the number 30 spot on the Forbe's 400 richest Chinese list (his stake in Yingli was worth almost $2 billion at the time). In 2012 and 2013, it was the world's leading producer of solar panels.[4]

Before entering the solar power business, Yingli was a trading company for consumer goods. In 1996 Miao Liansheng saw the opportunities in government-invested solar projects in the rural parts of Western China, and

[2] The RESA of 2000 was a modification to the 1990 Feed-In Law adopted by Germany, which had established a market for renewable power in Germany using price controls and development goals, each contributing to a booming wind power and to a lesser extent solar power market (see e.g. Lauber et al. 2006). The 2000 RESA set new long-term, above-market prices for renewable technologies according to their current and expected cost performance, with planned rate cuts. It proved to be a boon for solar power.

[3] In this chapter, Chinese names are written family name first, and given name after.

[4] Solarbuzz, "Top 10 Module Producers 2013," <http://www.solarbuzz.com/resources/articles-and-presentations/top-ten-module-suppliers-in-2013>.

decided to enter solar cell manufacturing as one of the first non-state solar PV firms in China. In 1998, Yingli, as a solar startup in Baoding, Hebei, was selected by the state to set up a solar cell production demonstration project. The demonstration project required the installation of a 3 MW solar cell production line, nearly as much capacity as existed in China at that time (Liu 2009).

Yet even with the state contract in hand, Yingli found it difficult to raise the 500 million RMB (approximately $60.4 million in 1998) required to finance the production line. To raise funds for the startup, the ownership of Yingli was split between the Yingli Group (owned by founder Miao), and the Baoding Gaoxin District Development Company, an investment firm owned and controlled by the city of Baoding. Operating within a high-tech development zone, Yingli benefitted from a number of subsidies common to Chinese high-tech firms, such as favorable tax treatment (Yingli SEC F-1 June 7, 2013: 80).

From 1998 until 2005, the Baoding Tianwei Group acted, like the city of Baoding, as a seed investor to Tianwei Yingli, supplying $15 million in no-interest loans and helping it secure $126 million in short-term commercial funding in exchange for a 51 percent controlling interest (SEC F-1 June 7, 2013: 53). A former state-owned manufacturer of electric transformers, the Baoding Tianwei Group, invested in Tianwei Yingli as part of its green energy segment development, which included wind turbine and solar PV production. The financial support from the Tianwei Group allowed Yingli to continue operations until the German market began its rapid growth in 2004, by which time Yingli was launching a second round of expansion in production capacity. Yingli financed a portion of this growth by going public, like Suntech, through the New York Stock Exchange (NYSE) with a $319 million Initial Public Offer (IPO) in 2007. By the early 2010s, Yingli supplied close to a quarter of the solar panels sold for China's Golden Sun Program—one of the largest state sponsored solar PV development initiatives in the world. It also operated with access to billions in credit provided through the China Development Bank.

Suntech: Returnee Firm Formation

While Yingli Green Energy was one of China's pioneer indigenous solar companies, it was the success of Suntech Power (尚德电力) that had the broadest and strongest impact on industry formation. Suntech was founded in 2001 by Shi Zhengrong, a returnee scientist who in his own words, "got in to the solar industry . . . by accident" (Batson 2006). Shi earned a Ph.D. in electrical engineering in 1992 from the University of New South Wales (UNSW), Australia, under the guidance of Professor Martin Green, an internationally recognized solar PV expert (Flannery 2006). At UNSW, Shi contributed to cutting-edge thin-film solar PV cell research, and gained solar technology commercialization experience by working at UNSW's joint venture with Pacific Power (Flannery 2006).

In 2000, Shi returned to China with technology developed at UNSW and a business plan to look for commercialization opportunities. He believed that large-scale manufacturing was more important for China at that time than R&D in emerging solar technology, given the opportunities to use existing technologies with expired patents and pursue incremental improvements using state of the art technologies in the future (Knight 2011). By 2001, the city of Wuxi, Jiangsu, agreed to invest $6 million in Shi's new venture, Wuxi Suntech, in exchange for 75 percent ownership (Batson 2006; Ahrens 2013). The investment was made through seven city-connected firms.[5] Li Yanren, the city's former director of Economic and Trade Commission became Suntech's chairman of the board and played a key role in making sure the financial needs of Suntech were met by the city of Wuxi. Shi obtained a 25 percent share of the new venture by contributing $400,000 of his personal savings and his research (He 2006).

By August 2002, Suntech launched its 15 MW solar cell production line, assembled in part from equipment acquired from the bankrupt US-based Astropower (Flannery 2006). Suntech also acquired parts and equipment from multiple Chinese and Japanese companies, as well as an Italian lab. The capacity of Suntech's first production line exceeded China's total production capacity of solar cells over the previous four years. Aided by a strategy of tweaking imported technology to fit local conditions of lower labor costs (Batson 2006), Suntech continued to expand with a second and a third line through 2004.[6] At the end of 2005 Suntech was embraced by investors on the NYSE, which poured $396 million into the firm during its IPO.

The "Suntech Effect"

Following the lead of Suntech, the Chinese solar PV industry grew from a handful of companies in the mid-2000s to approximately 100 firms in 2008, before surging to over 500 by 2011 (Niu and Li 2011). The rush into the solar business was less an outcome of China's environmental and renewable energy policy goals (Liu 2013), and more a response to the perceived success

[5] Among the seven companies, four were investment arms of the city (Wuxi High Tech Venture Capital Co., Wuxi Guolian Trust & Investment Company, Wuxi City Innovation Investment Company, and the Wuxi Keda innovation Investment Company), and three were city-connected State Owned Enterprises (SOEs)—(Jiangsu Little Swan Group, Wuxi Shuixing Group, and the Wuxi Shanhe Group).

[6] Later, as the company grew, Suntech invested more heavily in automation, reducing the number of workers per MW from 4 in 2008 to 1.49 in 2010 (Ahrens 2013).

of Suntech, which created "push" and "pull" effects. As Shi attracted international attention by becoming the richest man in China following Suntech's IPO (Flannery 2006; Batson 2006), his success "pulled" numerous returnees with technical and managerial experience acquired abroad back to China to seek opportunities in the solar business.

For example, Zhao Jianhua, Shi's colleague who at UNSW had achieved a world record of 25 percent conversion efficiency for a C-Si solar cell, returned to China to establish Sunergy as a subsidiary of China Electric Equipment Group (CEEG) of Nanjing, Jiangsu, in 2004. Zhao was an associate professor and deputy director at the Centre of Excellence for Advanced Silicon Photovoltaics and Photonics at UNSW through 2006, and also a senior fellow since 1991.[7] China Sunergy (CSUN) became a medium-sized manufacturing company with a technology lead among Chinese solar companies.[8]

On the "push" side, the Suntech experience legitimized the use of public funds to support the growth of solar startups, whose success produced financial returns and fulfilled development goals for high-tech industry. State investors in Suntech, convinced by Shi to divest their stakes before the IPO, still received returns 13.3 times their original investments (Ahrens 2013: 5). Without question, the City of Wuxi's success in nurturing a world-class solar startup and subsequent creation of high-tech jobs encouraged more local governments to fund solar startups.

Peng Xiaofeng, a young businessman in his late 20s persuaded the city of Yuxing, Jiangxi, to invest 200 million RMB (approximately $24.1 million) in a solar wafer production plant in 2004. The investment represented 17 percent of Yuxing's 1.2 billion RMB (approximately $145 million) tax revenues in 2004. Peng's company was named LDK Solar, which grew into the world's largest silicon wafer manufacturer and one of the largest vertically integrated solar PV companies in China. Even though Peng's success was unusual, it provides a reason why, since 2004, more than 100 city governments in China established industrial parks for solar PV companies, multiplying public support for firm formation (Wu 2010).

Within the Jiangsu province, solar PV clusters have emerged. Trina Solar, based in Changzhou, Jiangsu, began as a solar PV installer in 1998 and began manufacturing in 2004. Sunergy began operations in Nanjing, Jiangsu, in 2004. Canadian Solar, a solar startup registered in Canada and founded by a Chinese scientist, established all of its manufacturing operations in Suzhou, Jiangsu. Linyang Group, another electric equipment manufacturer from Nantong, Jiangsu, began manufacturing solar cells in 2004, and was later

[7] Sunergy FY2010 Annual Report, <http://media.corporate-ir.net/media_files/irol/21/211846/AR/AR2010.pdf>. Wang Aihua of China Sunergy's R&D center had also worked at UNSW.
[8] Author's interview with senior managers at China Sunergy, July 18, 2012.

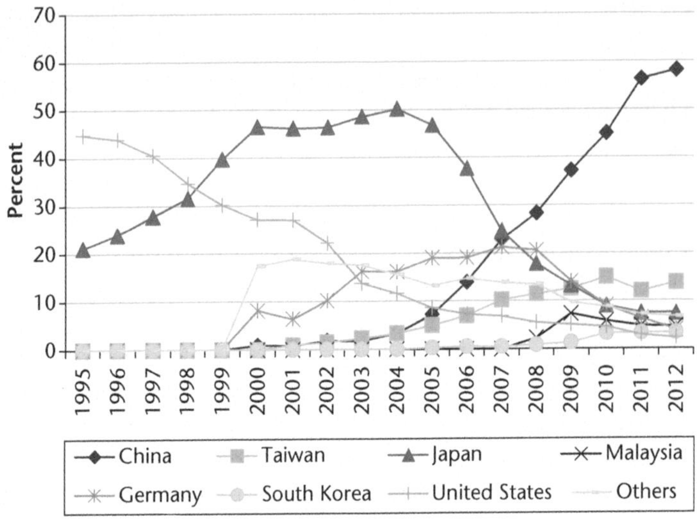

Figure 12.1. Producers of Solar PV by Percent of Annual Total, Selected Nations (1995–2012)

acquired by the Korean chemical conglomerate Hanwha in 2010 to become Hanwha Solarone.[9]

By 2008, China surpassed Japan as the world's largest producer of solar PV products. As shown in Figure 12.1, China was supplying 11 GWs or 45 percent of the global solar PV market two years later. Chinese businesses had captured market share from such powerful firms as Sharp (Japan), Kyocera (Japan), Q-Cells (Germany), and BP Solar (United Kingdom).

Supply Chain Development

The rise of Chinese solar firms went hand in hand with the formation of a complete supply chain in China. Initially, Chinese solar manufacturers concentrated on downstream segments of solar cell and module production, where inexpensive energy and low-cost labor could provide a competitive advantage (Liu and Goldstein 2013: 423–4; Tour et al. 2011: 463). But labor costs are less relevant in the more capital-intensive upstream production of silicon ingot and wafer production (USITC 2011; Ahrens 2013). China's labor cost advantage is even less important when factoring in shipping costs to

[9] Other solar companies setting up operations in Jiangsu include Changzhou Skypower Solar Energy Industry (founded 1998), EGing Photovoltaic Technology (2003), GD Solar (2010), and Hareon Solar (2004).

overseas markets (Goodrich et al. 2011). In the mid-2010s, leading Chinese firms aggressively invested in upstream and downstream segments of the solar PV value chain, in an effort to lower the cost and reap the benefits of vertical integration. Table 12.1 provides a stylized presentation of solar PV supply chain.

Simon Tsuo, chairman of Motech Industries, a major Taiwanese solar PV manufacturer, said in 2009 that it was "control of polysilicon supply and government loans [that] allowed China to grow vertically integrated champions like Suntech Power Holdings... and Yingli Green Energy" (as quoted in Pevzner 2009). Tsuo added that, in contrast, Taiwan was constrained by its dependence on foreign countries for its upstream solar PV supply chain inputs. China's transition from dependency on foreign producers for raw solar-grade silicon to becoming the world's leading supplier in under a decade is the clearest illustration of the effectiveness of Chinese supply chain development. It was also pivotal in enabling Chinese firms to exploit the efficiencies of vertical integration, and control pricing of the primary raw input of C-Si solar panels.

In the early 2000s, the world supply of raw silicon was dominated by a handful of chemical companies from advanced economies, including Wacker Chemical AG (Germany), MEMC (USA), Hemlock (USA), and M. Setek (Japan). Barriers to the raw silicon production industry are high, given that the raw silicon refinery process requires substantial capital investment and proprietary technology. The dominant suppliers were able to reap large shares of industry profits from China's booming solar PV manufacturing, which grew so large that it became vulnerable to price shocks or disruption in the global supply of raw silicon.

Policymakers in Beijing became aware of the need to lower the cost of raw silicon and improve value capture by the Chinese solar PV industry in the 2000s. The central government of China assumed the role of encouraging domestic investment in raw silicon production and diffusing key technologies. Developing silicon refinement technology was written into the Ministry of Science and Technology's 11th Five Year Plan (2006–10), which set a research agenda and directed funding priorities through national R&D programs (i.e. the 863 Program). In 2005, three members of the Chinese Academy of Sciences (Liang Junwu, Zhou Lian, Que Duanlin) wrote a joint proposal to the CCP Central Committee pledging state support for raw silicon production. State demonstration projects for large-scale silicon production were then set up using domestically manufactured equipment (Liu 2009). Chinese patenting for raw silicon purification activities intensified over the same period (Tour et al. 2011).

The 2007 "Major Projects on the Industrialization of High-Purity Silicon Material Technology" by China's economic planning agency, the National

Table 12.1. A Styled Description of the Solar PV Value Chain

Supply Chain	Upstream Components			Downstream components	
Sub-Industry	Silicon	Ingots	Wafers	Cells	Modules
Main Processes	Quartz Sand Melted into silicon. Additives influence + or – charged silicon.	Silicon processed into Round, Square, or Flat (String Ribbon) Shapes	Saws or other cutting tools slice ingots into Wafers	+ and – charged wafers sandwiched, forming a P-N Junction.	Cells joined to metal backplates, soldered, wired, framed, encased in glass or other durable, then baked
Industry Structure	Small number of firms			Large number of firms	

Supply Chain	Installation and Development	
Sub-Industry	Balance of System Components	Solar Projects
Main Processes	Batteries Inverters Racks and frames Tracking system	Site identification and assessment Project finance Technology consumption Building integrated, roof-mounted, or utility-scale
Industry Structure		Large number of firms

Source: Compiled by the authors.

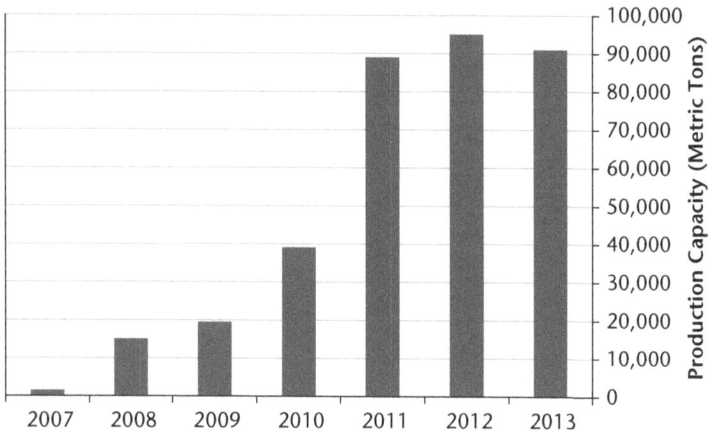

Figure 12.2. Reported Raw Silicon Production Capacity, Selected Firms (2006–13)

Development and Reform Commission (NRDC), was a mission-oriented program pushing for industry involvement in the technology development process. The program formalized support for firm entry into raw silicon production and encouraged bank lending for silicon production projects. About 70 billion RMB (or $10 billion) was invested between 2006 and 2008, half originating from Chinese firms and half from state-owned banks (Wu 2010). Induced by the rising price of raw silicon and access to capital, more than 50 Chinese companies raced into production. Large companies such as LDK Solar, Jiangsu Zhongneng Polysilicon Technology Development Co., and the Tianwei Group each invested about 1 billion RMB or $100–140 million in individual projects competing for the chance to become a leading producer.[10]

In 2005, China could produce only 80 tons of silicon annually (Dingding, 2013). In 2007, as shown in Figure 12.2, Chinese companies could produce about 1,500 metric tons (MT) of raw silicon annually and by the end of 2011 GCL-poly, LDK, Renesola, and Yingli Green Energy had a combined production capacity of 89,000 MT. China rather suddenly became one of the largest raw silicon-producing regions in the world.[11] By investing heavily in domestic production of solar-grade silicon, LDK, Renesola, and GCL-Poly asserted control over the supply and price of raw silicon used for domestic solar PV

[10] Jiangsu Zhongneng was formerly the largest silicon producer in China and was acquired by GCL-Poly in 2009. GCL-Poly is part of the state-owned conglomerate Poly Group, whose main business was electric equipment before entering raw silicon production through acquisition of Jiangsu Zhongneng.

[11] In similar fashion as with downstream components, however, some of the capacity brought online in China made use of production equipment sourced from outside the country. LDK Solar e.g. uses furnaces by GT Solar (US) and Sunways AG (Germany).

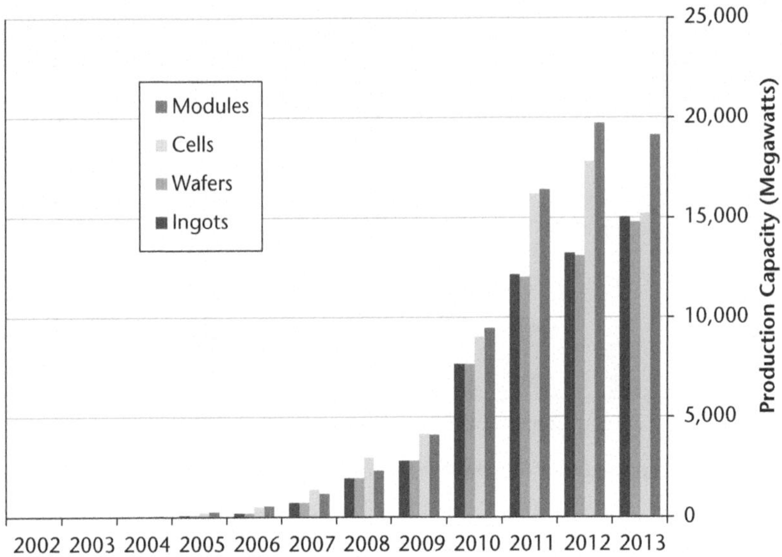

Figure 12.3. Reported Manufacturing Capacity, Selected Firms (2005–13)

manufacture. GCL-Poly aggressively reduced its costs of raw silicon from $66/kg to $36/kg during 2008–9 (Wu 2010).

Yet by October of 2012, only seven of China's polysilicon producers were still in business, as almost all of China's small- and medium-sized polysilicon firms had failed (Dingding 2013). Smaller firms, often competing with older production equipment, could not maintain profitability in a polysilicon market where prices had plummeted to $24/kg in the summer of 2012; nor were they expected to be able to meet scale, efficiency, and environmental standards put in place in 2011 (Dingding 2013; *China Chemical Reporter* 2011).[12]

Indigenous control over the domestic supply of raw silicon complemented broad investment in vertical integration being made by leading Chinese solar PV firms. Figure 12.3 shows the rapid expansion of production capacity into other areas of the supply chain made by China's leading firms. As presented, the timeline masks the fact that many major Chinese firms moved from downstream into upstream manufacturing activities.

[12] One consequence of the consolidation of China's polysilicon industry was a call for anti-dumping investigations into its US, Korean, and German trading partners. Dow Corning CEO Robert Hansen had described Chinese expansion as threatening Hemlock Semiconductor sales and ongoing development of the industry (see Hansen 2012). It should be noted that Chinese raw silicon production was mainly "solar grade," and not the high-purity required for high-yield semiconductor production.

Technology Development and Innovation

Becoming the dominant regional supplier of C-Si solar PV technologies was not just a matter of rapid growth, it required increasing technological competence that could exploit existing technologies and approach the current technological frontier. Evidence from several aspects of industry development indicate that the Chinese solar industry is striving to compete on the basis of innovation, not simply cost—including rising R&D expenditures on behalf of the state and leading firms; development of indigenous equipment supply; linkages between firms, universities, and government agencies; and improvement in the manufacturing process such that higher quality and lower-cost products are made.

Improved Performance and Process Innovation

Two key indicators of innovation in solar PV technology are the performance and cost of solar PV cells. Figure 12.4 shows how solar PV technology has become less expensive over time (as expressed in the dollars/watt ratio). A number of factors contribute to these cost reductions overtime, including the investments of R&D made by other firms. Cost reduction began to accelerate around 2008, or at the approximate time that most major Chinese manufacturers had gone public and established significant production capacity.

One aspect of the cost reduction relates to the performance of solar PV cells, indicated by their relative efficiency levels. Incremental gains in solar cell efficiency have proportionally large impacts. For example, a solar cell with 20 percent conversion efficiency can, under ideal conditions, produce

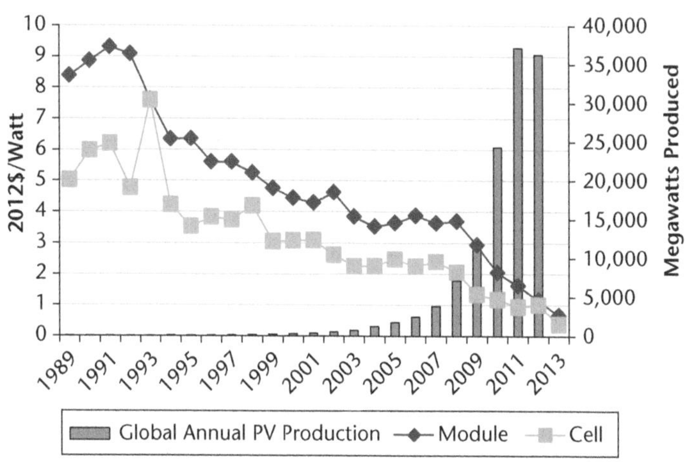

Figure 12.4. Solar PV Cell and Module Prices (1989–2013) (2012$/Watt)

33 percent more power than a solar cell rated at 15 percent efficiency. Likewise, a solar cell with 21 percent efficiency produces 5 percent more power than one rated at 20 percent, and so on. More productive solar cells increase the energy yield of solar projects from which the cost of solar electricity is set.

Mono-crystalline solar cells are made from a purer silicon base than multi-crystalline solar cells, and generally reach higher efficiency levels. In 2013, leading producers from the US (such as SunPower) claimed over 24 percent efficiency in their leading cells. Hybrid cells produced in Japan by Panasonic achieve efficiencies over 25 percent. In 2006, most Chinese manufacturers reported cell efficiencies in the 15–16 percent range, and by 2013 many claimed efficiencies of approximately 20 percent.

Likewise, state of the art multi-crystalline solar cells reached efficiencies of approximately 20 percent in 2012. Chinese firms reported efficiencies in the 15 percent range in 2007, rising to 17–18 percent in 2013.

Rising solar cell efficiencies are one indication that China's leading firms are engaged in technological learning and rapidly upgrading the performance of their products. Many also offer comparable long-term warranties on their modules to promote confidence in product quality, though it will be some time before it is clear whether Chinese panels are as durable as those manufactured by firms that have been in business for longer periods.

The second aspect of innovation is the reduction of unit costs achieved through constant improvement of the production process. As the production scale of the Chinese solar firms increased in the 2000s, multiple sources for technology learning and development occurred. Learning by doing, collaborative R&D, and access to support provided by domestic equipment makers all contributed to performance and cost improvements. First, improved performance of solar panels and cost reduction can be achieved through better manufacturing process, such as reducing waste of raw silicon, elimination of some manufacturing steps, and materials substitution.

Second, large-scale manufacturing encouraged Chinese firms to seek implementation of process innovations identified through accumulation of production experience (Tour et al. 2011). Such process innovation requires the development of engineering capabilities that can absorb foreign inventions and integrate them into the production process (such as by modifying designs to fit manufacturing capability). China has developed unique manufacturing capabilities in this respect (Nahm and Steinfeld 2014; Pisano and Teece 2007; Pisano and Shih 2009). For example, Jinko Solar, JA Solar, Yingli Green Energy, and Hanwha Solar each license technology from Innovalight of California, incorporating its "silicon ink" technology in their solar cells to boost efficiency (see also Nahm and Steinfeld 2014). Ancillary to Innovalight's successful launch and commercialization of its technology through Chinese manufacturers was its success in getting acquired by DuPont in 2011.

R&D Investment

Technological innovation requires significant investment, and in China both state and business investments in solar technology R&D are strong. On the state side, solar R&D makes up part of the total investment made in energy research. About 10 percent of funding to China's national 973 and 865 R&D programs are allocated to energy research, which has grown from about 127 million yuan in 1996 ($15.3 million) to about 2 billion yuan ($242 million) in 2000, before leaping to 9.9 billion yuan in 2007 ($1.4 billion) and 6.1 billion yuan in 2008 ($890 million) (Huang et al. 2012: 123–4). The funding supports a diversity of energy initiatives, but the proportion allocated to renewable energy began to increase in particular in the year 2000 (Huang et al. 2012: 125).

Local governments providing matching funds to national programs also play an important role, with significant proportions devoted not only to the development of technology, but its commercialization and export (Tan 2010: 2924). China's investment in energy technologies is made with a view to ensuring that business leads in innovation in the long term. Chinese solar companies conduct substantial and growing amounts of R&D, which for the top 10 firms we studied peaked at $282 million in 2010. The 2006 Medium and Long-Term Plan (the "Indigenous Innovation" campaign) significantly boosted firm R&D expenditure, since it provides access to tax, finance, and procurement support for R&D and innovation (Tan 2010).

Indigenous Equipment Manufacturing

Chinese solar PV manufacturers relied heavily on imported solar PV manufacturing equipment and technology in the 2000s (GTM 2011; PCT 2011). As China's firms built downstream capacity they created a strong demand for indigenous upstream equipment makers that could provide substitutes for more expensive imported manufacturing equipment. For example, China lacked the indigenous capability to produce solar PV production equipment until the year 2000. Collaboration between the Institute of Solar Energy at Shanghai Jiaotong University and GoFly Green Energy, financed in part under the 9th Five-Year Plan, succeeded in solar PV equipment production by a Chinese firm (Yang et al. 2003: 4).

By 2008, the competitive pressure to lower costs encouraged Chinese solar PV firms to look for lower cost domestically made equipment, resulting in a rapid expansion of a solar PV equipment sector. By 2011, leading Chinese solar equipment makers, such as Beijing Seven Star Electronics and the 48th Research Institute of China Electronics Technology Group (CETC-48), supplied more than half of the Chinese market. Their products typically cost

70 percent less than comparable imported equipment, creating substantial advantages for Chinese solar firms (Marigo 2007; author's interviews).

University Linkages and Collaboration

Linkages with foreign as well as domestic universities have provided another source of technology transfer for Chinese solar companies. Suntech did much of its upgrading, for example, by incrementally incorporating University of New South Wales's PERL technology into their PLUTO line of solar PV products (Bullis 2011).[13] Suntech also gave funding to the Swinburne University of Technology (launched in April 2009) to develop nanoplamonic solar cells in a $12 million collaboration funded by the Australian government, Suntech, and Swinburne, which led to the construction of the Victoria-Suntech Advanced Solar Facility in 2010 (Kivivali 2012; Ladiges 2010). Suntech also maintained relationships with the Universities of Zhongshan, Shanghai Jiaotong, Zhengzhou, Nanjing Aeronautic, and Jaingnan of China (Ahrens 2013).

As a second example, Yingli's high efficiency PANDA technology was an end result of collaboration begun in 2009 (project PANDA) with the Energy Research Center of the Netherlands. Yingli then worked with Tempress Systems (a subsidiary of the Netherlands' Amtech Systems) and integrated US-based DuPont technologies into their solar cells before bringing their results to the market (Osborne 2012).[14] University collaborations have produced significant achievements. Lab-controlled and confirmed efficiency records for mono C-Si modules, from which future product designs benefit, have occurred out of research done at the University of Australia New South Wales which achieved 22.9 percent cell efficiency and Suntech which reached 21.4 percent, for example (Green et al. 2013).[15]

Industry Sustainability and Public Finance

We noted earlier the importance of local government seed funds in underwriting the early growth of Suntech and Yingli Green Energy. It is important

[13] Jianhua Zhao, head of Sunergy's (CSUN) R&D department since 2004 was a long-time Professor and Researcher for UNSW's PV research center. CSUN's researchers worked on C-Si cells that achieved record in-lab performance: <http://www.csun-solar.com/company-r-d.html>.

[14] Yingli's CTO, Dr Dengyuan Song, also worked at the ARC Photovoltaics Center at UNSW and earned his doctorate there.

[15] Controlled lab testing environments provide a standardized performance report, but not necessarily "real world" results. Included in Green et al. are "Notable Exceptions" which are confirmed records which are not records for a class of solar cells. Panasonic's (Japan) HIT C-Si technology is one such exception at 23.9%, and Q-Cells reported poly-crystalline performance of 19.5%.

to note that Western stock exchanges played an important role in providing capital from which China's leading firms expanded production capacity. IPOs were done primarily on the NYSE, but also NASDAQ stock exchanges, notably on the basis that listing requirements for the NYSE are more stringent. Follow-on public equity issues and long-term debt issues supplemented that finance. Table 12.2 highlights how China's leading firms raised $2 billion through IPOs, with $2.5 billion in follow-on funding and $2 billion in debt issues during the relatively brief span of 2005 to 2010.

The funding is substantial given that the US leads in the provision of venture capital (VC) and private equity (PE) to clean technologies, which flow largely into solar PV companies (PCT 2014).[16] To further highlight the relatively modest funding typically flowing into solar startups, approximately $1.4 billion in IPOs were completed by US firms since the 1980s, from which just two firms, GT Solar (an equipment supplier) and First Solar (a leading manufacturer of thin-film solar PV panels) received a majority of the funding. A number of China's firms completed IPOs well in excess of the $150 million average typical of the clean technology sector (Hopkins and Lazonick 2013).

By 2012, the growth of the Chinese solar PV industry had contributed to a crisis in global solar PV markets (where falling prices squeezed or undermined the profit margins of many competitors). As a result China's continued access to markets in Europe and the US was challenged by competitors (in particular by Solarworld AG, a German C-Si producer). The tremendous scale achieved by China's firms had made even the booming European markets too small to sustain the continued growth of all the firms in the industry. The continuing development of China's industry has largely been translated into two public policy issues: the first is whether public finance should sustain the operations of large solar PV firms as they mounted increasing losses, the second is whether the Chinese government should support the formation of a large domestic solar market.[17] By the end of 2012, the Chinese government responded positively to both issues.

[16] The PEW charitable trusts report does not disaggregate its published data. The US Department of Energy's *Renewable Data Book*, in 2011, found 10.5 billion ($2011) in VC and PE for solar companies between 2003 and 2011. Peak year for investment occurred in 2007, at $2.1 billion. The IPOs of Ja Solar, Yingli, and LDK may have represented a substantial share of those funds.

[17] As noted, a third issue to consider is environmental. New regulations enacted in 2011 respond to the need to control pollution forced consolidation of polysilicon manufacturers. The regulations came with an economic cost in that sense, but were also conducive to improving waste management practices that lower industry costs and bolster efficiency by recycling raw materials back into the supply chain.

Table 12.2. IPO, Follow-On, and Debt Financing for Selected Companies ($US m)

Name of Company	Exchange	Date IPO Completed	Completed IPO Value	Date Follow-On Funding Completed	Completed Follow-On value	Date Debt Issue	Completed Debt Amount	Total Funding Raised
Canadian Solar	NASDAQ	11/8/2006	116	11/7/2009	103	6/17/2008	75	294
JinkoSolar	NYSE	5/13/2010	64	11/5/2010	126			190
Renesola	NYSE	1/29/2008	130	10/5/2009	71			388
				6/23/2008	187			
LDK Solar	NYSE	5/31/2007	469	9/23/2008	200	4/14/2009	400	1,424
				2/1/2011	164	2/2/2011	191	
Yingli Green Energy	NYSE	6/8/2007	319	6/23/2009	227	12/12/2007	150	720
				12/12/2007	24			
JA Solar Holdings	NASDAQ	2/6/2007	225	10/11/2007	266	5/15/2008	400	891
Hanwha SolarOne	NASDAQ	12/20/2006	150	1/29/2008	135	1/29/2008	173	720
				11/12/2010	166			
				7/16/2008	97			
Trina Solar	NYSE	12/18/2006	98	7/21/2008	120	7/21/2008	138	680
				8/17/2009	149			
				3/24/2010	176			
Suntech Power	NYSE	12/14/2005	396	5/29/2009	277	9/30/2008	500	1,730
						12/31/2008	557	
Total		IPO	1,966	Follow-on	2,488	Debt	2,026	6,480

Sources: Company Filings, News.

Public Finance

The growth of the Chinese solar PV industry was heavily dependent on revenues that could be derived from Europe. The European states, particularly Germany, have provided large subsidies to spur creation of a solar power market. But such a strategy has its consequences. Table 12.3 shows that most major Chinese PV firms have at some point derived substantial revenues from sales to Europe. Reduced dependence on Europe for revenues since 2009 was mainly driven by growth of sales into the US market (which also subsidizes its domestic solar power market), and also partly due to the impact of the global financial crisis on European governments' reduced willingness to continue funding solar power subsidy programs.

Platzer (2012) showed that US imports of solar cells and modules from China expanded from $22 million in 2005 to $2.8 billion in 2011. China's disruption of the global solar PV market has not proceeded without some international backlash. The US and Europe have instituted new trade barriers and tariffs. While Chinese manufacturers significantly increased the accessibility of solar PV technology, their export-oriented growth imposed huge competitive pressures on both foreign and Chinese manufacturers. Since 2011 a number of less efficient US, European, and Chinese manufacturers have been driven out of the market, including Germany's Q-Cells (a market share leader in 2008 purchased by Korean conglomerate Hawaha in 2012) and Solyndra, a US firm with a highly publicized bankruptcy.

Table 12.3. Percent of Total Revenues Derived from Europe (2004–13)

Year	2013	2012	2011	2010	2009	2008	2007	2006	2005	2004
Canadian Solar	11	51	65	80	83	90	95	76	83	68
Hanwha Solar One	21	50	62	77	78	81	78	92	80	
Hareon Solar			70							
JA Solar	9	19	48	19	12	14				
JinkoSolar	8	41	71	52	7	13				
LDK Solar	17	34	29	36	20	37	0	2		
Renesola	26	27	31	34	10	6	0	2	66	
SunTech Power			45	66	74	78	89	70	71	89
Trina Solar	31	48	68	77	93	91	97	90	97	81
Yingli Green Energy	30	60	62	82	89	88	93	83	83	

Source: Company SEC 20-F filings. Companies selectively report regional sales, and not all report Europe. When there is no Europe-only figure, we add up all available European countries and report that figure.

JA Solar figures based on sales to Germany and Spain through 2010, Germany only thereafter. Overall, JA Solar exports close to 50 percent of sales.

For 2013, Jinksolar reported sales to Germany only. It is likely that a high percentage of sales were to European countries.

The inability of Chinese firms to continue to rely on exports for growth has been countered by Chinese policy. In 2010 the China Development Bank (CDB) made $43.2 billion available to 15 solar companies, to both buttress manufacturers against global uncertainty and also to provide an opportunity to upgrade capabilities through continued capacity expansion or acquisition (Aipeng and Sanderson 2012). After Chinese solar firms faced tariffs in the US market, in September 2012, the CDB did a second round of lending to 12 selected solar PV firms. Six firms were selected as large "scale" firms, which included LDK Solar, GCL Poly, Suntech Power, Yingli, Trina, and JA Solar; followed by six "technology" leaders: Canadian Solar, Jinko Solar, Sungrow Power, China Sunergy, ENN, and Renesolar.

Yingli Green Energy is an example of how large solar firms benefitted from state bank lending. The company received a $5.3 billion line of credit from CDB in July of 2010, and was extended new credit in September of 2012. The credit was several times what the firm raised through the NYSE, though it is unclear how much of the funding it may have drawn upon. Yingli's CFO Li Zongwei described the bank's funds as more restrictive and costlier than their Western finance. But it was "the volume of CDB lines of credit... that [gave] Chinese companies a leg up over their global competitors, allowing them to increase their scale above all else" (Sanderson and Forsythe 2013: 153).

Amid heavy losses during the global downturn of the solar PV market, Yingli survived with the help of a local government bailout by the city of Xinyu and also "thanks to CDB," which made it possible for "a company that is bleeding cash, whose core product is plunging in price, whose competitors across the globe are filing for bankruptcy, and whose share price was 82 percent below its IPO valuation [to] borrow money at 6 percent" (Sanderson and Forsythe 2013: 154). With approximately $1.3 billion in debt in 2012 (growing to $2.5 billion by May 2013), Yingli's plan was simply to roll over the debt into the new year, the majority of which being publicly financed with securities held by the Export-Import Bank of China, China Communications Bank, Bank of China, China Citic Bank, and China Development Bank (Ma 2012).

The public finance that has enabled the survival of Yingli Green Energy extends to the majority of China's leading solar PV firms. The process caused concern among critics, who dubbed the process of Chinese bank lending "extend and pretend" (Wei 2012). Yet there is no question that public finance has been a major competitive advantage for China's firms.

But public finance alone does not assure success. A 2011 *Xinhua news* report highlighted that 70 percent of China's 22 publicly listed solar firms were losing money at the time, and that small and medium-sized enterprises (and non-profitable firms) were unlikely to survive the industry consolidation in China as a result of industrial policy and competitive pressure (Niu and Li 2011). In particular, growth in China's production capacity outpaced European demand for solar panels which had fallen as a result of both policy revision and the European debt crisis. Shortly thereafter Suntech was driven into bankruptcy by its public backers, after failing to make payment on a convertible bond (it had suffered a loss of over $1 billion in 2011), and was acquired by Shunfeng in April 2014. LDK, a second firm to file bankruptcy (in 2014), survived on $300 million extended by 11 public banks and a restructuring effort (Bloomberg 2014).

A Domestic Market (Demand-Side Support)

We have noted some of China's supply side policies. Demand-side policies (that support domestic market formation) have taken many forms. For example, China has set targets to cut energy consumption, carbon intensity, reliance on coal power, and set a working target of getting 15 percent of its energy from non-fossil sources by 2020 (Lui and Liang 2013). It has also instituted energy subsidies favorable to solar power, and set capacity goals to boost investment in solar power capacity.

It should be pointed out that China's electric energy grid is second only to the United States in sheer size—with about 1,100 GW of generating capacity of different types currently deployed. China's grid reached this massive scale in a few decades, while the US did so over a century of technological change and population growth. Additionally, China's installed renewable electric capacity grew at 12 percent annually between 2000 and 2011, or four times the rate of the US, making over one-quarter of its installed electric capacity renewable-based. The size of China's electric grid provides a massive long-term opportunity to its solar industry, particularly given that the majority of its existing generating facilities rely on fossil technologies. Providing support for accelerated domestic solar power market development is an important means of mitigating the uncertainties of foreign markets as well as meeting development and environmental goals.

China has begun large-scale deployment of domestic solar PV power late relative to other countries. Until 2008, all of China's solar PV deployment was for rural off-grid applications. The National Energy Administration of China changed this pattern by subsidizing solar PV utility projects with a 25-year Feed-in-Tariff (FiT) instituted in 2008 (Huo and Zhang 2012).[18] This was followed by project capital subsidies made available through the 2009 Golden

[18] A FiT sets a minimum price for solar PV power intended to make investment in solar power projects economically attractive.

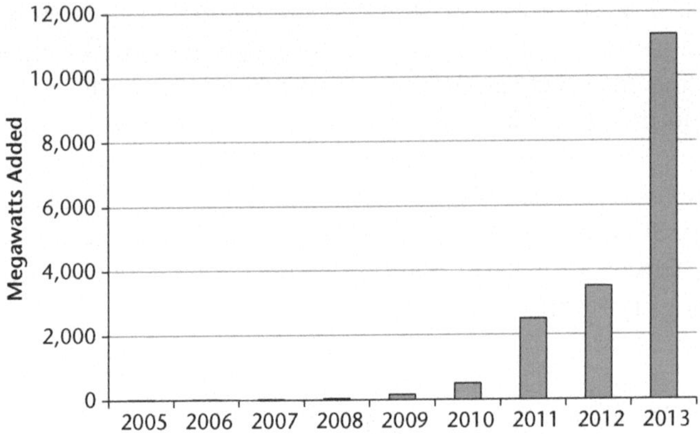

Figure 12.5. China's Annual Capacity Additions (2006–13)

Sun and Solar Roof programs, as well as region-specific subsidies started in, for example, the Jiangsu province.

Coupled to these price supports are aggressive targets for new installed solar power. Figure 12.5 shows that, after the FiTs were instituted in 2008, China's rate of solar PV capacity installation began rapid growth (EPIA 2012, 2013). In early 2013, China announced a goal of 35 GW of installed solar power by 2015 (Patton 2012; Choudhury 2013). In 2014, it revised the goal to set a target of 70 GW by 2017, with carve-outs for both large-scale and distributed projects (Parnell 2014).[19]

China's current solar power goals will likely create the largest market for solar PV power in the world.[20] That said, the "national policy aim" of China has been less to "address environmental and policy goals" and more to produce "world-class, export competitive" companies (Liu and Goldstein 2013: 424). Already, however, emphasis on export competitiveness appears to be shifting in favor of creating stable long-term opportunity for leading solar firms. Should China succeed in continuing to rapidly develop its domestic market, it could soon complete a feat similar to its foray into wind energy, in which it rapidly established itself as both a leading global market and a leading wind turbine manufacturer over a short period of time (as mentioned in Chapter 11 of this book) (Bradsher 2010; Zhou et al. 2012).

[19] Distributed projects typically mean those that are installed next to a point of consumption, such as a group of panels placed on a home or business.

[20] Germany had approximately 32 GW of solar PV installed at the end of 2012, Italy 16.2 GW. At the end of 2012, China had 8.3 GW installed, the US 7.6 GW, and Japan had 6.7 GW.

Discussion and Conclusions

Solar PV innovation in China is providing an example of how countries (through their leading firms) compete by developing superior manufacturing capabilities. Superior capabilities are a result of organizational learning that can occur in collaboration with foreign companies, universities, or emerge as outcomes of government programs. It is clear that technological development and cost reductions of Chinese solar PV products have followed changes in manufacturing technology and process improvement, or what is called learning by doing.

Scale was achieved using an import-substitution strategy—making use of available foreign equipment with a view to exploiting low-cost comparable indigenous equipment in the future. The significant capital requirements and technological barriers were overcome with the combined resources of willing government and foreign investment, with government acting as both a seed investor in early stages and as a guardian of industry investment when leading firms faced insolvency in the face of uncertainty. In that way, the government has provided the patient capital required to support continued firm investment in a process of innovation.

Markets for solar power were to a large extent a creation of European policies subsidizing and promoting them, and later became a central location for China's firms to export. Yet China has a long history of deploying solar technology within its own borders and will now aggressively pursue a large domestic market of its own, complete with aggressive targets and European-style feed-in tariffs. Whether their market will be open to foreign competition remains to be seen. The investments of the state provide a critical opportunity and a stabilizing force for its indigenous firms going forward, which can no longer rely on low-resistance access to foreign solar power markets. Given the complexity of the global solar PV trade, and the need for China to substitute its coal power for renewable sources of energy, we see a growing domestic solar PV market in China as a positive development and outcome of fierce competition.

Central to our argument has been the role of business enterprises and developmental states working in concert to produce innovation in energy technologies. A lesson from China is that the growth of a successful solar PV manufacturing base requires coordination and integration of research and development (R&D), manufacturing, and deployment activities related to solar PV technologies. Long-term stability and growth of the industry will relate in part to changing dynamics of global competition (such as the introduction of new competitors or solar PV technologies), but also the ability of solar energy to be generated at costs capable of matching or undercutting legacy fossil technologies. Progress in each of the three activities involves confronting uncertainty, as uncertainty is paramount to the innovative development process (Lazonick 2007). There are technological uncertainties concerning which technology to invest in (e.g. crystalline-silicon or emerging alternatives), competitive uncertainties concerning the ability to compete with incumbent or new manufacturers, and market uncertainties concerning the formation of durable and efficient solar markets whether through public policy or business initiatives.

To address these uncertainties, national governments around the world have established policy frameworks to catalyze innovation, business development, and national markets over decades as matters of environmental and economic imperative. Riding on this tide, Chinese innovative enterprises (and their foreign peers) employed combinations of strategy, organization, and finance to overcome the uncertainties of solar technology development and to develop superior productive capabilities (Lazonick 2007; Hopkins and Lazonick 2013). Thanks to the "patient capital" of the Chinese state, leading Chinese firms were able to sustain their development process while in many cases their foreign counterparts, lacking equivalent support mechanisms, could not. In short, we argue that the rise of the Chinese solar PV industry is an outcome of the formation of innovative business enterprises as well as supportive developmental states around the globe. Leading solar PV producers today command substantial labor and capital resources, but it is clear at this early stage of industrial development that where there has been absence of effective government policy, there is an absence of innovative solar technology, manufacturers, or solar power (Hopkins and Lazonick 2013).

China demonstrates that solar PV businesses can include vertically integrated large-scale operations that deliver solar technologies with increasing quality and rapidly declining costs in the face of tremendous uncertainty—strong evidence that innovation, and not just the presence of state subsidy or factor-input cost advantages, explains China's path to success. That said, China's firms, like the firms of other countries, benefit from state investments made in other countries around the world. We also cannot focus solely on the industry's central entrepreneurs, as competitiveness is an organizational, not individual outcome. The test for China's solar industry moving forward will be to transform its manufacturing lead into a technological lead. Given the variety of solar technologies available, and their adaptability to numerous applications (such as when integrated into buildings rather than installed on roofs), the future is very open to debate. We are at the beginning stages of a broader energy transition that will unfold over decades, not years. Is China becoming an innovation nation? Insofar as the early experience of the solar PV industry demonstrates, China is building innovative capabilities that will contribute toward its broader economic performance and transition toward a clean energy economy.

Acknowledgements

The authors would like to thank William Lazonick and Yu Zhou for their comments and suggestions during the development of this paper. Research for this chapter has been funded by the Ford Foundation (Financial Institutions for Innovation and Development project, directed by William Lazonick), Institute for New Economic Thinking (Impatient Capital in High-Tech Industries, directed by William Lazonick) and National Science Foundation (grant SES-0964907, directed by Dan Breznitz). Yin Li would extend gratitude to those supporting field research, especially Dan Breznitz, Michael Murphree, Xielin Liu, and Jinzhong Zhu.

References

Ahrens, Nathaniel (2013), "China's Competitiveness: Myth, Reality, and Lessons for the United States and Japan," Center for Strategic and International Studies, Jan., <http://csis.org/files/publication/130129_competitiveness_Suntech_casestudy_Web.pdf>.

Batson, Andrew (2006), "News in Depth: Rising Son: Chinese Entrepreneur Finds Rich Opportunities at Home—State's Willingness to Back Start-ups Energized Dr. Shi's Solar-Power Company upon his Return," *Wall Street Journal Asia*, Oct. 12.

Bloomberg News (2014), "LDK Solar Gets 2 Billion Yuan Bank Loans After Bond Default," May 26, <http://www.bloomberg.com/news/2014-05-26/ldk-solar-gets-two-billion-yuan-of-bank-loans-after-bond-default.html>.

Bradsher, Keith (2010), "To Conquer Wind Power, China Writes the Rules," NYTimes.com, Dec. 14, <http://www.nytimes.com/2010/12/15/business/global/15chinawind.html?_r=2&ref=windpower&>.

Bullis, Kevin (2011), "Chinese Solar Companies Thrive on Manufacturing Innovation," technologyreview.com, July 6, <http://www.technologyreview.com/news/424585/chinese-solar-companies-thrive-on-manufacturing>.

China Chemical Reporter (2011), "China Sets out New Rules for Polysilicon Industry," (Mar. 21), 22(6): 4.

Choudhury, Nilima (2013), "China Increases Solar Target by 67% – ," pv-tech.org, Feb. 1, <http://www.pv-tech.org/news/china_increases_solar_target_by_67_yet_again>.

Dingding, Zhang (2013), "A Hard Year for China's Polysilicon Industry," *China Chemical Reporter* (Jan. 6), 24(1): 16–17.

Hansen, Robert (2012), "Dow Corning President and CEO Robert D. Hansen's Response to China Ministry of Commerce Proceedings on U.S. Polysilicon," *Business Wire*, July 20.

European Photovoltaic Industry Association (EPIA) (2012), "Global Market Outlook for Photovoltaics until 2016," <http://www.epia.org/news/publications>.

European Photovoltaic Industry Association (EPIA) (2013), "Global Market Outlook for Photovoltaics 2013–2017," <http://www.epia.org/fileadmin/user_upload/Publications/GMO_2013_-_Final_PDF.pdf> accessed Sept. 25.

Flannery, Russel (2006), "Sun King; Photovoltaics Vendor Zhengrong Shi is worth only $2.2 billion. If he Could Just Make Solar Power Cost-Effective, he Could Be Really Rich," *Forbes Asia* (Mar. 27), 2: 5.

Goodrich, Alan, Ted James, and Michael Woodhouse (2011), "Solar PV Manufacturing Cost Analysis: U.S. Competitiveness in a Global Industry," <www.nrel.gov/docs/fy12osti/53938.pdf>.

Green, Martin, et al. (2013), "Solar Cell Efficiency Tables (Version 41)," *Progress in Photovoltaics: Research and Application*, 21(1): 1–11, <http://onlinelibrary.wiley.com/doi/10.1002/pip.2352/pdf>.

Greentech Media (GTM) (2011), "U.S. Solar Energy Trade Assessment 2011: Trade Flows and Domestic Content for Solar Energy-Related Goods and Services in the United States," Aug., <http://www.seia.org/research-resources/us-solar-energy-trade-assessment-2011>.

He, Yifan (2006), "The Suntech Model of Indigenous Innovation"(自主创新的"尚德模式"), *China Entrepreneurs*, Mar.

Hopkins, Matt, and William Lazonick (2013), *Soaking up the Sun and Blowing in The Wind: Clean Tech Needs Patient Capital*. AIR Working Paper Series #13-08/01, <http://www.theairnet.org/v3/publications-2>.

Huang, Cui, Jun Su, XiaoyuanZhao, Jigang, Sui, Peng Ru, HanweiZhang, and Xin Wang (2012), "Government Funded Renewable Energy Innovation in China," *Energy Policy*, 51: 121–7.

Huo, Mo-Lin, and Dan-wei Zhang (2012), "Lessons from Photovoltaic Policies in China for Future Development," *Energy Policy*, 51: 38–45.

Kivivali, Lea (2012), "Australian Innovators in Solar World First," swinburne.edu.au, Feb. 15, <http://www.swinburne.edu.au/chancellery/mediacentre/alumni/news/2012/02/australian-innovators-in-solar-world-first>.

Knight, Eric (2011), "The Sun King," *The Monthly*, June, <https://www.themonthly.com.au/issue/2011/june/1316152663/eric-knight/sun-king>.

Ladiges, Crystal (2010), "Advanced Solar Facility Opens in Melbourne," swinburne.edu.au, June 1, <http://www.swinburne.edu.au/chancellery/mediacentre/alumni/news/2010/06/advanced-solar-facility-opens-in-melbourne>.

Lauber, Volkmar, and Lutz Mez (2006), "Renewable Electricity Policy in Germany, 1974 to 2005," *Bulletin of Science, Technology and Society*, 26(2): 105–20.

Lazonick, William (2007), "Varieties of Capitalism and Innovative Enterprise," *Comparative Social Research*, 24: 21–69.

Lazonick, William, and Matt Hopkins (2011), "There Went the Sun: Renewable Energy Needs Patient Capital," Huffingtonpost.com, Sept. 23, <http://www.huffingtonpost.com/william-lazonick/there-went-the-sun-renewa_b_978572.html>.

Lian, Ryan (2014), "Top 10 PV Module Suppliers in 2013," *Solarbuzz.com*, Jan. 8, <http://www.solarbuzz.com/resources/articles-and-presentations/top-ten-module-suppliers-in-2013>.

Liu, Dawei, and Hideaki Shiroyama (2013), "Development of Photovoltaic Power Generation in China: A Transition Perspective," *Renewable and Sustainable Energy Reviews*, 25: 782–92.

Liu, Hengwei, and Dapeng Liang (2013), "A Review of Clean Energy Innovation and Technology Transfer in China," *Renewable and Sustainable Energy Reviews*, 18: 486–98.

Liu, Jialu, and Don Goldstein (2013), "Understanding China's Renewable Energy Technology Exports," *Energy Policy*, 52: 417–28.

Liu, Xia (2009), "China Solar: From Mono-Crystalline Silicon to Amorphous Silicon," *Keji Ribao, Science and Technology Daily*, Sept. 29, <http://www.stdaily.com/special/node_3114.htm>.

Ma, Wayne (2012), "Yingli to Roll Over Majority of $1.3 Bln Debt Due in 2012." *Dow Jones Newswires*, Sept. 3.

Marigo, Nicoletta (2007), "The Chinese Silicon Photovoltaic Industry and Market: A Critical Review of Trends and Outlook," *Progress in Photovoltaics: Research and Applications*, 15: 143–62.

Nahm, Jonas, and Edward Steinfeld (2014), "Scale-up Nation: China's Specialization in Innovative Manufacturing," *World Development*, 54: 288–300.

Niu, Huizhe, and Xiaohui Li (2011), "Analysis: China's Solar PV Industry to Speed up Restructuring." *Xinhua's China Economic Information Service*, Nov. 3.

Osborne, Mark (2012), "DuPont and Yingli Green Disclose N-Type 'Panda' Technology Collaboration," pv-tech.org, Jan. 23, <http://www.pv-tech.org/news/dupont_and_yingli_green_disclose_n_type_panda_technology_collaboration>.

Parnell, John (2014), "China Adopts 'More Aggressive' 70GW PV Target for 2017," pv-tech.org, May 16, <http://www.pv-tech.org/news/china_adopts_more_aggressive_70gw_pv_target_for_2017>.

Patton, Dominique (2012), "Further Huge Boost to Solar Target 'Not on China's Agenda'," rechargenews.com, Sept. 12, <http://www.rechargenews.com/business_area/politics/article322558.ece>.

Perlin, John (1999), *From Space to Earth: The Story of Solar Electricity*. Ann Arbor, MI: Aatec.

Pevzner, Alex (2009), "Taiwan Solar Cells Sector May Need to Consolidate," *Dow Jones Chinese Financial Wire*, Nov. 24.

Pew Charitable Trusts (PCT) (2011), *Who's Winning the Clean Energy Race? 2011 Edition*, <http://www.pewtrusts.org/en/research-and-analysis/reports/2012/04/11/whos-winning-the-clean-energy-race-2011-edition>.

Pew Charitable Trusts (PCT) (2013), "Advantage America: The U.S. Clean Energy Trade Relationship in 2011," Mar. 6, <http://www.pewenvironment.org/news-room/reports/advantage-america-the-us-china-clean-energy-technology-trade-relationship-in-2011-85899456253>.

Pew Charitable Trusts (PCT) (2014), *Who's Winning the Clean Energy Race? 2013 Edition*, <http://www.pewtrusts.org/en/research-and-analysis/reports/2014/04/03/whos-winning-the-clean-energy-race-2013>.

Pisano, Gary, and Willy Shih (2009), "Restoring American Competitiveness," *Harvard Business Review*, 87(7–8): 114–25.

Pisano, Gary, and David Teece (2007), "How to Capture Value from Innovation: Shaping Intellectual Property and Industry Architecture," *California Management Review*, 50(1): 278–96.

Platzer, Michaela (2012), "U.S. Solar Photovoltaic Manufacturing: Industry Trends, Global Competition, Federal Support," *Congressional Research Service*, June 13, <http://www.fas.org/sgp/crs/misc/R42509.pdf>.

Sanderson, Henry, and Michael Forsythe (2013), *China's Superbank: Debt, Oil and Influence—How China Development Bank is Rewriting the Rules of Finance*. Singapore: John Wiley & Sons.

Soo, Aipeng, and Henry Sanderson (2012), "China Development Bank Renews Solar Support, Journal Says," *bloomberg.com*, Sept. 25, <http://www.bloomberg.com/news/2012-09-25/chinese-lender-supports-solar-companies-securities-journal-says.html>.

Tan, Xiaomei (2010), "Clean Technology R&D and Innovation in Emerging Countries—Experience from China." *Energy Policy*, 38: 2916–26.

Tour, Arnaud, Matthieu Glachant, and Yann Ménière (2011), "Innovation and International Technology Transfer: The Case of the Chinese Photovoltaic Industry," *Energy Policy*, 39: 761–70.

US International Trade Commission (2011), "Crystalline Silicon Photovoltaic Cells and Modules from China," usitc.gov, Dec., <http://www.usitc.gov/publications/701_731/pub4295.pdf>.

Wang, Qiang (2010), "Effective Policies for Renewable Energy—The Example of China's Wind Power—Lessons for China's Photovoltaic Power," *Renewable and Sustainable Energy Reviews*, 14: 702–12.

Wei, Lingling (2012), "WSJ: Chinese Banks Make it Easy to Roll Over Loans," *Dow Jones Newswire*, Sept. 9.

Wu, Ying (2010), "China's Leap Forward in Solar Energy" (太阳能中国式跃进), *Cixin*, Apr. 1, <http://business.sohu.com/20100401/n271265393.shtml>.

Xinhuanet (2012), "China Development Bank to Support 12 Solar Firms" (in Chinese), xinhuanet.com, Sept. 26, <http://news.xinhuanet.com/finance/2012-09/26/c_123762053.htm>.

Yang, Hong, He Wang, Huacong Yu, JianpingXi, RongqiangCui, and Guangde Chen (2003), "Status of Photovoltaic Industry in China." *Energy Policy*, 31: 703–7.

Zhou, Yuanchun, Bing Zhang, Ji Zou, Jun Bi,and Ke Wang (2012), "Joint R&D in Low-Carbon Technology Development in China: A Case Study of the Wind-Turbine Manufacturing Industry," *Energy Policy*, 46: 100–8.

Acronym Glossary

3G	Third Generation Mobile Phone
4G	Fourth Generation Mobile Phone
AAMA	Asia America Multi-Technology Association
AC	alternating current
AIM	Alternative Investment Market, London Stock Exchange
AMC	American Motors Corporation
AMD	Advanced Micro Devices
AMEX	American stock exchange, third largest stock market in the U.S.
AMT	American Maglev Technology
ARM	Acorn RISC (reduced instruction set computing) Machine
ASMC	Advanced Semiconductor Manufacturing Corporation
AVIC	Aviation Industry of China
AVL	Anstalt für Verbrennungskraftmaschinen List (Institution for Combustion Engine Craft)
BAT	Baidu, Alibaba, and Tencent
BBC	British Broadcasting Corporation
BJ-SH	Beijing Shanghai Passenger Dedicated
BJ-TJ	Beijing-Tianjin
BMW	Bayerische Motoren Werke (Bavarian Motor Works)
BOP	bottom of the pyramid
BREW	Binary Runtime Environment for Wireless
BYD	Build Your Dreams, Chinese car company
CA	California
CAD	computer-aided design
CAS	Chinese Academy of Sciences
CASS	Chinese Academy of Social Sciences
CCP	(Central Committee of the) Chinese Communist Party
CCTV	China Central Television
CDB	China Development Bank
CDMA2000	Code Division Multiple Access 2000, 3G mobile phone standard
CEEG	China Electric Equipment Group
CEO	chief executive officer
CETC-48	The 48th Research Institute of China Electronics Technology Group
CGC	China General Certification
ChiNext	China second-board stock market, widely seen as modeled after Nasdaq

CIC	China Investment Corporation
CKD	Complete-Knocked Down
CMTTBA	China Machinery Tool & Tool Builders' Association
CNAJC	China National Automobile Joint Company
CNY	Chinese Yuan
COSTIND	Commission of State Technology Industry and Defense
COTS	commercial off-the-shelf
CPC	Communist Party of China (see CCP)
C-Si	Crystalline-silicon
CSIC	Corporation Haizhuang Windpower Equipment
CSMC	Central Semiconductor Manufacturing Corporation
CSR	China South Locomotive & Rolling Stock Corporation Limited
CSRC	China Securities Regulatory Commission
CSUN	China Sunergy
CTO	chief technology officer
DB	Deutsche Bahn
DEC	Dongfang Electric Corporation
DFG	Deutsche Forschungsgemeinschaft (German Research Foundation)
DMTF	Distributed Management Task Force
DOT	Department of Transport
DRAM	dynamic random access memory
DSL	Digital Subscriber Line
DSP	digital signal processing
EE	electronic engineering
EMU	Electric Multiple Units
ESOP	Employee stock ownership plan
EU	European Union
EV-DO	Evolution-Data Optimized
FAW	First Automobile Works
FDI	foreign direct investment
FFE	foreign funded enterprise
FiT	Feed-in-Tariff
FYP	five year plan
GAO	Government Accountability Office
GDP	gross domestic product
GDT	Great Dragon Telecommunications
GM	General Motors
GPA	Government Procurement Agreement
GPN	global production network
GPRS	general packet radio service
GRI	government research institution
GSM	Groupe Spécial Mobile (Global Standard for the second generation of Mobile Communications)
GVC	global value chain
GW	Gigawatt

HQ	Headquarter
HHNEC	Huahong Nippon Electric Company, Chinese joint-venture for integrated circuits foundry
HPM	Hydraulic Press Manufacturing
HS	Harmonized (Commodity Description and Coding) System
HSR	High Speed Rail
HTC	High-Tech Computer Corporation, Taiwan mobile phone company
IBM	International Business Machines
IC	Integrated Circuits
ICCT	International Conference on Computer Technology
ICE	Inter City Express
ICT	information and communications technology
ID	identification
IDC	International Data Corporation
IDF	International Data Group
IDM	integrated device manufacturer/ing
IDT	Integrated Device Technology
IEEE	Institute of Electrical and Electronic Engineers
IMEC	Interuniversity Microelectronics Centre
IMF	International Monetary Fund
IP	Internet protocol
IP	also intellectual property
IPO	initial public offering
IPR	intellectual property rights
IPU	integrated processor unit
IT	information technology
ITRI	Industrial Technology Research Institute
ITT	International Telephone and Telegraph
ITU	International Telecommunication Union
J2ME	Java 2 Platform, Micro Edition
JV	joint venture
K	kilobyte
KMPH	kilometers per hour
KOITA	Korean Industrial Technology Association
LP	limited partner
LRT	Light Rail Transit
LSI	large-scale integration
LTE	Long-Term Evolution
M&A	mergers and acquisitions
Maglev	Magnetic Levitation
MEI	Ministry of Electronics Industry
MetroPCS	Personal Communications Services
MI	Michigan
MIIT	Ministry of Industry and Information Technology
MIT	Massachusetts Institute of Technology

MIUI	Mobile Internet User Interface
MLP	The National Medium- and Long-Term Plan for the Development of Science and Technology
MMI	Ministry of Mechanical Industry
MNC	multinational corporation
MOE	Ministry of Education
MoR	Ministry of Railroad
MOST(or MoST)	Ministry of Science and Technology
MPH	miles per hour
MPT	Ministry of Posts and Telecommunications
MSI	medium-scale integration
MT	Metric tons
MVNO	Mobile Virtual Network Operators
MW	Megawatt
NASDAQ	National Association of Securities Dealers Automated Quotations, the second largest US stock market, mostly for start-ups.
NBS	National Bureau of Statistics
NBSC	National Bureau of Statistics of China
NDRC	National Development and Reform Commission
NEC	Nippon Electric Company
NIIP	national indigenous innovation product,
NIS	national innovation system
NV	Naamloze vennootschap (Limited/Ltd.)
NYSE	New York Stock Exchange
OECD	Organization for Economic Cooperation and Development
OEM	Original Equipment Manufacturing
OS	operating system
PATAC	Pan-Asian Technical Automotive Center
PC	personal computer
PDSS	public digital switch system
PE	private equity
PE	Private Equity
PMT	Passenger Miles Travelled
PRC	People's Republic of China
PRD	Pearl River Delta
PV	photovoltaic
PWHC (or PwC)	Price Waterhouse Coopers
QRD	Qualcomm Reference Design
R&D	Research and Development
RAM	random access memory
RDA	remote database access
RMB	Renminbi (Chinese Yuan)
ROW	right of way
S&T	science and technology
SAIC	Shanghai Automobile Industrial Corporation

SAW	Second Automobile Works
SBW	Shenyang Blower Works (Group) Co., Ltd.
SECI	Socialization-Externalization-Combination-Internalization
SETC	State Economic and Trade Commission
SIM	subscriber identity/identification module
SKD	Semi-Knocked Down
SKS	Shinkansen
SME	small and medium-sized enterprises
SMIC	Semiconductor Manufacturing International Corporation
SNEC	Shanghai New International Expo Center
SOE	State-owned Enterprise
SPV	Special Purpose Vehicle
SSI	small-scale integration
SSMEB	Shenzhen Small and Medium Enterprise Board
SSTC	State Science and Technology Committee (now MoST)
STEM	Science, Technology, Engineering, and Mathematics
TD	Time Division (Synchronous Code Division Multiple Access)
TDD	Time Division Duplex
TD-SCDMA	Time Division Synchronous Code Division Multiple Access
TGV	Train à Grande Vitesse
TI	Texas Instruments
TMFT	trading market for technology
TNC	transnational corporation
TSA	United States Transportation Security Administration
TSMC	Taiwan Semiconductor Manufacturing Company
TV	television
UK	United Kingdom
UMC	United Microelectronics Corporation
UN	United Nations
UNSW	University of New South Wales
US	United States
USD	United States dollar
USDOT	United States Department of Transportation
USPTO	United States Patent and Trademark Office
VAT	value-added-tax
VC	Venture Capital
WCC	World Computer Congress
WCDMA	Wideband Code Division Multiple Access
WIPO	World Intellectual Property Organization
WTO	World Trade Organization
YoY	year-over-year
ZCTT	Beijing Zhong Chuang Telecom Test Co., Ltd.
ZTE	Zhongxing Telecommunication Equipment Corporation

Index

Absorptive capacity 157
Academy of Agriculture Science China 36
Academy of Sciences Institute of Semiconductors, China 195
Advanced Micro Devices (AMD) 258
Aerodyn 296, 298
air travel, China 177
Alcatel-Lucent 219
ALD Vacuum Technology, Germany 123
Alibaba 11, 12, 16, 86, 90, 93, 268, 270, 271, 276
Alipay 90
Alstom Wind 284
Amazon 90
American Motors Corporation 138
American Recovery and Reinvestment Act (2009) 178
American Superconductor 296
AMSC-Windtec 298
An Conghui 8
Android platform (Google) 267, 268, 270, 273, 276, 277, 279, 280
angel funds, venture capital 68, 82–3
Anhui Anke Biotechnology Co. Ltd 39
A-Power 296, 298
App economy 267–8, 271, 273, 275
Apple 53, 258, 267, 268, 271, 272, 275–6, 278
apprenticeships, 'German model' 111
Asia–America MultiTechnology Association (AAMA) 84–5
AsiaInfo 76
Association of Mechanical Engineering Companies, China 115, 118
Audi-100 model 147, 147 n24
automobile industry 5, 7–8, 133–63, 135 n4, 137 Tab. 5.1
  Complete-Knocked Down (CKD) 138
  effective learning system of organization 158
  exports (2013) 7
  imitation to innovation 134
  import-assimilation-absorption 140–1, 141 Tab. 5.2
  indigenous brands 159–61
  indigenous innovation, rise of 149–58, 150 Fig. 5.2, 151 Tab. 5.5
  integrating external knowledge 154–8
  IP rights 157
  and the Japanese car industry 6–7
  new enterprise origins 151–2
  new entrants learning pattern 152–3
  'Old Three' 135, 135 n6, 150
  R&D centres 144, 146–7, 154
  recent changes 158–61
  Semi-Knocked Down (SKD) 138
  smuggling 137, 137 n7
  and state intervention 13
  strategic control and new entrants 157
  technological capabilities, building the basic core 153–4
  technology borrowing model 134
  Trading Market for Technology (TMFT) 135, 136–49
  turn-key projects 140
Automotive Promotion Act, Korea 134

Baidu 86, 91, 268, 270, 271, 276, 279
Ballon, P. 270
Baoding Gaoxin District Development Company 309
Baoding Tianwei Group 309
Baojun, Shanghai Auto 160
BAT internet 86
BeiDai conference (1987) 134
Beijing-AMC 148
Beijing Auto 138, 159
Beijing Beizhong 296
Beijing-Chrysler (formerly Beijing-AMC) 144
Beijing IC Industry Equity Investment Funds 209
Beijing Olympics 52, 187
'Beijing' platforms 148
Beijing Seven Star Electronics 319
Beijing-Shanghai Passenger Dedicated (BJ-SH) HSR 166
  Development Program (1990) 178
  Project Office 167
Beijing Shuanglu Pharmaceutical Co. Ltd 39
Beijing-Tianjin HSR 178, 186, 187
Beijing Vimicro 254
Bell Laboratories, US 307
Berlin-Hamburg Maglev Corridor Project 169
Bin Rao 137, 137 n8, 138
biopharmaceutical industry, US 10

Blackberry 268
'Blue Arrow' model 171, 185
BMW 6
Bombardier, Canada 183
Breznitz, D. 23
British Telecom 228
BYD 8, 9, 152, 155, 162

Caixin website 226
Canadian Solar 311
Cao, C. 114
CAS institutes *see* Chinese Academy of Sciences (CAS)
Catch up 217, 232
CDMA2000 W-CDMA wireless network, US 52, 264
censorship, internet 268, 277
Central Commission of National Planning, China 35–6
Central Semiconductor Manufacturing Corporation (CSMC) 197, 204
Centrotherm, Germany 123
Chaebol 198
Chang, Morris 192
Chang, Richard 201
ChangAn Auto 152
'Changbai Mountain' model 171
Changhong computers 248
Chartered Semiconductor, Singapore 202, 240
Chen, C. 274
Chen Tong 122
Chengdu Cension Semiconductor Manufacturing Corporation 206–7
Chery 8, 9, 42, 133 n1, 152, 153–4, 155–7, 156 n34, 159, 161, 161 n39, 163
  and AVL 156 Tab. 5.6
  external technical projects 156 Tab. 5.7
China
  banking laws 76
  centralized finance 9
  developing governance institutions 11–16
  Employment institution 16–19
  Framework for Development of the Integrated Circuit Industry 209
  Golden Sun and Solar Roof programs 325–6
  Government Procurement Agreements (GPA) 51
  government research institutes (GRIs) 36–7, 38–9, 47
  Guidelines for Medium and Long-tern Plans for Science and Technology Development (2006–2020) 2, 45–6, 113–14, 115, 116, 135–6
  'indigenous innovation' policy 45–50, 162
  as an innovation nation, conditions for 4–11
  Industrial eco-system 29, 258, 263, 266–7
  Investment institution 19–22
  labor market 17–18
  markets 192, 192 n1, 196
  Mechanical Engineering Research Laboratories 106 Fig. 4.3
  Medium to Long Term Railway Development Plan (2004) 178–9
  National Development and Reform Commission (NRDC) 36, 47, 151, 151 n28, 183
  National integrated circuit (IC) Industry Lead Group 209
  national indigenous innovation products (NIIP) 50–1
  Natural Science Fund 45
  '1000 Young Talents Program' 18, 253
  Outline of National Industrial Policy (1994) 140
  public procurement law (2002/2006) 50–1
  rapid growth of 1–4
  Renewable Energy Law (2005) 291, 304
  Science and Technology (S&T) System 34–45
  state and infrastructure investments 20–1
  Traditional industry 3
  'Twelve-Year Science and Technology Development Long-Term Program' 242
  see also state policies on innovation, China
China Academy of Telecommunication Research 270
China Administration of Foreign Exchange 78
China Development Bank 9, 15, 22, 209, 324
China Electric Equipment Group (CEEG) 311
China-European Union Standards project 266
China Investment Corporation (CIC) 208
China Mobile 52–3, 225, 230, 251, 252, 262, 270–1, 278
  and TD-SCDMA standard 226
China Netcom 224
China New Technology Venture Investment Corporation 75
China Railway Corporation 12, 20–1, 167
China Securities Regulatory Commission (CSRC) 73, 78
'China Star' HSR train model 171, 178, 185
China Telecom 224, 251
China Unicorn 224
China Wireless 273
China Young Angel Investor Leader Association 83
Chinese Academy of Sciences (CAS) 36, 126
  and integrated circuits 194–5
ChiNext 77, 92, 93
  board 79–81
Christensen, C. M. 264
Chuan Shi 122
Circular 18, 201, 202, 204
Clark, K. B. 114, 136

Co-evolution of technology and market 4, 263, 278
Cold War 34, 37
Commission of State Technology Industry and Defense (COSTIND), China 36
Computer and Large Scale IC Lead Group, China 196, 243
continuous improvement (kaizen) 6
Coolpad 268, 269, 272, 275
corruption 60, 61
Cortex processor cores 246–7
cross-shareholding 6

Da Tang Telecom Technology 52
Dalian University of Technology 122
Danneels, E. 114
Datang 216, 222, 223, 226, 229–32, 235, 247, 258, 262
Datang Group 272
Datang Telecom 208
Deng Zhonghan 254
Department of Transportation, China 167
Deutsche Bahn (DB), German 169
DeWind 296
Distributed Management Task Force (DMTF) 228, 228 n2
domestic VC 70–1, 73–5, 77–8, 80, 86, 88, 92–3
Dongfang Electric Corporation (DEC) 296
Dongfeng Auto 160
DongFeng-Nissan 146, 149, 154, 154 n32
Drucker, P. F. 134
DuPont 318
Dynamic support 221

eBay 90, 91
employee ownership 15
Enercon, Germany 284
Energy Research Centre, Netherlands 320
entrepreneurial habitats 81, 83–4
Ericsson 22, 219
Ernst, D. 44, 127
European Union (EU) 51
executives and value extraction, US 14
Export promotion 40

Facebook 16, 90, 268
family ownership in the "German model" 112
FAW-Volkswagen 146, 147, 147 nn25, 26, 148, 149, 153, 159
  First Automobile Works (FAW) China 141–2
foreign direct investment (FDI), 1990s–2000s 40, 41
feed-in tariffs (FiT) 13, 325–6
Feng Lu 195 n2
Fifteen year National Medium to Long-Term Plan for the Development of
  Science and Technology (2006–2020) 159
financing innovative enterprise equitably 11
First Automobile Works (FAW), China 141–2
First Solar 321
'530 Program', Wuxi 253
Five Year Plans
  6th (1981–1985) 196, 243
  7th 'plan outline' (1986) 134
  8th (1991–5) 143, 196
  9th (1996–2000) 185, 243, 290, 291, 319
  11th (2005–10) 102–3
  11th of Science and Technology (2006–10) 291, 313
  12th (2011–2015) 117
foreign VC 70–1, 73, 76, 78, 80, 82–3, 89
Founder, IT company 38, 49
4G LTE mobile phone standards 53, 225, 227, 230, 252, 258
Freeman, C. 263
Fuhrlander 296
Fujimoto, T. 136
Fukang car model 150
Fuller, D. B. 199

GalaxyCore Inc 254, 255
Gamesa, Spain 284
Geely 8–9, 42, 152, 153, 155, 158, 159, 161, 163
  acquisition of Volvo 8
General Electrics (GE), US 284, 286
General Motors (GM), US 134
Germany
  apprenticeships 111
  automobile industry 7, 8
  family ownership models 112
  mechanical engineering industry 99
  model of mechanical engineering and innovation 110–18
  Renewable Energy Sources Act (2000) 308, 308 n2
Give-Me-Five 275
global financial crisis (2007–8) 55, 78, 219
global production networks (GPNs) 22–4, 44–5, 61, 133, 141, 142
  Strategic control 24–5
global supply chains, and integrated circuit foundry development 205–9
global value chains 18
  semiconductors 241
GoFly Green Energy 319
Gogo technology 229, 235
Golden Shield, China 91
Golden Sun and Solar Roof programs, China 325–6
Goldwind 285, 295
'good-enough innovation' 262, 264–5, 274
  integrated circuit (IC) design industry 245–6, 254–5

'good-enough innovation' (*cont.*)
Integrated circuit (IC) design houses, or companies 244–7, 250
Google 16, 91, 267
Android platform 267, 268, 270, 273, 276, 277, 279, 280
governance institutions
developing 11–16
and 'rent-seeking' tendencies 14
Government Procurement Agreements (GPA), China 51
government research institutes (GRIs), China 36–7, 47
marketization of 38–9
Grace 21, 204, 208
Grace Semiconductor 201–2
Great Dragon 223
Great Firewall, China 91
Green, Martin 309
Green Mountain IC 207–8
Guangdong province 270
guanxi networks 88–9
Guojinmintui 55

H&Q Asia Pacific 76
HaFei 151–2, 152 n29, 154, 155, 159
HaFei-Pinifarina cooperative project 155
Haier 248
Hainan State government 137
Hangzhou Silan 254
Hanwha Solarone 312
Henderson, R. M. 114
Hewind 298
High Speed Rail (HSR) development 12, 20–1, 165–89
adaptation period (2004–2010) 171–3, 173 Tab. 6.3
assimilation period (2011–present) 173–4, 175 Tab. 6.4
average speeds 170 Fig. 6.2, 170–1
characteristics of technology transfer 181–7, 188
continuous research and development 184–7
evaluation period (1998–2004) 168–71, 169 Tab. 6.2, 170 Fig. 6.2
exploration period (1991–1887) 166–8, 167 Tab. 6.1
financial capabilities 177 Fig. 6.4, 177–8
historical development of railways 166–74, 167 Fig. 6.1, 181–2, 182 Fig. 6.5
key factors for success 174–80
Maglev/steel wheel on steel rail debate 168
market with central authority 182–4
operating speeds 186
and political will 178–9
positive return on investment 179–80, 180 Tab. 6.6
rail travel demand 174–7, 176 Fig. 6.3, 176 Tab. 6.5
Railway Construction Fund 177–8
Railway Enterprise Bonds 178
right of way (ROW) 179
technologies 171–2
track stability 186
higher education system 19
highway system, China 176, 176 Tab. 6.5
Hirukawa, M. 85
Hisense 248
HiSilicon Technologies 247, 249, 258, 272, 278
Hobday, M. 127
Honda 5
Hong Kong 40, 56, 58, 70
stock exchange 10
venture capital firms 82
'HongTa' car model 148
Hou Weigui 231
Hsinchu Science Park 244
HTC 268
Hu Jintao 216
Hu Qili 197, 198
Huahong 198, 199, 200, 204, 208
Huahong-NEC 204
Huali Microelectronics Corporation 208
Huanjing Group 197
Huarun 244
HuaTai 163
Huawei Technologies 12, 13, 14–15, 22, 47, 53–4, 215–16, 219, 223, 225, 226, 227–8, 231–2, 235, 248, 249, 251–2, 262, 263, 268, 269
patent applications 227–8
Hydraulic Press Manufacturing (HPM), US 120
hypermobility of labour 17–18

ICT industries 21
IDG 89
Import substitution 40
India
automobile industry 7–8
Chinese imports 108
'indigenous innovation' policy, China 45–50, 162
and globalization 50–4
Top-down 4, 13, 262
Bottom-up 4, 13, 262
Industrial ecosystem 209
industrial standards, China 51–2
Infiniti 6
infrastructural knowledge 69
Innovalight, California 318
innovation management, in a global production system 22–6
'flagship' corporations 23
Inspecting and confirming right 144, 157

integrated circuit (IC) design industry 240–58
'good-enough' products 254–5
history (1960s–late 1990s) 242–3
Project 909 243–4, 249–50, 257
returnees role 252–5
rise, post-2000 243–8, 244 Fig. 9.1, 245 Fig. 9.2, 245 Tab. 9.1, 246 Tab. 9.2
salaries 255
state, market and development of 248–53, 249 Tab. 9.3
value chain 248
vertical disintegration of global value chain 255
integrated circuit foundry development 21, 192–212
6-inch fabs 197, 197 n5, 198
8-inch fabs 198–9, 201, 205, 206, 250
12-in fabs 201, 205, 206
central planning regime 195, 195 n2
Circular 18 200–1, 201 n7, 202, 204–5
Computer and Large Scale IC Lead Group 196
financing the non-state industry 205–9
global supply chains 202–3
and the integrated circuit (IC) design sector 194
joint ventures (JV) 135, 138, 139, 140, 141, 142, 143, 144, 145, 146, 147, 148, 149, 150, 152, 154, 160, 162
local government initiatives 206–8, 209
non-state companies 200–5
policy initiatives 193
'pure-play foundry' model 202
repatriate executives 204
SMIC and employee stock option plans 203
state, market and business enterprise 209–11
state industries 194–200
strategic control 200
Integrated Device Technology (IDT) 253–4
Integrated Device Manufacturing Model (IDM), Semi-conductor 241
Intel 241, 258
Intel Hynix-Numonyx 250
intelligent antennas technology 230, 230 n5
Inter City Express (ICE), Germany 168
International Data Group (IDG) 76
Internal Development of Technology 231–2
International Telecommunications Union (ITU) 52
internet 91
boom 76–7, 86
censorship 268, 277
Interuniversity Microelectronics Centre (IMEC), Leuven 199
investment institutions 19–22
human capital 20, 20 n9
infrastructure projects 20–1
venture capital role 21–2
iPad 266, 276
iPhone 53, 266, 276
iPod 276
international property rights (IPR)
protection 115, 117–18
promotion strategies 51–4
'Iron Rice Bowl' 17
ITT, Belgium 196

Jacobs 295
Japan
automobile industry 5–7, 134
and Chinese HSR development 168
'social conditions of innovative enterprise' 6–7
venture capital (VC) firms 71
Jetta 150
A2 (1983) model 141–2, 142 Tab. 5.3
Jiang Mian-heng 201
Jiang Zemin 201
Jiangsu Zhongneng Polysilicon Technology Development Company 315, 315 n10
Jinghan Railway 181
joint ventures (jVs) *see* individual industries and sectors
Jun Shi 122
just-in-time inventory systems 6

Kalkowski, P. 111
Keller, W. W. 37
Kenney, M. 85
Kim, L. 44, 217
Kingsoft, Spain 82
Konka 248
Korea *see* South Korea
Korean War (1951–3) 34
Kortum, S. 85
Kushuda, K. 53

labor market, China 17–18
labor mobility, international 18–19
Latecomer 216, 218, 232, 234, 235
Lazonick, W. 14, 50, 85, 217, 220
LDK Solar 306, 311, 315, 325
Leadcore 254, 278
Leader without followers, Japan 53
Lee Kai-fu 82
Lei Jun 82, 270
Lenovo 12, 38, 47, 49, 234, 248, 268, 269
Lerner, J. 85
Lexus 6
Li Shufu 153, 158
Li Yanren 310
Li Zongwei 324
LiNian 160
Linyang Group 311–12
Liu, X. L. 114

Liu, Z. 186
Liu Zhijun 60, 172, 183, 187
local government, China
  financing industrial development 9–10
  roles in innovation 54–5
Localization 172, 182, 186
Locke, Gary 292
London AIM 10
Lotus 155
Low-carbon energy technology sector 283
Lu, Q. 39, 49
Lubo model, Lotus 155
Lucent Technologies 197, 221
Lundin, N. N. 114
Lyu, G.-Q. 107

Macau 56, 58
Made 290
Maglev HSR technology 168–70, 169 Tab. 6.2
Mali graphics processor cores 246–7
Manhattan Project, US 47
Manske, F. 111
Mao era, Science and Technology System 34–8, 35 Fig. 2.1
Market diversity 264
mechanical engineering industries 98–129
  apprenticeships 111
  cluster development 117
  development of a national system (1949–1978) 100–2
  disruptive innovation vs. incremental innovation 114
  export destinations 108–9
  export/import unit values 109
  export-led growth promotion 102
  family ownership 112
  FDI and domestic knowledge base 114
  firm-level approaches to innovation 118–26
  five-year plan (2005–10) 102–3
  foreign equipment and technology (1979–2005) 102
  German model and indigenous innovation 110–18, 127
  growth and spatial structures 100–3, 101 Fig. 4.1, 104 Fig. 4.2
  import countries 108–9
  indigenous innovation 99, 113–15
  indigenous innovation (since 2006) 102–3
  innovation-related structures 115–18
  innovation systems approach 112
  instructed innovation 111
  IPR protection 115, 117–18
  low-cost/high-tech innovation 114–15, 121, 125
  machinery sector growth 103, 103 n1
  output figures 107–8, 108 Fig. 4.1
  patents 105, 106–7
  Propower case study 121–4
  radical innovation 114, 125
  regional systems 102
  state procurement 117
  technological capabilities 103–10, 106 Fig. 4.3
  universities 113
  Yizumi case study 119–21
  Yizumi/Propower cross-case analysis 124 Fig. 4.4, 124–6
Mediatek 252, 262, 265, 272–3, 280
Medium and Long-Term Plan for Science and Technology (2006), China 2
Medium to Long-Term Railway Development Plan (2004), China 178–9
Mercedes-Benz 6
merchant venture capital *see* venture capital (VC)
MetroPCS 275
Miao Lingsheng 308
Ministry of Education (MOE), China 36
Ministry of Electronics Industry (MEI), China 196–7, 198
Ministry of Finance, China 47, 75, 291
Ministry of Industry and Innovation Technology (MIIT), China 221–2, 225
Ministry of Railroads (MOR), China 165, 180, 183–4, 188
  first round of high-speed EMU tenders 184
  technology studies and scientific research 185
Ministry of Science and Technology (MOST), China 36, 45, 46 Fig. 2.3, 47, 49, 75, 123
  and HSR development 185
  and wind power technologies 291
Minyang 284
Mitsubishi 155
mobile phone industry evolution 261–80
  3G mobile communications 267–73, 269 Fig. 10.3, 271 Fig. 10.4, 272 Fig. 10.5
  BOP innovations 264
  discussion and conclusions 274–80, 276 Fig. 10.6, 277 Fig. 10.7
  internet censorship 268, 277
  and internet connections growth 219, 220 Fig. 8.2
  good enough innovation 274
  platformization 262, 275, 278, 280
  production 220 Fig. 8.2
  'Shanzai' (guerrilla) handsets 261, 261 n1, 262, 264–5, 266, 267, 270, 279
  smartphone industry 262, 263–7, 265 Fig. 10.1, 266 Fig. 10.2
  'techno-economic paradigm' 263
Mobile Virtual Network Operators (MVNO) 277
Motech Industries 313

Motorola 219, 220, 221
Motorola Mobility 269
Murphree, M. 23

Nanjing-Hangzhou Passenger Dedicated Line 187
NASDAQ, New York 10, 321
National Development and Reform Commission (NRDC), China 36, 47, 151, 151 n28, 183
National Framework for Development of the Integrated Circuit Industry, China 209
National integrated circuit (IC) Industry Lead Group, China 209
national indigenous innovation products (NIIP) 50–1
National Mechanical Engineering Research Laboratories, China 106 Fig. 4.3
National Natural Science Fund, China 45
NEC, Japan 196, 198–9, 200
Netease 76, 86
Netscape 76
Neuer Markt, German 92
'New Economy' business model, US 11, 16
New Wave (Xintao) Science and Technology 253–4
New York Stock Exchange (NYSE) 10
Nie Rongzhen, Marshal 34, 35
9/11 terrorist attacks 171
Ningbao Zhongwei 207, 208
Nissan 5, 6
'no exit, no entry' policy 93
Nokia 221, 275, 276–7
Nokia-Siemens 219
Nonaka, I. 147
Nordex, Germany 284, 290
Nortel, Canada 196, 219
Northwest University, Xi'an 122
Norwin, Denmark 296, 298

offshoring 77–8
Off-the-shelf technology 171, 172
oil crisis (1973–4) 5
OmniVision 209
'1000 Young Talents Program' 18, 253

Pacific Development Venture Fund 76
Pan-Asian Technical Automotive Center (PATAC) 146
Parsons Binkerhoff Inc. 181
Partnership Enterprise Law 78–9
patents
  applications 59 Fig. 2.9, 60
  Huawei 227–8
  mechanical engineering 105, 106–7
  Propwer 122
  universities 107
  wind power technology sector 300, 301 Fig. 11.7
  ZTE 229
'patient capital' 9, 19
Patton, D. 85
Paypal 90
'Peacock Program', Shenzhen 253
Peighambari, A. 127
Peng Xiaofeng 311
Perez, C. 263
Permanent Select Committee on Intelligence, US 215
Philips 196, 249
photovoltaic (PV) industry *see* solar photovoltaic industry
Pininfarina 155
'Pioneer' model 171
Platzer, M. 323
Pony Ma 76–7
Power, J. D. report (2011) 160–1
Prahalad, C. K. 264
private equity (PE) 68, 70, 86
private equity (PE) investment 68
private equity firms 70
Product innovation 148, 150
Production localization 139, 140, 141, 142, 144, 145, 146, 147, 158
Proprietary platform 267
Project 908 197, 198, 204, 257
Project 909 198, 208
Propower case study 121–4, 127
  Yizumi cross-case analysis 124–6
Public Digital Switch Systems (PDSS) 232, 233, 235
public procurement law (2002/2006), China 50–1

Q-Cells 323
Qihoo 270, 271
Qinhuangdao-Shenyang Passenger Dedicated Line 185
Qualcomm 23, 52, 221, 252, 258, 266, 273

Rao Yi 49
RDA Microelectronics 247
'red chip'/Sina model 77–8
'RedFlag' car model 147, 148
Reliance Webstore 275
Ren Zhengfei 231
Renesola 315
Renewable Energy Law (2005), China 291, 304
Repower 296
Research Alliance for the Purification of Poly-silicon, China 122
Reverse-engineering 197, 224, 227
Richardson, G. B. 144
Rover 159
Russia *see* Soviet Union

Saab 159
SaiMa project 155
Samsung 23, 240, 247, 268
Samuels, R. G. 37
Sangyong 159
'SanKouLe' project 148, 154
'Santana' model 141, 142, 142 Tab. 5.3, 143, 150
  2000 model 144
Science and Technology (S&T) System, China 34–45
  Basic research 48
semiconductor industry *see* integrated circuit (IC) design industry
Semiconductor Manufacturing International Corporation (SMIC) 86, 201–5, 241
Sewind 284, 298
Shanghai 17
  Pu Dong International Airport 169
  Jiaotong University 122
Shanghai Automobile Industrial Corporation (SAIC) 138, 152, 159, 160
  venture with Volkswagen 41, 139, 139 n12, 142, 143–4, 145, 149
Shanghai Belling Semiconductor 249
Shanghai Huahong Group 198–9, 198 n6, 200
  NEC Electronics Co. Ltd (HHNEC) 241
'Shanghai' platforms 148
Shanghai to Hangzhou HSR 170
Shanghai Wireless Electronics Factory 242
'Shanzhai' (guerrilla) handsets 261, 261 n1, 262, 264–5, 266, 267, 270, 279
Shanzhen 55, 77
shareholder value, 'maximising' and innovation 15–16
Sheng Guangzu 187
Shenyang Sunshine Pharmaceutical Co. Ltd. 39
Shenzhen 17, 204, 205, 205 n10, 270
  industrial cluster 18–19
  Small and Medium Enterprise Board (SSMED) 78, 80–1, 92
Shi Yigong 49
Shi Zhengrong 309–10, 311
Shinkansen (SKS) HSR, Japan 166, 168
Shroder, Gerhard 169
Shuguang 39
Shunfeng 325
Siemens, Germany 184, 247, 284
  and China's HSR development 185–6
Silicon Valley 10, 77, 83, 84, 85
Simon, Denis Fred 195
Sina 76, 86
Singapore 40
  Stock Exchange 10
  VC firms 71
Sinovel 284, 295–6, 298
Semi-conductor Manufacturing International Corporation (SMIC) 21, 208, 209, 210, 244
  investment and capital costs 206
  local government investment 206–7
  School 203
SNEC 'Terawatt Diamont Award' 121–2
Snowden, Edward 215
Social condition of innovative enterprises 6–10
  Japan 6–7
Sohu 76, 86
solar photovoltaic industry 10, 306–29
  Brightness Program (1996) 308
  China Development Bank 324
  domestic market (demand-side support) 325–6, 326 Fig. 12.5
  feed-in tariffs (FiT) 325–6
  formation and local initiatives 307–12, 312 Fig. 12.1
  Golden Sun Program 309
  improved performance and process innovation 317 Fig. 12.4, 317–18
  indigenous equipment manufacturing 319–20
  Jiangsu Province 54, 311–12
  mono-crystalline solar cells 318
  multi-crystalline solar cells 318
  nanoplamonic solar cells 320
  Propower 121–4
  public finance 323 Tab. 12.3, 323–5
  R&D investment 319
  raw silicon production 313–16
  rural electrification projects 307–8
  Suntech 306, 308, 309–10, 311
  Suntech effect 310–12
  supply chain development 312–16, 314 Tab. 12.1, 315 Fig. 12.2, 316 Fig. 12.3
  sustainability 320–1, 321 n16, 322 Tab. 12.2
  technology development and innovation 317–20
  unit cost reduction 318
  university linkage and collaboration 320, 320 n15
  Yingli Green Energy 308–9
Solar PV
  organizational learning 327
    silicon refinement technology 313
  solar cell efficiency 318
  solar startups 309, 311, 321
  vertical integration 313, 316
Solyndra 323
South Korea 40, 42–3
  automobile industry 7, 8
  Automotive Promotion Act 134
  electronics industries 198
  venture capital (VC) firms 71

Soviet Union
automobile industry 7–8
foreign aid to China 37
Special Purpose Vehicles (SPV) 77
Spreadtrum Communications 53, 54, 252–3, 254, 257, 258, 273, 278
state capitalism model 12
state capitalism/liberal capitalism debate 12
State Economic and Trade Commission (SETC), China 36
state investment, top-down 13–14
state policies on innovation 33–62
3G wireless standard TD-SCDMA 52–3
centrality of (2000 onwards) 45–61
consortia 47
defense-led techno-nationalism (1950–1980) 34–8, 35 Fig. 2.1
FDI 40, 41
global production networks (GPNs) 44–5
government research institutes (GRIs) 36–7
Funding 45
Marketization 38, 39
Task 36, 37, 38
Technology standard 52 high-tech exports 40–1
imported technologies 37
indigenous 45–50
indigenous, and globalization 50–4
industrial parks and start-up incubators 55
IPR promotion strategy 51–4
local government roles 54–5
marketization of research institutes and competitive technological enterprises (1980s and 1990s) 38–9
mega-projects 47–8
public procurement 50–1
task-led model 35
trading market for technology (1990s to 2000s) 40–5, 43 Fig. 2.2
SOEs and non-state firms 55–61, 56 Fig. 2.4, 57 Fig. 2.5, 57 Fig. 2.6, 58 Fig. 2.7, 59 Fig. 2.8, 59 Fig. 2.9
top-down national research projects 45–50, 46 Fig. 2.3, 48 Tab. 2.1, 48 Tab. 2.2
steel industry 9, 21
stock buyback 10
stock markets 14
and initial public offerings (IPOs) 15
stock-option culture 18
strategic control 26, 145, 149, 154, 157
and the Chinese automobile industry 8, 9
and Japanese automobile industry 6
Strategic intent, 152
students studying abroad 19
Suling Chen 122
Sun 266
Sunenergy Company 311
Suntech Power 306, 308, 309–10, 311, 313, 320, 325
Suzhou 55
Suzlon, India 184
Swinburne University of Technology, Australia 320
switch systems and technology 227, 228, 232, 233, 235

Taiwan 40, 56, 58
electronics industries 198
VC firms 71
Taiwan Semi-conductor Manufacturing Corporation (TSMC) 23, 202, 205, 207, 240, 250
Tang-Xu Railway 181
Taobao 90, 91
Taobao Alipay 273
TD-LTE 4G standard TD-SCDMA wireless network development 230, 242, 252–8, 225–7, 230–1, 235, 236, 247, 251, 252–3, 254, 255, 258, 261, 263, 263 n2, 268, 271, 275, 276, 277, 279
Technological capabilities 145, 153, 154
technology assimilation 40–5, 43 Fig. 2.2
technology localization 43
technology transfer 43, 50
and HSR development 165–89
telecommunication equipment industry 232–3
wind power technology, and firm-level models 289–91, 293–5, 295 Fig. 11.4
Techno-nationalism, (also techno-nationalist policy, techno-nationalists)
Definition 33
Mao Era p. 34–39
Technology standards 52, *see also* TD-SCDMA
telecommunication equipment industry 215–36
competition 224–5
Customer-Producer Coordination Conference 222, 224
financialization strategies 220
fixed-line connections growth 218–19
government support for domestic innovation 221–3
Huawei patents 227–8
from imitation to innovation 234–5
influencing local firms' participation in international standards 225–6
innovation, Chinese firms vs. MNCs 233–4
innovation capability, developing 216–18, 222–3, 227–35
internal development of technology vs. technology transfer 232–3
intervening in demand condition, government 223–5

telecommunication equipment industry (*cont.*)
joint ventures 220–1, 222
market and infrastructure 218–21, 219 Fig. 8.1, 220 Fig. 8.2
mobile phone and internet connections growth 219, 220 Fig. 8.2
national science and technology programs 222
role of government 218, 221–7
rural telecom markets 227
social conditions of innovation enterprises 217–18
switch equipment and technology 222, 223, 224
TD-SCDMA development 225–7, 230–1
technology capability ladder 217
ZTE patents 229
Tencent 12, 77, 86, 90, 91, 93, 268, 271, 272, 278
'321 Program', Nanjing 253
3G mobile standard and technology 19, 47, 52–3, 216, 222, 225, 226, 230, 247, 252, 262, 263, 265, 266, 274, 275
Japanese 53
see also TD-SCDMA wireless network development
Thun, E. 139 n13
Tianyu 265
Tianwei Yingli 309, 315, 318, 320, 324
Tjinnova 155
Torch program, China 121
Toshiba 196
Toyota 5, 6, 159
Trading Market for Technology (TMFT) 8, 13, 23, 40–5, 43 Fig. 2.2, 61, 158
activities sequence 148–9, 149 Tab. 5.4
learning pattern 140–8, 140 n16, 141 Tab. 5.1, 142 Tab. 5.3
origins of policy 135, 136–9, 138 n10, 150, 152, 154, 162
Train à Grande Vitesse (TGV), France 168
Transnational talent circulation 203
Transrapid International, Inc. 168, 169, 170
Trina Solar 306
Tsinghua Unigroup 247, 258
Tsinghua University 36
Tsuo, Simon 313
Tulum, Ö. 85
'Twelve-Year Science and Technology Development Long-Term Program', China 242
Twitter 90, 268
2G mobile standard and technology 265, 266

Ueda, M. 85
United Microelectronics Corporation (UMC) 23
United Power 284
United States of America (US) 51
automobile industry 7, 8
Department of Transportation (USDOT) 169
exports to China 108
Transportation Security Administration (TSA) 171
universities
and the "German model" 113
holding patents 107
University of New South Wales (UNSW), Australia 309–10, 311, 320
UTStarcom 76

value chains, modularized 18
value creation, and value extraction 15–16, 22
Vensys Energiesysteme GmbH 295, 298
venture capital 68–94
angel funds 68, 82–3
Asia America MultiTechnology Association (AAMA) 84–5
China 10–11
and the Chinese internet 91
cycles 75–81, 79 Fig. 3.4, 80 Tab 3.2
e-commerce 90
foreign firms 76
formation of networked entrepreneurial habitats 81 Fig. 3.5, 81–5, 84 Fig. 3.6
globalization 69
and the Great Firewall 91
Hong Kong 82
and the internet boom 76–7
investment syndication 83–4
and national banking laws 76
'no exit, no entry' 75
offshoring 77–8
Partnership Enterprise Law 78–9
private equity firms 70
guanxi networks 88–9
regulation (from 2005) 78
'red chip'/Sina model 77–8
rise of 71 n3, 71–4, 72 Fig. 3.1, 73 Fig. 3.2, 74 Fig. 3.3, 74 Tab. 3.1
Shenzhen Small and Medium Enterprise Board (SSMED) 78, 80–1
as a spur to innovation 85–92, 87 Fig. 3.7, 88 Fig. 3.8
technology sector 82
Vestas 284, 286, 294, 298
Victoria-Suntech Advanced Solar Facility 320
Vodafone 228
Volkswagen 138, 141, 142
Beetle 5
venture with Shanghai Auto 41
Volvo 161

Wacker Chemie, Germany 123
Walden International 76

Walled garden 267 277–9
Wan Li 196, 243
Wang, Winston 201
Wang Yongqin 201
WCDMA (Europe) 52
Wei, S. 40
Weifeng Zong 122
Weixin 90
Welkener 143
Wen Jiabao 47
Wenzhou City train crash (2011) 187
Wind energy
  Concession program 288, 290
  Joint development or co-developing, 293, 297–8
  Local content 289–90, 292
  Mergers and acquisitions (M&A) 294–5, 304
  Preferential tax policies 291
  Technology License 293–4, 296–7
  Unified pricing policy 288–9
wind power technology sector development 283–304
  Access Conditions for Wind Power Equipment Industry (2010) 292–3
  Double Increase Program (1997) 289–90
  Enterprise Income Tax Law (2005) 291
  feed-in tariffs (FiT) for renewable electricity 287, 289
  firm-level innovation models 295–9, 296 Fig. 11.5, 297 Tab. 11.1
  firm-level models of technology acquisition 293–9
  firm-level technology transfer models 293–5, 295 Fig. 11.4
  German engineering design firms 294–5
  global context 284–6, 285 Fig. 11.1, 285 Fig. 11.2
  industrial and trade policies 292 n2, 292–3
  joint ventures (JV) 290, 294
  learning and technology costs 301–3, 302 Fig. 11.8
  national framework policy 287–8
  National Renewable Energy Law 286, 287, 304
  Notice on Abolishing the Localization Rate Requirement (2009) 292
  Notice on the Relevant Requirements (2005) 290–1
  patents 300, 301 Fig. 11.7
  policies and institutions 286–93, 287 Fig. 11.3
  Policy to Improve Grid-Connected Power Pricing (categories I–IV) 288–9
  pricing policies 288–9
  Provisions for Grid-Connected Wind Farm Management 288
  R&D facilities 298
  Renewable Energy Industrial Development Guidance Catalog 291
  Ride the Wind Program (1997) 290
  science and technology policies 291–2
  technology size 299–300, 300 Fig. 11.6
  technology transfer, and localization policies 289–91, 293–95
Wind World India 284
Windey 296, 298
Windtec 296
Womack, J. 136
World Trade Organization (WTO) 13, 169, 200
Wuhan Xinxin Semiconductor Manufacturing Corporation 206–7
Wuhan-Guangzhou HSR 179
Wuxi Jiangnan Wireless Device Factory 196, 196 n4, 243

XEMC 284
Xi Jingping 209
Xi'an 290
Xiaomi 12, 26, 82, 269, 270, 271, 272, 274, 275, 276, 279
'XiaoWangZi' car model 148
Xu, B. 40
Xu Xiaoping 83

Yahoo 76
Yang Ning 83
YangFeng Zhu 159, 159 n36
Yangtze River Delta mechanical engineering industries 107
Yin Tongyue 152–3
Yingli Green Energy 306, 308–9, 313, 315, 320, 324
Yituo 90
Yizumi case study 119–21, 127
  Propower cross-case analysis 124–6
Yoo, Y. O. 266, 267

Zero2IPO Database 70
Zhang, Charles 76
Zhang Shugan 187
Zhangjiang 55
Zhao Jianhua 311
Zhejiang Institute of Mechanical and Electrical Engineering 298
Zhengfei Ren 15
Zhongguancun, Beijing 17, 38, 55
'ZhongHua' model, Brilliant Auto 157
ZhongTai 163
ZhongYi prototype 155
Zhou, Y. 39, 49
Zhou Enlai 34
Zhou Huan 231

Zhou Quan 88–9
Zhu Rongli 141, 141 n17, 169
Zhuhai Actions 254
Zizhu Chuangxin (indigenous innovation) strategy 2–3
Zhongxing Telecommunication Equipment Corporation (ZTE) 12, 13, 215–16, 219, 223, 226, 228–9, 231–2, 235, 248, 251–2, 263, 268–9
patents 229

The manufacturer's authorised representative in the EU for product safety is Oxford University Press España S.A. of El Parque Empresarial San Fernando de Henares, Avenida de Castilla, 2 - 28830 Madrid (www.oup.es/en or product.safety@oup.com). OUP España S.A. also acts as importer into Spain of products made by the manufacturer.
Printed and bound by CPI Group (UK) Ltd, Croydon, CR0 4YY
14/09/2026
02216304-0010